Wnt Signaling

METHODS IN MOLECULAR BIOLOGY™

John M. Walker, Series Editor

475. Cell Fusion: *Overviews and Methods*, edited by *Elizabeth H. Chen, 2008*
474. Nanostructure Design: *Methods and Protocols*, edited by *Ehud Gazit and Ruth Nussinov, 2008*
473. Clinical Epidemiology: *Practice and Methods*, edited by *Patrick Parfrey and Brendon Barrett, 2008*
472. Cancer Epidemiology, Volume 2: *Modifiable Factors*, edited by *Mukesh Verma, 2008*
471. Cancer Epidemiology, Volume 1: *Host Susceptibility Factors*, edited by *Mukesh Verma, 2008*
470. Host-Pathogen Interactions: *Methods and Protocols*, edited by *Steffen Rupp and Kai Sohn, 2008*
469. Wnt Signaling, Volume 2: *Pathway Models*, edited by *Elizabeth Vincan, 2008*
468. Wnt Signaling, Volume 1: *Pathway Methods and Mammalian Models*, edited by *Elizabeth Vincan, 2008*
467. Angiogenesis Protocols: *Second Edition*, edited by *Stewart Martin and Cliff Murray, 2008*
466. Kidney Research: *Experimental Protocols*, edited by *Tim D. Hewitson and Gavin J. Becker, 2008*
465. Mycobacteria, Second Edition, edited by *Tanya Parish and Amanda Claire Brown, 2008*
464. The Nucleus, Volume 2: *Physical Properties and Imaging Methods*, edited by *Ronald Hancock, 2008*
463. The Nucleus, Volume 1: *Nuclei and Subnuclear Components*, edited by *Ronald Hancock, 2008*
462. Lipid Signaling Protocols, edited by *Banafshe Larijani, Rudiger Woscholski, and Colin A. Rosser, 2008*
461. Molecular Embryology: *Methods and Protocols, Second Edition*, edited by *Paul Sharpe and Ivor Mason, 2008*
460. Essential Concepts in Toxicogenomics, edited by *Donna L. Mendrick and William B. Mattes, 2008*
459. Prion Protein Protocols, edited by *Andrew F. Hill, 2008*
458. Artificial Neural Networks: Methods and Applications, edited by *David S. Livingstone, 2008*
457. Membrane Trafficking, edited by *Ales Vancura, 2008*
456. Adipose Tissue Protocols, Second Edition, edited by *Kaiping Yang, 2008*
455. Osteoporosis, edited by *Jennifer J. Westendorf, 2008*
454. SARS- and Other Coronaviruses: *Laboratory Protocols*, edited by *Dave Cavanagh, 2008*
453. Bioinformatics, Volume II: *Structure, Function and Applications*, edited by *Jonathan M. Keith, 2008*
452. Bioinformatics, Volume I: *Data, Sequence Analysis and Evolution*, edited by *Jonathan M. Keith, 2008*
451. Plant Virology Protocols: *From Viral Sequence to Protein Function*, edited by *Gary Foster, Elisabeth Johansen, Yiguo Hong, and Peter Nagy, 2008*
450. Germline Stem Cells, edited by *Steven X. Hou and Shree Ram Singh, 2008*
449. Mesenchymal Stem Cells: *Methods and Protocols*, edited by *Darwin J. Prockop, Douglas G. Phinney, and Bruce A. Brunnell, 2008*
448. Pharmacogenomics in Drug Discovery and Development, edited by *Qing Yan, 2008*
447. Alcohol: *Methods and Protocols*, edited by *Laura E. Nagy, 2008*
446. Post-translational Modification of Proteins: *Tools for Functional Proteomics*, Second Edition, edited by *Christoph Kannicht, 2008*
445. Autophagosome and Phagosome, edited by *Vojo Deretic, 2008*
444. Prenatal Diagnosis, edited by *Sinhue Hahn and Laird G. Jackson, 2008*
443. Molecular Modeling of Proteins, edited by *Andreas Kukol, 2008.*
442. RNAi: Design and Application, edited by *Sailen Barik, 2008*
441. Tissue Proteomics: *Pathways, Biomarkers, and Drug Discovery*, edited by *Brian Liu, 2008*
440. Exocytosis and Endocytosis, edited by *Andrei I. Ivanov, 2008*
439. Genomics Protocols, Second Edition, edited by *Mike Starkey and Ramnanth Elaswarapu, 2008*
438. Neural Stem Cells: *Methods and Protocols*, Second Edition, edited by *Leslie P. Weiner, 2008*
437. Drug Delivery Systems, edited by *Kewal K. Jain, 2008*
436. Avian Influenza Virus, edited by *Erica Spackman, 2008*
435. Chromosomal Mutagenesis, edited by *Greg Davis and Kevin J. Kayser, 2008*
434. Gene Therapy Protocols: Volume II: *Design and Characterization of Gene Transfer Vectors*, edited by *Joseph M. LeDoux, 2008*
433. Gene Therapy Protocols: Volume I: *Production and In Vivo Applications of Gene Transfer Vectors*, edited by *Joseph M. LeDoux, 2008*
432. Organelle Proteomics, edited by *Delphine Pflieger and Jean Rossier, 2008*
431. Bacterial Pathogenesis: *Methods and Protocols*, edited by *Frank DeLeo and Michael Otto, 2008*
430. Hematopoietic *Stem Cell Protocols*, edited by *Kevin D. Bunting, 2008*
429. Molecular Beacons: *Signalling Nucleic Acid Probes, Methods and Protocols*, edited by *Andreas Marx and Oliver Seitz, 2008*
428. Clinical Proteomics: *Methods and Protocols*, edited by *Antonia Vlahou, 2008*
427. Plant Embryogenesis, edited by *Maria Fernanda Suarez and Peter Bozhkov, 2008*
426. Structural Proteomics: *High-Throughput Methods*, edited by *Bostjan Kobe, Mitchell Guss, and Huber Thomas, 2008*
425. 2D PAGE: *Sample Preparation and Fractionation*, Volume II, edited by *Anton Posch, 2008*
424. 2D PAGE: *Sample Preparation and Fractionation*, Volume I, edited by *Anton Posch, 2008*
423. Electroporation Protocols: *Preclinical and Clinical Gene Medicine*, edited by *Shulin Li, 2008*
422. Phylogenomics, edited by *William J. Murphy, 2008*
421. Affinity Chromatography: *Methods and Protocols*, Second Edition, edited by *Michael Zachariou, 2008*

METHODS IN MOLECULAR BIOLOGY™

Wnt Signaling

Volume 1
Pathway Methods and Mammalian Models

Edited by

Elizabeth Vincan, PhD

University of Melbourne, Parkville, Victoria, Australia

Humana Press

Editor
Elizabeth Vincan
University of Melbourne
Parkville, Victoria VIC 3010
Australia

Series Editor
John M. Walker
School of life Sciences
University of Hertfordshire
Hatfield, Hertfordshire
AL10 9AB, UK

ISBN: 978-1-58829-912-3 e-ISBN: 978-1-59745-249-6
ISSN: 1064-3745 e-ISSN: 1940-6029
DOI: 10.1007/978-1-59745-249-6

Library of Congress Control Number: 2008936263

Cover illustration:

Printed on acid-free paper

9 8 7 6 5 4 3 2 1

springer.com

Preface

Since their discovery some 20 years ago, Wnt signaling molecules have been shown to control key events in embryogenesis, maintain tissue homeostasis in the adult and, when aberrantly activated, promote human degenerative diseases and cancer. Elucidation of Wnt signaling mechanisms has relied on both biochemical methodologies and vertebrate and invertebrate model systems. Therefore, I felt that an issue dedicated to Wnt signaling had to include both assays (biochemical readout) and model systems (functional readout) of Wnt signaling. It is not an exhaustive catalog, but rather a point of reference to current molecular protocols and the diverse model systems employed to study this signaling pathway. The issue is divided into two volumes. The first volume includes assays to measure activation of the diverse Wnt signaling pathways as well as models and strategies used to study mammalian Wnt/FZD function (from protein–protein interaction and simple cell line models to organoid cultures and mouse models). The second volume is dedicated to the diverse vertebrate and invertebrate models that have shaped the Wnt signaling field. It provides an entry point for the novice and an overview of the unique properties of each organism with respect to studying Wnt/FZD function (for example asymmetric cell division in *Caenorhabditis elegans*, epithelial morphogenesis in *Dictyostelium* and so on). Given the collective expertise and knowledge of the contributors, I anticipate that this two-volume issue will be an invaluable resource.

The Wnt field advances at an exceptionally rapid rate for several reasons. First, diverse fields of research converge on this pathway. Second, the Wnt community is very generous: reagents, knowledge, and ideas are shared freely. This is facilitated by informative web sites and regular Wnt meetings that are packed back-to-back with cutting-edge research. The "no-frills" approach to these meetings means that the whole community, including students, can participate. Equally important is the elusive nature of the Wnt pathway itself, which continues to intrigue and fascinate both novice and veteran researchers alike. This book is a testament to all these. It was steered by the generosity and enthusiasm of contributors from diverse fields. I thank them all. Special thanks to Randall Moon and Stefan Hoppler; their suggestions for authors and chapters helped shape this issue.

On a personal note, I would also like to take this opportunity to acknowledge Bill Boyle for being an inspirational mentor during my formative years; his infectious enthusiasm for research set me on this exciting and rewarding career path. I am indebted to Bob Thomas and Rob Ramsay for generously supporting my research into FZD7 function in colon cancer when funding in Australia for the Wnt field was scarce in the early years. Most importantly, I thank my very patient and accommodating children for allowing me to indulge myself!

I thank Tony Goodwin, Scott Bowden, and the University of Melbourne—without their assistance this book would not have been possible. John Walker and all at Humana Press, especially David Casey and Amina Ravi, for their generosity and for the opportunity to edit this issue—a truly rewarding experience.

Contents

Contributors

HELEN E. ABUD, Ph.D. • *Department of Anatomy and Developmental Biology, Monash University, Clayton, Victoria, Australia*

STEPHANE ANGERS, Ph.D. • *Leslie Dan Faculty of Pharmacy, University of Toronto, Toronto, Ontario, Canada*

KOJI AOKI, Ph.D. • *Department of Pharmacology, Graduate School of Medicine, Kyoto University, Yoshida-Konoé-cho, Sakyo, Kyoto, Japan*

NICK BARKER, Ph.D. • *Hubrecht Institute for Developmental Biology and Stem Cell Research, Utrecht, The Netherlands*

TRAVIS L. BIECHELE, B.Sc. • *Howard Hughes Medical Institute and Department of Pharmacology and Institute for Stem Cell and Regenerative Medicine, University of Washington School of Medicine, Seattle, WA, USA*

DANIEL D. BILLADEAU, Ph.D. • *Division of Oncology Research, Mayo Clinic College of Medicine, Rochester MN, USA*

MAAIKE VAN DEN BORN. • *Hubrecht Institute for Developmental Biology and Stem Cell Research, Utrecht, The Netherlands*

THOMAS BRABLETZ, M.D., Ph.D. • *Department of Surgery, University of Freiburg, Freiburg, Germany*

MARIANA BROCARDO, Ph.D. • *Westmead Millennium Institute, The University of Sydney, Westmead, NSW, Australia*

ADAM R. COLE, Ph.D. • *Pathology and Neurosciences, University of Dundee, Ninewells Hospital, Dundee, Scotland*

WILLIAM H. CONRAD, Ph.D. • *Howard Hughes Medical Institute, Department of Pharmacology, and Institute for Stem Cell and Regenerative Medicine, University of Washington School of Medicine, Seattle, WA, USA*

SAMUDRA K. DISSANAYAKE, Ph.D. • *Laboratory of Immunology, National Institutes of Health, National Institute on Aging, Gerontology Research Center, Baltimore, MD, USA*

JOHN D. DORAN, Ph.D. • *Protein Biochemistry, Vertex Pharmaceuticals, Cambridge, MA, USA*

YOSHIMI ENDO, M.D., Ph.D. • *National Cancer Institute, Bethesda, MD, USA*

MAREE C. FAUX, Ph.D. • *Ludwig Institute for Cancer Research, Parkville, Victoria, Australia*

CHRISTINA M. FREISINGER, Ph.D. • *Department of Biology, University of Iowa, Iowa City, IA, USA*

AMANDA D. FRENCH, B.Sc. • *Laboratory of Immunology, National Institutes of Health, National Institute on Aging, Gerontology Research Center, Baltimore, MD, USA*

LYNN E. HEASLEY, Ph.D. • *Department of Medicine, University of Colorado Health Sciences Centre, Denver, Colorado, USA*

BERIC R. HENDERSON, Ph.D. • *Westmead Millennium Institute, The University of Sydney, Westmead, NSW, Australia*

RANDALL F. HOLCOMBE, M.D. • *Division of Hematology/Oncology, University of California–Irvine, Orange, CA, USA*

DOUGLAS W. HOUSTON, Ph.D. • *Department of Biology, University of Iowa, Iowa City, IA, USA*

MARC D. JACOBS, Ph.D. • *Structural Biology, Vertex Pharmaceuticals, Cambridge, MA, USA*

RICHARD G. JAMES, Ph.D. • *Howard Hughes Medical Institute, Department of Pharmacology, and Institute for Stem Cell and Regenerative Medicine, University of Washington School of Medicine, Seattle, WA, USA*

JENNIFER M. JONES, Ph.D. • *Department of Otolaryngology, Washington University School of Medicine, St. Louis, MO, USA*

MICHAEL KÜHL, Ph.D. • *University of Ulm, Institute for Biochemistry and Molecular Biology, Ulm, Germany*

POLOKO D. LEOTLELA, Ph.D. • *Laboratory of Immunology, National Institutes of Health, National Institute on Aging, Gerontology Research Center, Baltimore, MD, USA*

MIREILLE MONTCOUQUIOL, Ph.D. • *Equipe Avenir, Development Neurosciences, INSERM U862, Institut Francois Magendie, Université Bordeaux II, Bordeaux Cédex, France*

RANDALL T. MOON, Ph.D. • *Howard Hughes Medical Institute, Department of Pharmacology, and Institute for Stem Cell and Regenerative Medicine, University of Washington School of Medicine, Seattle, WA, USA*

DONALD F. NEWGREEN, Ph.D. • *Murdoch Children's Research Institute Royal Children's Hospital, Parkville, Victoria, Australia*

MICHAEL P. O'CONNELL, Ph.D. • *Laboratory of Immunology, National Institutes of Health, National Institute on Aging, Gerontology Research Center, Baltimore, MD, USA*

ANDREI V. OUGOLKOV, M.D., Ph.D. • *Division of Oncology Research, Mayo Clinic College of Medicine, Rochester, MN, USA*

PETRA PANDUR, Ph.D. • *University of Ulm, Institute for Biochemistry and Molecular Biology, Ulm, Germany*

MARINA PLANUTIENE, Ph.D. • *Division of Hematology/Oncology, University of California–Irvine, Orange, CA, USA*

KESTUTIS PLANUTIS, Ph.D. • *Division of Hematology/Oncology, University of California–Irvine, Orange, CA, USA*

JEFFERY S. RUBIN, M.D., Ph.D. • *National Cancer Institute, Bethesda, MD, USA*

NATHALIE SANS, Ph.D. • *Equipe Avenir, Molecular Neurobiology, INSERM U862, Université Bordeaux II, Bordeaux Cedex, France*

OTTO SCHMALHOFER, M.Sc. • *Department of Surgery, University of Freiburg, Freiburg, Germany*

DIANE C. SLUSARSKI, Ph.D. • *Department of Biology, University of Iowa, Iowa City, IA, USA*

SIMONE SPADERNA, Ph.D. • *Department of Surgery, University of Freiburg, Freiburg, Germany*

CALUM SUTHERLAND, Ph.D. • *Pathology and Neurosciences, University of Dundee, Ninewells Hospital, Dundee, Scotland*

MAKOTO M. TAKETO, M.D., Ph.D. • *Department of Pharmacology, Graduate School of Medicine, Kyoto University, Yoshida-Konoé-cho, Sakyo, Kyoto, Japan*

ELIZABETH VINCAN, Ph.D. • *Cancer Biology Laboratory, Department of Anatomy and Cell Biology, University of Melbourne, Victoria, Australia*

ASHANI T. WEERARATNA, Ph.D. • *Laboratory of Immunology, National Institutes of Health, National Institute on Aging, Gerontology Research Center, Baltimore, MD, USA*

ROBERT H. WHITEHEAD, M.Sc., Ph.D. • *Departments of Medicine, Cancer Biology and Cell and Developmental Biology, Vanderbilt University, Nashville, TN, USA*

KARL H. WILLERT, Ph.D. • *Cellular and Molecular Medicine, University of California–San Diego, La Jolla, CA, USA*

ROBERT A. WINN, M.D. • *Veterans Affairs Medical Center, Denver, Colorado, USA*

VLADIMIR WOLF, M.D. • *National Cancer Institute, Bethesda, MD, USA*

TERRY P. YAMAGUCHI, Ph.D. • *Cancer and Developmental Biology Laboratory, Center for Cancer Research, National Cancer Institute–Frederick, the National Institutes of Health, Frederick, MD, USA*

HEATHER M. YOUNG, Ph.D. • *Department of Anatomy and Cell Biology, University of Melbourne, Parkville, Victoria, Australia*

Contents of Volume 2

Part I

Wnt Signaling Pathway Methods

Section A

Canonical Wnt / FZD Signaling

Chapter 1

The Canonical Wnt/β-Catenin Signalling Pathway

Nick Barker

Abstract

Embryonic development of multicellular organisms is an incredibly complex process that relies heavily on evolutionarily conserved signalling pathways to provide crucial cell–cell communication. Typically, secreted signalling proteins such as Wnts, BMPs, and Hedgehogs released by one cell population will trigger concentration-dependent responses in other cells located some distance away. In adults, the same signalling pathways orchestrate tissue renewal in organs such as the intestine and skin, and direct tissue regeneration in many organs following injury. Strict regulation of these signalling pathways is critical, with insufficient or excess activity having catastrophic consequences including severe developmental defects or, later in life, cancer. This chapter deals with the β-catenin-dependent branch of Wnt signalling (also referred to the canonical pathway).

Key words: Wnt, Morphogen, β-catenin, Groucho, Tcf, Target gene, Constitutive activation, Stem cell, Colon cancer.

1. Introduction

The Wnt pathway derives its name from the Drosophila (fruit-fly) *W*ingless gene and the mouse I*NT*-1 gene. The mouse Wnt1 gene (originally named Int1), was identified in 1982 as a gene inappropriately activated by integration of the Mouse Mammary Tumor Virus in virally induced breast tumors *(1)*. This established the first link between mis-expression of Wnt genes and cancer (branding Wnt-1 a proto-oncogene) and revealed Wnt-1 to be a secreted cysteine-rich protein with the potential to act as a signalling molecule. However, the real breakthrough in linking this gene to a signalling pathway was made by fly geneticists, who

Elizabeth Vincan (ed.), *Wnt Signaling, Volume I: Pathway Methods and Mammalian Models, vol. 468*

Book doi: 10.1007/978-1-59745-249-6

demonstrated that the Drosophila Wnt1 counterpart, Wingless (Wg) was a crucial component of a novel signal transduction pathway controlling body patterning during larval development *(2)*. This pathway was subsequently found to be highly conserved in frogs, where it orchestrates proper induction of the body axis during embryonic development. A particularly graphic example of the influence of the Wnt pathway on vertebrate development was provided by McMahon and Moon in 1989, when they demonstrated that forced activation of the Wnt pathway at the ventral (anterior) side of early frog embryos by injection of Wnt1 messenger RNA (mRNA) caused a complete duplication of the body axis and, as a consequence, the development of two-headed embryos *(3)*. More recently, altered Wnt signalling activity in adults has been linked to a wide range of human diseases, including cancer, bone defects, schizophrenia, and arthritis *(4, 5)*.

This chapter aims to provide a simplified overview of how Wnt proteins are produced and secreted and how they subsequently activate the canonical Wnt signalling pathway in recipient cells to effect changes in cell growth, movement, and cell survival.

2. The Wnt Family

Since the identification of Wnt1, genome sequencing has revealed the existence of another 18 Wnt genes in mammals, which can be divided into 12 highly conserved subfamilies on the basis of sequence similarity *(6)*. Many of these subfamilies are highly conserved in early multicellular organisms such as the sea anemone, highlighting the crucial role of the Wnt pathway in driving body patterning throughout the animal kingdom. All Wnt proteins share common features that are essential for their function, including a signal peptide for secretion, many potential glycosylation sites and multiple cysteine residues responsible for ensuring proper folding and secretion. With a few exceptions (Wg, Wnt3/5, and Wnt4), Wnt proteins are generally around 350 amino acids long and have an approximate molecular weight of 40 kDa *(7)*.

3. Wnt Secretion and Delivery

The recent success in developing cell-based systems for expressing and purifying biologically active Wnt proteins has provided invaluable insights into how Wnts are converted from immature precursors in the cell into secreted signalling molecules capable of interacting with specific receptor complexes on target cells located up to 20–30 cell diameters distant *(8, 9)*.

Wnts start out as precursors containing an N-terminal hydrophobic signal peptide that directs the immature protein to the endoplasmic reticulum (ER). In the ER, the signal peptide is cleaved off by a resident protease and the Wnt protein is extensively modified by the addition of sugars and lipids to ensure efficient secretion, intercellular delivery, and to maximize biological activity on target cells. Probably the most significant modification to occur is the attachment of a palmitate moiety to a conserved cysteine residue on the Wnts, thereby converting them into hydrophobic proteins. This modification was shown to be essential for the biological activity of Wnt proteins produced in cell lines, when inhibition of essential acyltransferase enzyme activity (typically required for lipid modifications) or mutation of the conserved Wnt cysteine modification site resulted in a protein that was neither hydrophobic nor active *(8, 10)*. It has been proposed that this lipid modification serves to anchor the Wnt proteins in the vicinity of the oligosaccharyl complex (OST) at the ER membrane, thereby facilitating their efficient *N*-linked glycosylation at conserved asparagine residues *(11)*. Alternatively, the palmitate moiety may prevent the modified cysteine residue from forming disulphide bonds that would otherwise result in mis-folding and retention of the Wnt protein in the ER. The ER-resident acyltransferase enzyme believed to perform this lipid modification in vivo is encoded by the *Drosophila porcupine* gene and its homologues (*mom-1* in worms) *(9, 10, 12)*.

More recently, it has become increasingly clear that efficient secretion of Wnt proteins is a complex process requiring the concerted actions of several conserved genes (**Fig. 1.1**). This is highlighted by the discovery of another gene termed *Wntless/eveness interrupted*, which, like *Porcupine*, is also indispensable for Wnt secretion in *Drosophila (13, 14)*. *Wntless* encodes a seven-pass transmembrane protein that is highly conserved across species from worms (mom-3) to man (hWLS). Inactivation of this gene in Wnt-producing cells short-circuits the Wnt secretion process and leads to retention of the Wnt protein inside the cell. Although the precise function of *Wntless* remains elusive, it largely resides in the Golgi apparatus, where it physically interacts with the Wnt proteins. This has prompted speculation that it may act as a chaperone to guide the Wnt proteins through other post-translational modifications necessary for their efficient secretion and/or regulate the intracellular trafficking of Wnt between different cell compartments en route to its release into the extracellular space.

Genetic screens in *Caenorhabditis elegans* have revealed yet more proteins involved in controlling the fate of Wnt as Wnt passes through the cell secretion machinery. These proteins reside

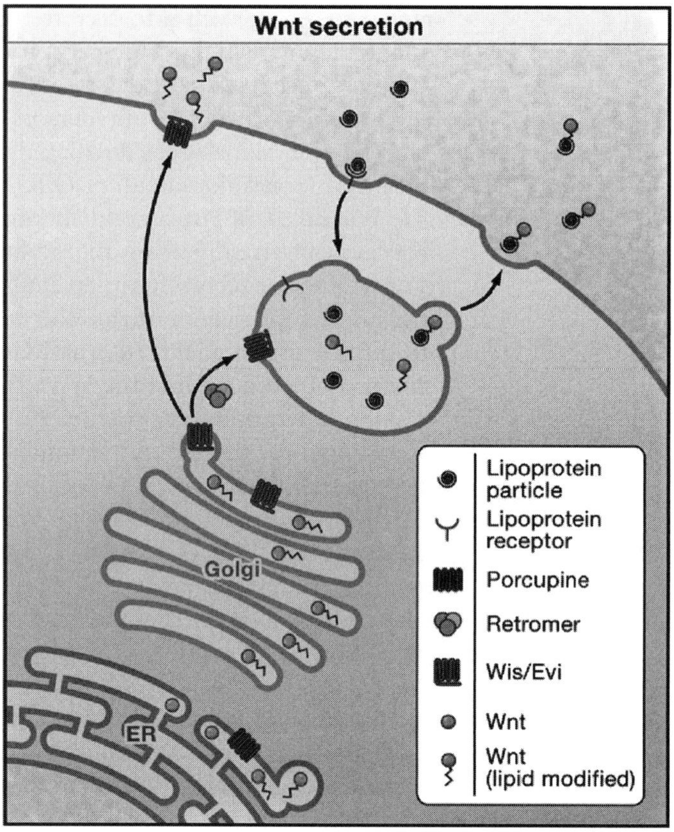

Fig. 1.1. Wnt secretion. To facilitate their secretion, Wnt proteins must first be palmitoylated in the endoplasmic reticulum (ER) by the actions of Porcupine. Wnt proteins must also complex with Wntless (Wls/Evi) in the Golgi to be efficiently routed to the outside of the cell. Loading onto lipoprotein particles may occur in a dedicated endocytic/exocytic compartment. The retromer complex may shuttle Wls between the Golgi and the endo/exocytic compartment. Reprinted with permission from ref. *(4)*. Copyright Elsevier (2006).

in a complex called the retromer, which is involved in intracellular trafficking in many species ranging from yeast to man *(15)*. Unlike *Porcupine* and *Wntless* mutations, abrogation of retromer function in mammalian Wnt-producing cell lines does not substantially impair Wnt secretion. Instead, depletion of the retromer complex in worms and frogs appears to prevent long-range transport of secreted Wnts to their target cells. In contrast, delivery of the Wnt proteins to neighboring target cells (short-range signalling) remains largely intact. These observations have led to the proposal that retromers direct Wnt proteins from the Golgi apparatus into specialized intracellular compartments dedicated to long-range Wnt secretion.

Wnts are classic morphogens (long-range signalling molecules whose activity is concentration dependent) and must therefore form long-range concentration gradients capable of activating signalling on target cells up to 20–30 cell diameters distant from their source *(16)*. It is likely that the various post-translational modifications bestowed on Wnt whilst en route through the cell play an important part in setting up these concentration gradients. For instance, the addition of palmitate to the Wnts may facilitate the association of Wnts with lipoprotein particles involved in lipid transport (referred to as argosomes) *(17)*. This interaction may dislodge Wnts from cell membranes in the immediate vicinity of the Wnt source thereby allowing them to spread further afield. Alternatively, the argosomes may simply capture and concentrate the Wnt proteins to a level required for efficient activation of the pathway on distant target cells. Genetic studies in *Drosophila* indicate that interaction with heparin sulphate proteoglycans (HSPG) present on cell membranes and the extracellular matrix is also likely to play a role in transporting and stabilizing Wnts *(6, 18)*. Finally, the Wnts may be actively transported by cytonemes, which are long, thin, filopodial cell processes present in the extracellular matrix *(19)*.

Binding of these secreted Wnt proteins to specific receptor complexes on target cells activates one of three intracellular signalling pathways: the canonical (T-cell factor [Tcf]/β-catenin) pathway, the non-canonical (planar cell polarity) pathway and the Wnt/Ca^{2+} pathway *(20, 21)*. Each of these pathways delivers a very different set of instructions to the recipient cell by activating specific sets of target genes. This chapter focuses on the canonical pathway, which is better characterized and generally considered to be more relevant for cancer development.

4. The Canonical Wnt Pathway

The canonical Wnt pathway strictly controls the levels of a cytoplasmic protein known as β-catenin, which has crucial roles in both cell adhesion and activation of Wnt target genes in the nucleus *(4)*. In the absence of a Wnt signal, β-catenin is efficiently captured by a scaffold protein termed Axin, which is present within a protein complex (referred to as the destruction complex) that also harbors adenomatous polyposis coli (APC) and the protein kinases casein kinase (CK)-1 and glycogen synthase kinase (GSK)-3 (**Fig. 1.2**, left panel). APC is an essential component of the destruction complex, where it is thought to ensure the efficient recruitment and anchoring of β-catenin. The resident CK1 and GSK3 protein kinases sequentially phosphorylate conserved serine and threonine residues in the N-terminus of the trapped β-catenin *(22)*, generating a binding site for an

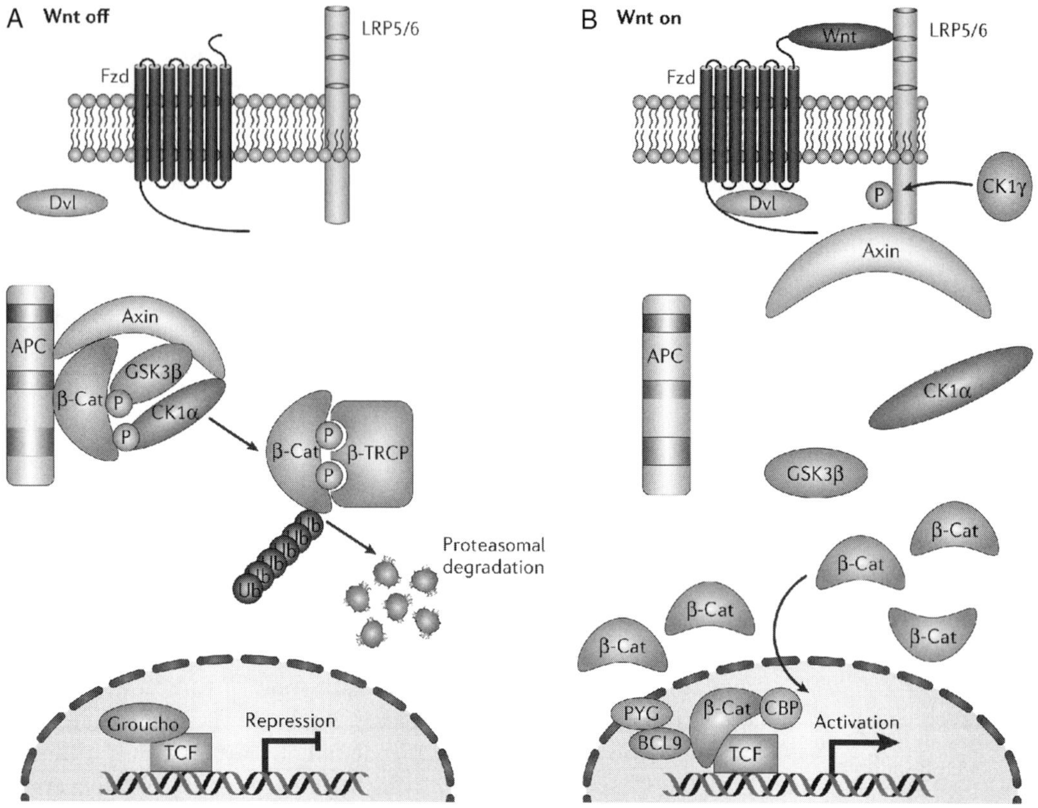

Fig. 1.2. Overview of the canonical Wnt signalling pathway. **A** *Wnt off*: in the absence of a Wnt signal, β-catenin is captured by APC and Axin within the destruction complex, facilitating its phosphorylation (*P*) by the kinases CK1α and GSK3. CK1α and GSK3 then sequentially phosphorylate a conserved set of serine and threonine residues at the N-terminus of β-catenin. This facilitates binding of the β-transducin repeat-containing protein (β-TRCP), which subsequently mediates the ubiquitylation (*Ub*) and efficient proteasomal degradation of β-catenin. The resulting β-catenin "drought" ensures that nuclear DNA-binding proteins of the Tcf/Lef transcription factor family (Tcf-1, Tcf-3, Tcf-4, and Lef-1) actively repress target genes by recruiting transcriptional co-repressors (Groucho/TLE) to their promoters and/or enhancers. **B** *Wnt on*: interaction of a Wnt ligand with its specific receptor complex containing a Frizzled family member and low-density lipid receptor (LRP)-5 or LRP6 triggers the formation of Dishevelled (Dvl)/Frizzled (Fzd) complexes. The resulting generation of LRP/Fzd/Dsh aggregates at the cell membrane induces the phosphorylation of LRP by CK1α, thereby facilitating relocation of Axin to the membrane and inactivation of the destruction box. This allows β-catenin to accumulate and enter the nucleus, where it interacts with members of the Tcf/Lef family. In the nucleus, β-catenin converts the Tcf proteins into potent transcriptional activators by displacing Groucho/TLE proteins and recruiting an array of coactivator proteins including CBP, TBP, BRG-1, Bcl9, Legless, Mediator, and Hyrax. This ensures efficient activation of Tcf target genes such as c-Myc, which instruct the cell to actively proliferate and remain in an undifferentiated state. Following dissipation of the Wnt signal, β-catenin is evicted from the nucleus by the APC protein, and Tcf proteins revert to actively repressing the target gene program. *APC*, adenomatous polyposis coli; *CK1α*, casein kinase 1α; *CBP*, CREB-binding protein; *Tcf*, T-cell factor; *Lef*, lymphoid enhancer factor; *Bcl9*, B-cell lymphoma-9; Dvl, ; β-cat, β-catenin; *PYG*, Pygopus. (For more details, I recommend the Wnt homepage: http://www.stanford.edu/~rnusse/Wntwindow.html). Reprinted with permission from ref. *(5)*. Copyright Nature Publishing Group (2006).

E3 ubiquitin ligase, which subsequently targets the β-catenin for rapid proteasomal degradation *(23)*. Such efficient suppression of β-catenin levels ensures that Groucho proteins are free to bind Tcf/lymphoid enhancer factor (Lef) proteins occupying the promoters and enhancers of Wnt target genes in the nucleus *(24, 25)*. These Tcf/Groucho complexes actively suppress the transcriptional activation of Wnt target genes such as c-Myc, thereby silencing an array of biological responses, including cell proliferation.

Rapid activation of the canonical pathway occurs when Wnt proteins interact with specific cell surface receptor complexes comprising members of the Frizzled family of seven-pass transmembrane proteins and the single-pass transmembrane proteins, low-density lipid receptor (LRP)-5 or LRP6 (**Fig. 1.2**, right panel). This triggers the phosphorylation of Dsh proteins and promotes their interaction with the Frizzled proteins *(26)*. The resulting Dsh/receptor complexes are thought to stimulate the formation of LRP6 aggregates at the membrane, which facilitates the phosphorylation of the LRP6 intracellular tails by the CK1γ. As a consequence, Axin is recruited to this receptor complex and the proteasomal degradation of β-catenin is blocked *(22, 27, 28)*. This allows β-catenin to accumulate and enter the nucleus, where it interacts with members of the Tcf/Lef family and converts them into potent transcriptional activators by recruiting co-activator proteins and ensuring efficient activation of Wnt target genes.

5. Regulation of Wnt Target Gene Activity

In the absence of Wnt signalling, Tcf proteins occupy target gene enhancers and promoters independently of β-catenin. There are four family members in vertebrates, Tcf-1, Tcf-3, Tcf-4, and Lef-1, which share a highly similar DNA-binding domain termed the high mobility group (HMG) box. This HMG box provides the target gene specificity of the Tcf proteins by ensuring that they exclusively bind DNA at a conserved motif defined as AGA/TA/TCAAAG *(29, 30)*. Interaction of Tcf proteins with this motif in enhancers and promoters of target genes causes the DNA to dramatically bend through more than 90 degrees. When β-catenin is absent from the nucleus, the Tcf proteins bound to the Wnt target genes act as transcriptional repressors by recruiting members of the Groucho/TLE protein family *(24, 25)*.

Relocation of β-catenin from the cytoplasm to the nucleus following its Wnt-induced stabilization is essential for achieving the efficient activation of Wnt target genes and ensuring the appropriate physiological response. Exactly how this is achieved remains somewhat of a mystery. Its nuclear import appears independent of the nuclear localization signal (NLS)/importin machinery; although β-catenin is itself related to the importin/karyophilin protein family and may gain access through direct interaction with the nuclear pores. There has also been speculation that β-catenin may shuttle to the nucleus as a complex with other proteins such as Pygopus/B-cell lymphoma (Bcl)-9 and the Tcf/Lef family, which are actively imported by virtue of their NLS. However, this is unlikely to be the major access route because β-catenin still localizes to the nucleus in the absence of these proteins *(31)*.

Following its entry into the nucleus, β-catenin binds to the N-terminus of Tcf and converts Tcf into a potent transcriptional activator *(32–34)*. The resulting transient activation of the Wnt target gene program signals the final step in the Wnt signalling pathway. β-catenin achieves this by displacing the Groucho/TLE co-repressor proteins from Tcf *(35)* and efficiently recruiting a variety of proteins capable of effecting changes in local chromatin structure to the Wnt target genes *(31)* (**Fig. 1.3**). Many of these co-activator proteins, such as the histone acetylase CREB-binding protein (CBP), Brahma-related gene (BRG)-1 (a component of the SWI/SNF chromatin remodelling complex), and Hyrax interact directly with the C-terminus of β-catenin. Another protein, Pygopus, indirectly binds the N-terminus of β-catenin via a common binding partner, Bcl9. The precise role of the β-catenin/Bcl9/Pygopus complex is somewhat controversial; one line of evidence suggests that it facilitates the nuclear import/retention of β-catenin *(36)*, whilst another study supports a direct role for this complex in enhancing the ability of β-catenin to activate Wnt target genes *(37)*.

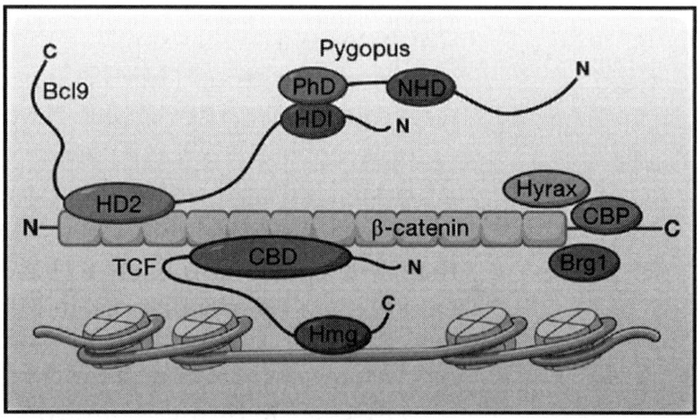

Fig. 1.3. Transactivation of Wnt target genes. The Tcf/β-catenin complex interacts with a variety of chromatin-remodelling complexes to activate transcription of Wnt target genes. The recruitment of β-catenin to Tcf target genes affects local chromatin in several ways. B-cell lymphoma (Bcl) 9 acts as a bridge between Pygopus and the N-terminus of β-catenin. Evidence suggests that this trimeric complex is involved in nuclear import/retention of β-catenin, but may also directly enhance the ability of β-catenin to activate transcription. The C-terminus of β-catenin also binds several co-activators, including the histone acetylase CREB binding protein (CBP), Hyrax, and Brahma-related gene (Brg-1). Reprinted with permission from *(4)*. Copyright Elsevier (2006).

6. Biological Consequences of Wnt Signalling

Activation of the Wnt pathway at the cell surface is ultimately translated into a biological response through activation of a select set of Tcf/β-catenin-responsive target genes. Of the currently estimated 400 target genes present in the mammalian genome, only a fraction of these are thought to be Wnt responsive at any given time in a particular cell type *(4)*. This allows the transcriptional output of the Wnt signal to be tailored to meet the specific needs of a given cell type. For example, in the epithelium of intestinal crypts, Wnt signalling drives both the proliferation of stem/progenitor cells *(38, 39)* and terminal differentiation of Paneth cells *(40)*. The advent of microarray technologies has facilitated the identification of many of these Wnt target genes *(41–43)*. This has provided important clues as to how Wnt signalling influences such diverse biological events as cell proliferation, cell fate specification, terminal differentiation, and cell migration. For example, c-Myc and cyclin D1 are considered to be potent activators of cell proliferation, whilst other target genes encoding the guidance receptors EphB2 and EphB3 are instrumental in cell positioning within tissues such as the intestine *(44)*.

Given the wide-ranging influence of Wnt signalling on such a diverse set of biological processes, it is perhaps not too surprising that loss of proper regulation of this signalling activity can have disastrous consequences for embryonic development and tissue renewal in adults *(4, 39)*. For proof of this, we only have to look at the large variety of severe phenotypes in multiple tissues and organs caused by artificially induced loss of Wnt signalling components in flies, frogs, fish, and mice *(7)*. In adults, Wnt signalling remains essential throughout life for driving tissue renewal in organs such as the intestine and skin *(39)*. In these rapidly self-renewing tissues, Wnt signalling is instrumental in maintaining proliferation of stem cell populations and driving expansion of new epithelial cell precursors. More recent evidence suggests that Wnt signalling is likely to have a more general role in maintaining stem cell populations in a variety of tissues, including the hematopoietic system *(39)*. This raises the attractive possibility of stem cells expressing unique Wnt target genes that could be used as specific markers for identifying and ultimately isolating these cells from a variety of tissues.

However, it is also becoming increasingly apparent that Wnt pathway mutations can occur. These mutations upset the homeostatic balance in self-renewing tissues and cause a variety of diseases including bone defects and cancer. In the early stages of colon cancer for example, mutations frequently occur in either APC or β-catenin that cause constitutive activation of the Wnt pathway and promote uncontrolled cell proliferation *(29, 45)*.

Deregulation of the Wnt pathway is also associated with several other types of human cancer and disease *(4, 46)*. This has fuelled efforts to try and develop specific inhibitors of the Wnt pathway for use as cancer therapeutics, although serious challenges remain to be overcome before this can become a reality *(5)*.

References

1. Nusse, R. and Varmus, H. E. (1982) Many tumors induced by the mouse mammary tumor virus contain a provirus integrated in the same region of the host genome. *Cell* **31**, 99–109.
2. Rijsewijk, F., Schuermann, M., Wagenaar, E., Parren, P., Weigel, D., and Nusse R. (1987) The Drosophila homolog of the mouse mammary oncogene int-1 is identical to the segment polarity gene wingless. *Cell* **50**, 649–657.
3. McMahon, A. P. and Moon, R. T. (1989) Ectopic expression of the proto-oncogene int-1 in Xenopus embryos leads to duplication of the embryonic axis. *Cell* **58**, 1075–1084.
4. Clevers, H. (2006) Wnt/beta-catenin signaling in development and disease. *Cell* **127**, 469–480.
5. Barker, N. and Clevers, H. (2006) Mining the Wnt pathway for cancer therapeutics. *Nat. Rev. Drug Discov.* **5**, 997–1014.
6. Coudreuse, D. and Korswagen, H. C. (2007) The making of Wnt: new insights into Wnt maturation, sorting and secretion. *Development* **134**, 3–12.
7. Miller J. R. (2002) The Wnts. *Genome Biol.* **3**, REVIEWS3001.
8. Willert, K., Brown, J. D., Danenberg, E., Duncan, A.W., Weissman, I. L., Reya, T., et al. (2003) Wnt proteins are lipid-modified and can act as stem cell growth factors. *Nature* **423**, 448–452.
9. Mikels, A. J. and Nusse, R. (2006) Wnts as ligands: processing, secretion and reception. *Oncogene* **25**, 7461–7468.
10. Zhai, L., Chaturvedi, D., and Cumberledge, S. (2004) Drosophila wnt-1 undergoes a hydrophobic modification and is targeted to lipid rafts, a process that requires porcupine. *J. Biol. Chem.* **279**, 33220–33227.
11. Tanaka, K., Kitagawa, Y., and Kadowaki, T. (2002) Drosophila segment polarity gene product porcupine stimulates the post-translational N-glycosylation of wingless in the endoplasmic reticulum. *J. Biol. Chem.* **277**, 12816–12823.
12. Hofmann, K. (2000) A superfamily of membrane-bound O-acyltransferases with implications for wnt signaling. *Trends Biochem. Sci.* **25**, 111–112.
13. Banziger, C., Soldini, D., Schutt, C., Zipperlen, P., Hausmann, G., and Basler, K. (2006) Wntless, a conserved membrane protein dedicated to the secretion of Wnt proteins from signaling cells. *Cell* **125**, 509–522.
14. Bartscherer, K., Pelte, N., Ingelfinger, D., and Boutros, M. (2006) Secretion of Wnt ligands requires Evi, a conserved transmembrane protein. *Cell* **125**, 523–533.
15. Coudreuse, D. Y., Roel, G., Betist, M. C., Destree, O., and Korswagen H. C. (2006) Wnt gradient formation requires retromer function in Wnt-producing cells. *Science* **312**, 921–924.
16. Logan, C. Y. and Nusse, R. (2004) The Wnt signaling pathway in development and disease. *Annu. Rev. Cell. Dev. Biol.* **20**, 781–810.
17. Panakova, D., Sprong, H., Marois, E., Thiele, C., and Eaton S. (2005) Lipoprotein particles are required for Hedgehog and Wingless signalling. *Nature* **435**, 58–65.
18. Lin, X. (2004) Functions of heparan sulfate proteoglycans in cell signaling during development. *Development* **131**, 6009–6021.
19. Hsiung, F., Ramirez-Weber, F. A., Iwaki, D. D., and Kornberg, T. B. (2005) Dependence of Drosophila wing imaginal disc cytonemes on Decapentaplegic. *Nature* **437**, 560–563.
20. Katoh, M. (2005) WNT/PCP signaling pathway and human cancer (review). *Oncol. Rep.* **14**, 1583–1588.
21. Kohn, A. D. and Moon, R. T. (2005) Wnt and calcium signaling: beta-catenin-independent pathways. *Cell. Calcium* **38**, 439–446.
22. Zeng, X., Tamai, K., Doble, B., Li, S., Huang, H., Habas, R., et al. (2005) A dual-kinase mechanism for Wnt co-receptor phosphorylation and activation. *Nature* **438**, 873–877.
23. Aberle, H., Bauer, A., Stappert, J., Kispert, A., and Kemler, R. (1997) Beta-catenin is a target for the ubiquitin-proteasome pathway. *Embo J.* **16**, 3797–3804.

24. Cavallo, R. A., Cox, R. T., Moline, M. M., Roose, J., Polevoy, G. A., Clevers, H., et al. (1998) Drosophila Tcf and Groucho interact to repress Wingless signalling activity. *Nature* **395**, 604–608.
25. Roose, J., Molenaar, M., Peterson, J., Hurenkamp, J., Brantjes, H., Moerer, P., et al. (1998) The Xenopus Wnt effector XTcf-3 interacts with Groucho-related transcriptional repressors. *Nature* **395**, 608–612.
26. Wallingford, J.B. and Habas, R. (2005) The developmental biology of Dishevelled: an enigmatic protein governing cell fate and cell polarity. *Development* **132**, 4421–4436.
27. Davidson, G., Wu, W., Shen, J., Bilic, J., Fenger, U., Stannek, P., et al. (2005) Casein kinase 1 gamma couples Wnt receptor activation to cytoplasmic signal transduction. *Nature* **438**, 867–872.
28. Bilic, J., Huang, Y. L., Davidson, G., Zimmermann, T., Cruciat, C.M., Bienz, M., et al. (2007) Wnt induces LRP6 signalosomes and promotes dishevelled-dependent LRP6 phosphorylation. *Science* **316**, 1619–1622.
29. Korinek, V., Barker, N., Morin, P. J., van Wichen, D., de Weger, R., Kinzler, K.W., et al. (1997) Constitutive transcriptional activation by a beta-catenin-Tcf complex in APC-/- colon carcinoma. *Science 275,* 1784–1787.
30. van de Wetering, M. and Clevers, H. (1992) Sequence-specific interaction of the HMG box proteins TCF-1 and SRY occurs within the minor groove of a Watson–Crick double helix. *Embo J.* **11**, 3039–3044.
31. Stadeli, R., Hoffmans, R., and Basler, K. (2006) Transcription under the control of nuclear Arm/beta-catenin. *Curr. Biol.* **16**, R378–385.
32. Behrens, J., von Kries, J. P., Kuhl, M., Bruhn, L., Wedlich, D., Grosschedl, R., et al. (1996) Functional interaction of beta-catenin with the transcription factor LEF-1. *Nature* **382**, 638–642.
33. Molenaar, M., van de Wetering, M., Oosterwegel, M., Peterson-Maduro, J., Godsave, S., Korinek, V., et al. (1996) XTcf-3 transcription factor mediates beta-catenin-induced axis formation in Xenopus embryos. *Cell* **86**, 391–399.
34. van de Wetering, M., Cavallo, R., Dooijes, D., van Beest, M., van Es, J., Loureiro, J., et al. (1997) Armadillo coactivates transcription driven by the product of the Drosophila segment polarity gene dTCF. *Cell* **88**, 789–799.
35. Daniels, D. L. and Weis, W. I. (2005) Beta-catenin directly displaces Groucho/TLE repressors from Tcf/Lef in Wnt-mediated transcription activation. *Nat. Struct. Mol. Biol.* **12**, 364–371.
36. Townsley, F. M., Thompson, B., and Bienz, M. (2004) Pygopus residues required for its binding to Legless are critical for transcription and development. *J. Biol. Chem.* **279**, 5177–5183.
37. Hoffmans, R., Stadeli, R., and Basler, K. (2005) Pygopus and legless provide essential transcriptional coactivator functions to armadillo/beta-catenin. *Curr. Biol.* **15**, 1207–1211.
38. Korinek, V., Barker, N., Moerer, P., van Donselaar, E., Huls, G., Peters, P. J., et al. (1998) Depletion of epithelial stem-cell compartments in the small intestine of mice lacking Tcf-4. *Nat. Genet.* **19**, 379–383.
39. Reya, T. and Clevers, H. (2005) Wnt signalling in stem cells and cancer. *Nature* **434**, 843–850.
40. van Es, J. H., Jay, P., Gregorieff, A., van Gijn, M. E., Jonkheer, S., Hatzis, P., et al. (2005) Wnt signalling induces maturation of Paneth cells in intestinal crypts. *Nat. Cell Biol.* 7, 381–386.
41. van de Wetering, M., Sancho, E., Verweij, C., de Lau, W., Oving, I., Hurlstone, A., et al. (2002) The beta-catenin/TCF-4 complex imposes a crypt progenitor phenotype on colorectal cancer cells. *Cell* **111**, 241–250.
42. Van der Flier, L. G., Sabates-Bellver, J., Oving, I., Haegebarth, A., De Palo, M., Anti, M., et al. (2007) The intestinal Wnt/TCF signature. *Gastroenterology* **132**, 628–632.
43. Sansom, O. J., Reed, K. R., Hayes, A. J., Ireland, H., Brinkmann, H., Newton, I. P., et al. (2004) Loss of Apc in vivo immediately perturbs Wnt signaling, differentiation, and migration. *Genes Dev.* **18**, 1385–1390.
44. Batlle, E., Henderson, J. T., Beghtel, H., van den Born, M. M., Sancho, E., Huls, G., et al. (2002) Beta-catenin and TCF mediate cell positioning in the intestinal epithelium by controlling the expression of EphB/ephrinB. *Cell* **111**, 251–263.
45. Morin, P. J., Sparks, A. B., Korinek, V., Barker, N., Clevers, H., Vogelstein, B., et al. (1997) Activation of beta-catenin-Tcf signaling in colon cancer by mutations in beta-catenin or APC. *Science* **275**, 1787–1790.
46. Polakis, P. (2000) Wnt signaling and cancer. *Genes Dev.* **4**, 1837–1851.

Chapter 2

Isolation and Application of Bioactive Wnt Proteins

Karl H. Willert

Abstract

Wnt proteins and their signaling cascades are involved in a wide variety of developmental processes, and deregulation of this pathway is frequently associated with tumorigenesis. Unlike many other growth factors, Wnts long eluded biochemical purification, in large part because of their hydrophobic nature, which is imparted by one or more lipid modifications *(1–3)*. Here I describe a complete protocol that outlines the purification process for Wnt proteins. While this protocol has not been applied to all known Wnt proteins, it has been successfully applied to the purification of a large subset of Wnts, including the very divergent Wnt protein, Drosophila Wnt8 (Dwnt8 or WntD), indicating that this protocol is likely applicable to all Wnts.

Key words: Wnt, Wnt3A, β-catenin, Purification, Blue Sepharose, Immobilized metal affinity chromatography (IMAC), Gel filtration, Heparin cation exchange.

1. Introduction

The protocol described here, based on a previous publication *(3)*, outlines the purification protocol of Wnt proteins, starting with a crude and dilute Wnt sample, usually in the form of conditioned medium (CM) *(4)*, to the final purified protein. In addition, protocols to assay Wnt activity are described.

Two key observations have made the purification of Wnts possible: 1) inclusion of detergent to facilitate solubility of the highly hydrophobic Wnt protein, and 2) fractionation over Blue Sepharose, which binds Wnts with high selectivity. The purification consists of four chromatography steps: Blue Sepharose,

Elizabeth Vincan (ed.), *Wnt Signaling, Volume I: Pathway Methods and Mammalian Models, vol. 468*

Book doi: 10.1007/978-1-59745-249-6

immobilized metal affinity chromatography (IMAC), gel filtration, and heparin cation exchange. Throughout the purification, samples are assayed for both the presence of the Wnt protein (using a Wnt immunoblot or a Coomassie- or silver-stained gel) and activity (β-catenin stabilization or activation of Wnt reporter constructs) to ensure optimal recovery of protein and associated activity.

This protocol describes the purification of Wnt3A specifically; modifications for other Wnt proteins are necessary (e.g., elution profiles may vary; for detection use the appropriate Wnt antibody; not all Wnts stabilize β-catenin; etc.). The resulting purified product is useful in a variety of assays, including signaling studies in established cell lines, explant manipulations, and in vivo experiments (for examples, see refs. *(3, 5–15)*).

2. Materials

2.1. Production of Wnt3A CM

1. Cell line: L-Wnt3A (ATCC, Manassas, VA; ATCC# CRL-2647) (*see* **Note 1**).
2. Cell culture medium: Dulbecco's minimum essential medium (DMEM) supplemented with 10% (v/v) fetal bovine serum (FBS), 1:100 dilution of penicillin–streptomycin solution with 10,000 U penicillin (base)/mL and 10,000 μg streptomycin (base)/mL in 0.85% NaCl (liquid) (*see* **Note 2**).
3. Dulbecco's phosphate-buffered saline (PBS).
4. Trypsin (0.25% liquid trypsin).
5. Cell culture dishes: 10- or 15-cm dishes or large surface area culture dishes (e.g., Corning® CellSTACK®).
6. CO_2 incubator and biosafety cabinet.
7. Filter bottle, 0.5–1 L, 0.2-μm pore size (Corning or equivalent).

2.2. Preparation of Wnt3A CM for Fractionation

1. 20% (v/v) Triton X-100.
2. 1 M Tris-HCl, pH 7.5.
3. 10% (w/v) NaN_3.
4. Filter bottle, 0.5–1 L, 0.2-μm pore size (Corning or equivalent).

2.3. Fractionation of Wnt3A CM

2.3.1 Step 1: Blue Sepharose

1. Sample: 1–4 L Wnt3A CM as prepared in **Section 2.2**.
2. Column: Blue Sepharose HP (Cibacron Blue F3G-A coupled to Sepharose), ~100 mL packed into an empty column with 100–200 mL of bed volume.

3. Binding Buffer: 1% (w/v) CHAPS, 150 mM KCl, and 20 mM Tris-HCl, pH 7.5, sterile filtered.
4. Elution Buffer: 1% (w/v) CHAPS, 1.5 M KCl, and 20 mM Tris-HCl, pH 7.5, sterile filtered.
5. Syringe (30–50 mL) with 0.2-μm pore size filter.

2.3.2. Step 2: IMAC

1. Sample: Pooled Wnt3A containing fractions from **Section 2.3.1**.
2. HiTrap™ Chelating, 1-mL column (GE Healthcare, Piscataway, NJ), loaded with Cu^{2+}.
3. Binding Buffer: 0.5 M NaCl, 20 mM Tris-HCl, pH 7.5, and 1% (w/v) CHAPS.
4. Elution Buffer: Binding Buffer with 100 mM imidazole, pH 7.5.
5. Syringe (10 mL) with 0.2-μm pore size filter.

2.3.3. Step 3: Gel Filtration

1. Sample: 5–10 mL pooled Wnt3A-containing fractions from **Section 2.3.2**.
2. HiLoad 26/60 Superdex 200 preparative grade.
3. Buffer: 1× PBS, 1% (w/v) CHAPS.
4. Syringe (30 mL) with 0.2-μm pore size filter.

2.3.4. Step 4: Heparin Cation Exchange

1. Sample: 20–40 mL pooled Wnt3A-containing fractions from **Section 2.3.3**.
2. 1 mL HiTrap heparin (GE Healthcare).
3. Binding Buffer: 1× PBS, 1% (w/v) CHAPS.
4. Elution Buffer: Binding Buffer, 1 M NaCl (adjust pH to that of Binding Buffer if necessary).
5. Syringe (10 mL) with 0.2-μm pore size filter.

2.4. Assays for Wnt Proteins

2.4.1. Wnt3A Immunoblots

1. Sample: Any fraction containing the Wnt protein of interest.
2. Protein Sample Loading Dye (4×): 250 mM Tris-HCl, pH 6.8, 8% (w/v) sodium dodecyl sulfate (SDS), 40% (v/v) glycerol, 20% (v/v) 2-mercaptoethanol, pinch of bromo phenol blue.
3. Suitable SDS-polyacrylamide gel electrophoresis (PAGE) and transfer setup (e.g., the BioRad Criterion Precast Gel System or equivalent).
4. Nitrocellulose or polyvinyl difluoride (PVDF) membrane.
5. Wnt3A antibody (available from R&D Systems, Minneapolis, MN).
6. Appropriate conjugated secondary antibody for detection.

2.4.2. β-Catenin Stabilization

1. Mouse L-cells (ATCC CCL-1.3 or CRL-2648).
2. Cell culture medium: Same as **Section 2.1.2**.
3. 96-well tissue culture plate (flat well).
4. Wnt3A samples (CM, fractions from **Sections. 2.3** and **2.4**).
5. PBS.
6. Lysis Buffer: 50 mM Tris-HCl, pH 7.5, 150 mM NaCl, and 1% (v/v) Triton X-100.
7. Suitable SDS-PAGE and transfer setup as in **Section 2.4.1**.
8. Protein Sample Loading Dye (4×) as in **Section 2.4.1**.
9. Nitrocellulose or PVDF membrane.
10. Mouse anti-β-catenin antibody (available from various vendors, including BD Transduction Laboratories, Franklin Lakes, NJ; R&D Systems; and Santa Cruz Biotechnology, Santa Cruz, CA).
11. Appropriate conjugated secondary antibody for detection.

3. Methods

3.1. Production of Wnt3A CM

1. Grow L-Wnt3A cells to confluency.
2. Wash with warm (37°C) PBS, trypsinize, and divide the L-Wnt3A cells into plates with a surface area 20 times greater than the original dish(es). Use about 10 mL of cell culture medium per 75-cm^2 surface area).
3. Incubate cells in a humidified CO_2 incubator at 37°C for 4 days.
4. Remove CM and filter through 0.2-μm filter. Add fresh media to all plates and incubate another 3 days.
5. Harvest a second batch of CM. Discard dishes. Filter the second batch and combine with first batch of CM. The CM media can be stored at 4°C for several months without appreciable reduction in activity, as assessed by β-catenin stabilization in L-cells, over the course of 1 year. However, a sensitive and quantitative assay may detect a change in specific activity.

3.2. Preparation of Wnt3A CM for Fractionation

1. To the filtered CM from **Section 3.1**, add Triton X-100 to 1% (v/v), Tris-HCl, pH 7.5, to 20 mM, and NaN_3 to 0.01% (w/v) (e.g., to 930 mL CM, add 50 mL of 20% (v/v) Triton X-100, 20 mL of 1 M Tris-HCl, pH 7.5, and 1 mL of 10% (w/v) NaN_3).
2. Filter through a 0.2-μm filter.

3.3. Fractionation of Wnt3A CM

This fractionation can be performed with any chromatography setup, such as the Äkta FPLC (GE Healthcare). If such an instrument is not available, the purification can also be performed using a peristaltic pump that can accurately control flow rates of 1–5 mL/min. The purification involves four steps: Blue Sepharose, IMAC, gel filtration, and heparin cation exchange, each of which is described in detail below. All buffers and the sample should be sterile filtered through a 0.2-μm filter (*see* **Note 3**).

3.3.1. Step 1: Blue Sepharose

Purpose: This step serves to recover a large fraction of Wnt3A protein from the CM. In an optimal experiment, the Wnt protein can be enriched 2,000- to 2,500-fold (*see* **Notes 4** and **5**).

1. Column Packing: Pour a 1:1 slurry of the Blue Sepharose into a clean and empty column, such as XK26-20 or XK50-20 (GE Healthcare). Allow resin to settle overnight; close the column from the top by lowering and tightening the plunger so that no air is trapped and no gap exists between the top of the resin bed and the plunger.
2. Column Equilibration: Using a suitable pump, wash the column with 2–3 column volumes (CV) of filtered and distilled water or 20 mM Tris-HCl, pH 7.5, at a flow rate of 1–5 mL/min, being sure not to exceed the allowable back pressure for the Blue Sepharose resin. Wash the column with 2–3 CV of binding buffer. If using a UV monitor, adjust it to zero. The column is now ready for sample application.
3. Sample Application: At a flow rate of 1–5 mL/min, apply the entire volume of the Wnt3A CM. The flow rate can be adjusted so that this step can be performed within the span of a few hours or overnight (e.g., a 1 L sample can be applied at 5 mL/min in 3 hours and 20 minutes or at 1 mL/min in 16 hours and 40 minutes). Collect the flow-through material, which can be examined for the presence of Wnt protein by immunoblotting to ensure depletion (*see* **Note 6**). During this loading step, the UV monitor may exceed its detection limit, which is in part due to the high protein content of the sample (10% FBS) and in part due to the Triton X-100, a nonionic detergent containing an aromatic group that exhibits intense UV absorbance.
4. Washing Column: Wash the column with Binding Buffer until the UV reading has stabilized near baseline. This may take four to five CVs. Washing can be done at a flow rate of up to 5 mL/min, as long as the maximal allowed back pressure tolerated by the resin is not exceeded.
5. Elution: Once the UV reading has established a stable baseline, start the elution by switching the buffer to Elution Buffer. Start collecting 10-mL fractions. A large protein peak

will emerge immediately coincident with the increase in salt concentration. Continue to collect fractions for about 100 to 200 mL beyond the main protein peak. Assay all fractions for Wnt3A protein by immunoblotting (*see* **Section 3.4.1**). It is important to note that a large portion of the Wnt3A protein trails behind the main protein peak. These trailing Wnt fractions (Pool 2 in **Fig. 2.1**) contain very little of the other proteins that bind to Blue Sepharose under these conditions. If this separation is not achieved and all the Wnt protein coelutes with the main protein peak, the following fractionation steps will serve to remove the majority of contaminating proteins. Combine the fractions containing the highest amounts of Wnt protein and sterile filter through a 0.2-μm syringe filter. Depending on the amount of starting material and the size of the column, the total volume of pooled eluate fractions will vary from 40 to 100 mL. Store the samples at 4°C before proceeding to the next step. Do not freeze the fractions.

6. Regeneration of column: Wash the column into distilled water and then into 20% (v/v) ethanol for long-term storage. To clean the column rigorously, wash the column in reverse flow

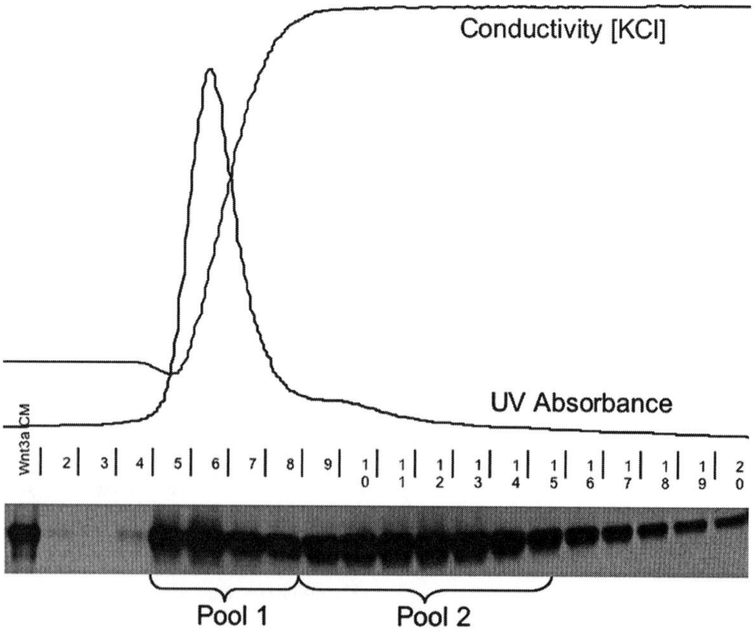

Fig. 2.1. Blue Sepharose fractionation of Wnt3a CM. This step leads to an approximate enrichment of the Wnt protein of 2,000- to 2,500-fold. A large proportion of Wnt3A, detected by immunoblotting, trails behind the major protein peak and consequently contains a higher specific activity of Wnt protein. While the eluate fractions containing Wnt3A can be combined and used for further purification, separating Pool 2 from Pool 1 will yield a purer final Wnt preparation. The Wnt3A protein continues to elute for quite a while, suggesting that in this elution buffer Wnt3A is sticking to Blue Sepharose.

with two to three CVs of 0.1–0.5 N NaOH. This treatment will remove some of the immobilized Cibacron Blue dye from the resin, so contact time should be minimized (*see* **Note 4**).

3.3.2. Step 2: IMAC

Purpose: This step serves to remove contaminating proteins and to concentrate the volume of Wnt3A-containing fractions.

1. Column Equilibration: At a flow rate of 1 mL/min, wash the column with 5 mL distilled and filtered water, and then with 5 mL Binding Buffer. The column is now ready for sample application.
2. Sample Application: Apply the pooled and filtered fractions from Step 1 at a flow rate of 1 mL/min, using a Superloop™ (GE Healthcare) or a sample pump. Collect the flow-through material. Once the entire sample has been applied, wash the column with 5 to 10 CVs of Binding Buffer or until the UV absorbance has established a stable baseline.
3. Elution: Bound protein is eluted by a combination of a step followed by a linear gradient elution (*see* **Fig. 2.2**): 1) step to 5% Elution Buffer/95% Binding Buffer and collect 1-mL fractions for 10 CV (note: a large protein peak should elute as shown in **Fig. 2.2**); 2) elute with a gradient from 5% Elution Buffer/95% Binding Buffer to 100% Elution Buffer over

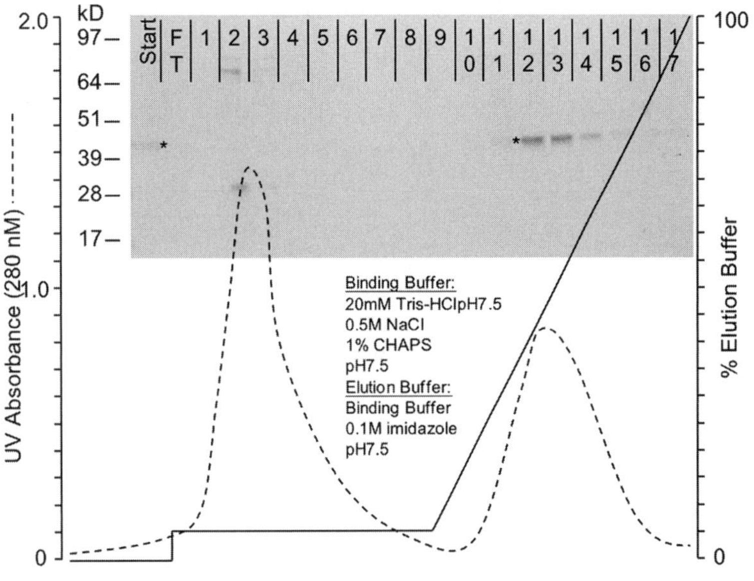

Fig. 2.2. IMAC fractionation of Wnt3A. This step separates Wnt3A from some of the contaminants that co-purify in the Blue Sepharose step. In addition, this step acts to concentrate the samples significantly. After this fractionation step, Wnt3A can be readily detected on a Coomassie-stained gel, which also serves to indicate the purity of the Wnt protein samples. The starting material (*Start*) is Wnt3A eluted from Blue Sepharose; the flow-through material is indicated by *FT*. The *asterisk* indicates the position of the Wnt3A band.

10 CV and collect 1-mL fractions; 3) collect an additional ten 1-mL fractions at 100% Elution Buffer. Verify the presence of Wnt3A by immunoblotting (*see* **Section 3.4.1**). The vast majority of Wnt3A should be present in fractions collected during the gradient from 5 to 100% Elution Buffer. The bound Wnt protein elutes in a broad peak, totaling about 5 mL, at a concentration of approximately 10–40 mM imidazole. If sufficient Wnt3A CM is fractionated (>1 L), a Coomassie stained gel should be sufficient to allow detection of the Wnt protein in the eluate fractions (as shown in **Fig. 2.2**). Combine fractions containing highest amounts of Wnt protein and sterile filter through a 0.2-µm syringe filter. The sample can be stored at 4°C before continuing with the purification. Do not freeze the fractions.

3.3.3. Step 3: Gel Filtration

Purpose: This step serves to separate low molecular weight Wnt from high molecular weight Wnt and some contaminating proteins. Additionally, it serves to exchange the buffer from high (0.5 M) to physiological (1× PBS) salt concentrations, which is important to perform Step 4 of this purification.

1. Column Equilibration: At a flow rate of 1–2.5 mL/min (be sure not to exceed the maximal back pressure permissible for this column), wash the column with one CV distilled and filtered water, then with two CV Buffer. The column is now ready for sample application.
2. Sample Application and Fractionation: Load the sample using a Superloop™ at a flow rate of 1 to 2.5 mL/min. If the sample exceeds the maximum volume that can be efficiently fractionated on this column (consult the vendor's recommendations for details), split the sample and perform two identical fractionations. Once the entire sample has been applied, collect 10-mL fractions for an entire CV. The majority of the Wnt protein should emerge at the same position as a molecular weight standard of 50 kD (*see* **Fig. 2.3**). Occasionally a small amount of Wnt protein elutes with the void volume of the column suggesting that it is part of a large complex or aggregate (*see* **Note 7**). The Wnt protein can be readily detected on a Coomassie-stained gel. Combine all Wnt-containing fractions and pass through a 0.2-µm syringe filter.

3.3.4. Step 4: Heparin Cation Exchange

Purpose: This fractionation step serves to concentrate the Wnt protein and remove remaining contaminating proteins, such as bovine serum albumin (BSA).

1. Column Equilibration: Wash the column with 5 mL distilled and filtered water at a flow rate of 1 mL/min, then with 5 mL Binding Buffer. The column is now ready for sample application.

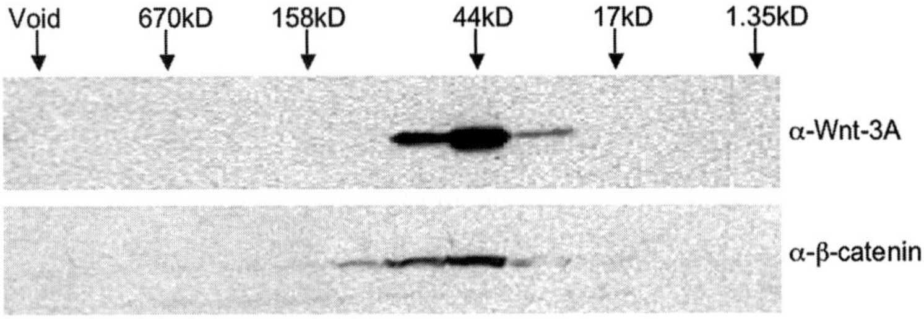

Fig. 2.3. Gel filtration of Wnt3A and β-catenin assay. The elution peak positions of molecular weight standards on the Superdex 200 column are indicated at the top. Wnt3A has an approximate molecular weight of 50 kD (*top panel*). Eluate fractions were diluted 1:100 in DMEM with 10% FBS and applied to mouse L cells for a β-catenin stabilization assay (bottom panel, *see* **Section 3.4.2**). Only fractions containing Wnt3A stimulate the stabilization of β-catenin.

2. Sample Application: Load the entire sample from **Section 3.3.3** and collect the flow-through material. Upon loading, wash the column with 5 to 10 CV or until UV absorbance has established a stable baseline.
3. Elution: Elute the bound Wnt protein by applying a linear gradient from 0 to 100% Elution Buffer. The Wnt protein elutes in a tight peak at approximately 200 mM NaCl (i.e., 20% Elution Buffer = PBS + 200 mM NaCl). Alternatively, the Wnt protein can be eluted in a single step from 0 to 100% Elution Buffer. The maximum concentration of Wnt that can be achieved in these buffer conditions is approximately 100 μg/mL (2.5 μM for Wnt3A). The Wnt protein can be readily detected on a Coomassie-stained gel. Combine all Wnt-containing fractions, filter through a 0.2-μm syringe filter, and store (*see* **Note 8**).

3.4. Assays for Wnt Proteins

There are three simple assays for Wnt3A: 1) Wnt3A immunoblot, 2) β-catenin stabilization in L-cells, and 3) Super-TOP-Flash reporter assays. Here I describe the Wnt3A immunoblot and the β-catenin assay. For details on the reporter assay, *see* **Chapter 8 (Volume 1)** of this book. Wnt proteins that do not lead to β-catenin stabilization, such as Wnt5A, can be assayed for their ability to inhibit Wnt3A-stimulated Super-TOP-Flash reporter assays *(11)*. Additional biochemical assays to monitor Wnt activity(ies) are described elsewhere.

3.4.1 Wnt3A Immunoblotting

1. Prepare samples by combining a small amount of a Wnt sample with 4× loading dye and water to give a final concentration of the loading dye of 1× in a total volume of 20 to 40 μL. Use the following amounts of Wnt per 20 μL sample: 1) Wnt3A CM: 1 μL; 2) Blue Sepharose fractions: 1 μL; 3) IMAC fractions: 1 μL for Wnt3A immunoblot (or 10 μL for

Coomassie-stained gel); 4) gel filtration fractions: 2 μL for Wnt3A immunoblot (or 15 μL for Coomassie-stained gel); 5) heparin fractions: 1 μL for Wnt3A immunoblot or Coomassie-stained gel.

2. Boil all samples for 5 minutes.
3. Perform SDS-PAGE (10% acrylamide gel) followed by transfer to a suitable membrane, such as nitrocellulose or PVDF using standard protocols.
4. Detect Wnt3A protein using a Wnt3A-specific antibody according to manufacturer's recommendations.

3.4.2. β-Catenin Stabilization

1. One or 2 days prior to the assay, seed a 96-well plate with L-cells (*see* **Note 9**). Seeding one tenth of the cells from a confluent 10-cm dish of L-cells into all 96 wells will give the desired cell density to perform the assay. Most of the following manipulations can be performed with 12-well multichannel pipetters.
2. Prepare samples to be tested in a separate 96-well plate: To 100 μL of cell culture media, add 1–2 μL of the Wnt3A-containing fractions (*see* **Note 2**).
3. Aspirate media from L-cells in 96 well plates.
4. Transfer diluted Wnt samples to the cells. For a positive control, use 100 μL of Wnt3A CM per well and for a negative control use cell culture medium (or CM from L-cells prepared as described for Wnt3A CM in **Section 3.1**).
5. Incubate cells for 2 hours at 37°C in a humidified CO_2 incubator.
6. Aspirate all control and test wells and wash once with 100 μL of PBS.
7. Add 30 μL of Lysis Buffer, incubate 1–2 minutes, and transfer lysates (do not try to dislodge nuclei by scraping) to tubes containing 10 μL of Protein Sample Loading Dye.
8. Boil and resolve protein samples by SDS-polyacrylamide gel electrophoresis.
9. Transfer protein from the gel to a suitable membrane and perform a β-catenin immunoblot as recommended by manufacturer. An example of this β-catenin stabilization assay is shown in **Fig. 2.3.**

4. Notes

1. This purification can be applied to other Wnt proteins: to date, Wnt3A, 5A, 7A, 16, Wingless, and Dwnt8/WntD have been successfully purified using this scheme *(3, 11)*. Aside from the

cells producing Wnt3A, ATCC makes available a similar cell line overexpressing Wnt5A (ATCC# CRL-2814).

2. Lowering FBS concentrations in the media will lead to a reduction of soluble Wnt3A in the CM. Since Wnt proteins are extremely hydrophobic, it is likely that the high lipid content of serum promotes the solubility of the Wnt protein. Upon fractionation of the Wnt CM, the requirement for serum is displaced by the presence of detergent. If purified Wnt3A is diluted into an aqueous buffer, either detergent or 10% FBS should be included. In fact, dilution of purified Wnt into a buffer lacking either detergent or serum will lead to precipitation and loss of activity.
3. The purification has been successfully performed at 4°C and room temperature. However, I have not carefully examined whether one of these temperatures yields a superior purified Wnt protein preparation. Given common wisdom of protein stability and temperature, I recommend that the purification is performed at 4°C and all samples are stored in the cold. I recommend against freezing any samples, from the starting material to the final purified protein fractions, as this may result in protein denaturation and precipitation.
4. While Blue Sepharose binds all the Wnts tested so far, I recommend that a given column is dedicated to the purification of a single Wnt protein. Wnt proteins are extremely sticky and are likely present at low levels even after washing the column extensively.
5. During the Blue Sepharose fractionation, the detergent is exchanged from 1% Triton X-100 to 1% CHAPS. This is done because Triton X-100 is significantly less expensive than CHAPS. However, Triton X-100 is not a suitable detergent for the entire purification because of its low critical micelle concentration (0.0155% w/v) relative to CHAPS (0.492–0.615% w/v) and its cellular toxicity. In addition, in the presence of Triton X-100, Wnt3A is incorporated into high molecular weight complexes (potentially micelles) as assessed by gel filtration, and is significantly less active.
6. The concentration of Wnt3A in CM is approximately 200 ng/mL. Blue Sepharose has a binding capacity of approximately 0.64 mg Wnt3A protein per 100 mL Blue Sepharose HP. This estimation is derived from these observations: the maximal amount of Wnt3A CM loaded onto a 100-mL Blue Sepharose column has been 4 L (~0.8 mg Wnt3A) and 80% of Wnt3A was depleted, indicating that 0.64 mg of Wnt3A was bound. Flow-through fractions collected at the end of the loading step contained significantly more Wnt3A protein than flow-through fractions collected at the beginning, suggesting

that the column was approaching saturation with respect to Wnt binding.

7. By gel filtration, the majority of Wnt3A should emerge as a protein whose molecular size corresponds closely to its monomeric or slightly larger size. A high molecular weight complex containing Wnt3A is often observed. This high molecular weight Wnt is most likely aggregated protein, because 1) it is less active than low molecular weight Wnt3A, 2) it cannot be readily dissociated with detergent, and 3) it can be pelleted by centrifugation (50,000×*g*).

8. The stability and shelf life of purified Wnt protein, specifically Wnt3A, is not known. However, it is clear that purified Wnt protein is unstable and looses activity over time. I have found that after 6 months of storage at 4°C, only 10% activity will remain. Therefore, it is best to use the Wnt protein shortly after purification. Preliminary studies indicate that the purified Wnt protein can be flash frozen in liquid nitrogen and stored at –150°C without appreciable loss in activity. While it has not been tested experimentally, it is likely that stored under such conditions, the Wnt protein will retain maximal activity indefinitely. Avoid repeated freeze–thaws by freezing down multiple small volume aliquots. Some Wnt proteins (Wnt3a, 5a, 5b, and 7a) are provided by R&D Systems in a lyophilized form, suggesting that lyophilization does not adversely affect the activity of the Wnt protein.

9. Mouse L cells lack cadherins and consequently have very little to no membrane-associated β-catenin. As a result, the Wnt-stimulated cytoplasmic accumulation of β-catenin is readily detectable. If cells with cadherins are used in place of L cells, the Wnt-stimulated cytoplasmic accumulation of β-catenin may be masked by the large amounts of β-catenin at the membrane. In this case, cells should be fractionated to separate cytosolic proteins from membrane-bound proteins (e.g., by lysis in hypotonic buffer without detergent).

References

1. Kurayoshi, M., Yamamoto, H., Izumi, S., and Kikuchi, A. (2007) Post-translational palmitoylation and glycosylation of Wnt-5a are necessary for its signalling. *Biochem J.* **402**, 515–523.
2. Takada, R., Satomi, Y., Kurata, T., Ueno, N., Norioka, S., Kondoh, H., et al. (2006) Monounsaturated fatty acid modification of Wnt protein: its role in Wnt secretion. *Dev Cell*, **11**, 791–801.
3. Willert, K., Brown, J. D., Danenberg, E., Duncan, A. W., Weissman, I. L., Reya, T., et al. (2003) Wnt proteins are lipid-modified and can act as stem cell growth factors. *Nature*, **423**, 448–452.
4. Shibamoto, S., Higano, K., Takada, R., Ito, F., Takeichi, M., and Takada, S. (1998) Cytoskeletal reorganization by soluble Wnt-3a protein signalling. *Genes Cells*, **3**, 659–670.

5. Bryja, V., Schulte, G., and Arenas, E. (2007) Wnt-3a utilizes a novel low dose and rapid pathway that does not require casein kinase 1-mediated phosphorylation of Dvl to activate beta-catenin. *Cell Signal*, **19**, 610–616.
6. Bryja, V., Schulte, G., Rawal, N., Grahn, A., and Arenas, E. (2007) Wnt-5a induces Dishevelled phosphorylation and dopaminergic differentiation via a CK1-dependent mechanism. *J Cell Sci*, **120**, 586–595.
7. Gadue, P., Huber, T. L., Paddison, P. J., and Keller, G. M. (2006) Wnt and TGF-beta signaling are required for the induction of an in vitro model of primitive streak formation using embryonic stem cells. *Proc Natl Acad Sci U S A*, **103**, 16806–16811.
8. Galli, L. M., Barnes, T., Cheng, T., Acosta, L., Anglade, A., Willert, K., et al. (2006) Differential inhibition of Wnt-3a by Sfrp-1, Sfrp-2, and Sfrp-3. *Dev Dyn*, **235**, 681–690.
9. Galli, L. M., Willert, K., Nusse, R., Yablonka-Reuveni, Z., Nohno, T., Denetclaw, W., et al. (2004) A proliferative role for Wnt-3a in chick somites. *Dev Biol*, **269**, 489–504.
10. Kishida, S., Yamamoto, H., and Kikuchi, A. (2004) Wnt-3a and Dvl induce neurite retraction by activating Rho-associated kinase. *Mol Cell Biol*, **24**, 4487–4501.
11. Mikels, A. J. and Nusse, R. (2006) Purified Wnt5a protein activates or inhibits beta-catenin-TCF signaling depending on receptor context. *PLoS Biol*, **4**, e115.
12. Reya, T., Duncan, A. W., Ailles, L., Domen, J., Scherer, D. C., Willert, K., et al. (2003) A role for Wnt signalling in self-renewal of haematopoietic stem cells. *Nature*, **423**, 409–414.
13. Schmitt, A. M., Shi, J., Wolf, A. M., Lu, C. C., King, L. A., and Zou, Y. (2006) Wnt-Ryk signalling mediates medial-lateral retinotectal topographic mapping. *Nature*, **439**, 31–37.
14. Schulte, G., Bryja, V., Rawal, N., Castelo-Branco, G., Sousa, K. M., and Arenas, E. (2005) Purified Wnt-5a increases differentiation of midbrain dopaminergic cells and dishevelled phosphorylation. *J Neurochem*, **92**, 1550–1553.
15. Yue, Z., Jiang, T. X., Widelitz, R. B., and Chuong, C. M. (2006) Wnt3a gradient converts radial to bilateral feather symmetry via topological arrangement of epithelia. *Proc Natl Acad Sci U S A*, **103**, 951–955.

Chapter 3

Purification and Wnt-Inhibitory Activities of Secreted Frizzled-Related Proteins

Vladimir Wolf, Yoshimi Endo, and Jeffrey S. Rubin

Abstract

Recombinant expression of secreted Frizzled-related proteins (sFRPs) in mammalian expression systems is a convenient source of these proteins for biological studies. Yields of protein vary; screening of clonal lines for high expression is usually worthwhile. Heparin affinity chromatography is an easy step that provides a major enrichment, particularly for sFRP-1 and sFRP-2. Alternatively, sFRP derivatives tagged with poly-histidine at their carboxyl termini are functional and can be readily isolated by chelating chromatography. Once purified, the proteins are stable indefinitely if stored frozen and they tolerate multiple rounds of freeze–thawing. Pre-incubation of Wnt samples with sFRP protein for 30 min at 37°C is sufficient to inhibit Wnt activity in various assays. The concentration of sFRP required to block Wnt signaling should be determined empirically, as it will vary with the Wnt preparation and cellular context.

Key words: sFRP, Recombinant expression, Protein purification, Heparin chromatography, Chelating chromatography, Wnt.

1. Introduction

Secreted Frizzled-related proteins (sFRPs) contain a Frizzled (Fzd)-type, cysteine-rich domain (CRD) that serves as a binding site for Wnts as well as other sFRPs and Fzds *(1)*. Much evidence indicates that sFRPs can function as Wnt antagonists, presumably by sequestering Wnts or interacting with Fzds to disrupt ligand-dependent receptor signaling *(1, 2)*. However, additional activities of sFRPs have been described *(2, 3)*, with one report suggesting that sFRP-1 itself is a ligand for Fzd-2 that regulates axonal extension *(4)*.

Elizabeth Vincan (ed.), *Wnt Signaling, Volume I: Pathway Methods and Mammalian Models, vol. 468*

Book doi: 10.1007/978-1-59745-249-6

Although some members of the sFRP family have been isolated from whole tissue or cell culture *(5, 6)*, relatively small quantities have been obtained in this manner. There are no reports of recombinant expression in prokaryotic systems, presumably because of the folding problems associated with production of disulfide-rich proteins in bacteria. A Fzd-type CRD has been successfully produced in yeast *(7)*; however, we are not aware of similar results for full-length sFRPs. Alternatively, mammalian expression systems have proven to be a satisfactory source of recombinant, biologically active sFRP proteins *(8, 9)*. Typically, yields can be optimized by screening clonally derived cell lines for maximal release of the desired protein into conditioned medium. While various purification strategies are feasible, heparin chromatography is a straightforward procedure that does not require sophisticated equipment and often provides purified material suitable for biological studies. Once isolated, sFRP proteins can be stored frozen for use over a period of several years.

2. Materials

The following description presupposes that the reader has access to complementary DNAs (cDNAs) encoding the sFRPs of interest or cell lines already transfected with an appropriate sFRP expression vector. We inserted sFRP coding sequences into pcDNA3.1 vectors (Invitrogen, Carlsbad, CA), which contain a promoter from cytomegalovirus and antibiotic resistance elements for selection. Information about other expression vectors is available from commercial sources and the scientific literature. For simplicity, we outline the scheme used to purify sFRP-1. Minor adjustments, as indicated below, also gave satisfactory results for sFRP-2. Experience with sFRP-3 and sFRP-4 is shared in the Notes.

2.1. Cell Culture for Conditioned Medium

1. sFRP-1/MDCK (Madin-Darby canine kidney) and sFRP-2/MDCK clonally derived, transfectant cell lines.
2. Dulbecco's modified Eagle's medium (DMEM) (GIBCO; Invitrogen), with and without 10% (v/v) fetal calf serum (Biosource; Invitrogen).
3. Phosphate-buffered saline (PBS): 1.54 mM KH_2PO_4, 155.17 mM NaCl, and 2.71 mM $NaHPO_4$, pH 7.2.
4. 0.4-μm Filter.

2.2. Heparin Chromatography

1. Hi-Trap heparin affinity column (1.0-mL bed volume, GE Healthcare, Piscataway, NJ).
2. Peristaltic pump (for instance, Econo Pump, Bio-Rad, Hercules, CA).

3. Solution A: 0.05 M sodium phosphate, pH 7.4 (4×). Prepare 0.2 M stock sodium phosphate monobasic solution (13.8 g $NaH_2PO_4 \cdot H_2O$ dissolved in 0.5 L distilled H_2O) and 0.2 M stock sodium phosphate dibasic solution (53.65 g $Na_2HPO_4 \cdot 7H_2O$ or 28.4 g of the anhydrous form dissolved in 1 L distilled H_2O); combine 57 mL of the former with 243 mL of the latter.
4. Solution B: 0.05 M sodium phosphate, pH 7.4, and 2.0 M NaCl. Dissolve 116.8 g NaCl in 1 L Solution A.
5. Elution buffers: NaCl stepwise elution buffers are generated by combining Solutions A and B in various ratios (for instance, 5% Solution B for 0.1 M NaCl; 15% Solution B for 0.3 M NaCl; 25% Solution B for 0.5 M NaCl; 35% Solution B for 0.7 M NaCl; 50% solution B for 1.0 M NaCl; and 60% solution B for 1.2 M NaCl).

2.3. Chelating Chromatography

1. Hi-Trap chelating affinity column (1.0-mL bed volume, GE Healthcare).
2. 0.1 M nickel sulfate ($NiSO_4$).
3. Equilibration solution: 0.05 M $NaPO_4$ and 0.01 M imidazole, pH 7.4.
4. Elution solutions: 0.05 M $NaPO_4$ and 0.05 M imidazole, pH 7.4; 0.05 M $NaPO_4$ and 0.1 M imidazole, pH 7.4.

2.4. Sodium Dodecyl Sulfate (SDS)-Polyacrylamide Gel Electrophoresis (PAGE)

1. Mini-gel electrophoresis apparatus and power supply.
2. Pre-poured SDS-polyacrylamide mini-gels (Bio-Rad or Invitrogen); store refrigerated.
3. Running buffer: 25 mM Tris, 192 mM glycine, and 0.1 % (w/v) SDS. Store at room temperature.
4. Pre-stained molecular mass markers: Kaleidoscope markers (Bio-Rad).
5. Coomassie Brilliant Blue (CBB) staining solution: 0.025 % (w/v) CBB (Pierce), 40 % (v/v) methanol, and 7 % (v/v) acetic acid. Dissolve CBB in methanol by stirring, followed by filtration. Then add acetic acid and bring to final volume with distilled water.
6. Destaining solution: 30% (v/v) methanol and 10% (v/v) acetic acid.
7. Gel drying solution: 50% (v/v) methanol and 10% (v/v) acetic acid.
8. Cellophane wrap.

2.5. Immunoblotting

1. Mini-gel transfer apparatus to match electrophoresis equipment.
2. Transfer buffer: 25 mM Tris, 192 mM glycine, and 10% (v/v) methanol, pH 8.3.
3. Gel blot paper.

4. Polyvinylidene difluoride (PVDF) membrane (Millipore, Billerica, MA).
5. Methanol.
6. Tris-buffered saline (TBS): 10 mM Tris-HCl and 137 mM NaCl, pH 7.4; add 0.025% (w/v) sodium azide for long-term storage at room temperature.
7. Blocking solution: 5% (w/v) non-fat dry milk in 0.05% (v/v) Tween-20 in TBS (T-TBS).
8. Primary antibodies to sFRPs: we generated multiple rabbit polyclonal sFRP-1 antisera directed against either the full length protein or peptide segments; alternatively, one can use commercially available reagents such as a goat polyclonal antiserum (sc-7425; Santa Cruz Biotechnology, Santa Cruz, CA).
9. Secondary antibody to use with goat polyclonal reagent: donkey anti-goat IgG-horseradish peroxidase (HRP) (sc-2056; Santa Cruz Biotechnology).
10. Secondary antibody to use with rabbit polyclonal reagent: donkey anti-rabbit IgG-HRP (#NA934V, GE Healthcare).
11. Enhanced chemiluminescent reagents (Femto Super Signal; Pierce, Rockford, IL).
12. X-Omat film (Kodak, Rochester, NY).

2.6. Wnt Activity Assays

1. β-catenin primary antibody: clone 4 (#610153; BD Biosciences, San Jose, CA).
2. Dishevelled (Dvl)-2 primary antibody: clone 10B5 (sc-8026; Santa Cruz Biotechnology).
3. Dvl-3 primary antibody: clone 4D3 (sc-8027; Santa Cruz Biotechnology).
4. Secondary antibody to use with mouse monoclonal primary antibodies: sheep anti-mouse IgG-HRP (#NA931V, GE Healthcare).
5. L929 fibroblasts (ATCC, Manassas, VA), RIMM-18 *(10)* and HEK293 cells (ATCC); Wnt-3a/L929 transfectant (ATCC line, conditioned medium routinely contains serum; we generated another Wnt-3a/L929 transfectant to collect serum-free conditioned medium *(11)*).

3. Methods

When expressing recombinant proteins, the choice of vector has a major impact on purification strategy, yield, and utility of the product. Vectors that link a poly-histidine tag to the recombinant protein typically simplify purification by enabling chelating (nickel) chromatography to serve as a major enrichment step. In addition, an epitope tag facilitates monitoring of the protein during purification, if antibody reagents that recognize the native sequence are not available. Epitope-tagged reagents are particularly useful for co-precipitation experiments, as an antibody to the foreign epitope is less likely to interfere with protein–protein interactions mediated by the native sequence. Moreover, incorporation of the same epitope into a series of recombinant proteins provides a way to normalize their concentrations when only partially purified preparations are available. However, the addition of an epitope would be problematic if it disrupted protein folding and/or interactions with other molecules.

With regard to sFRPs, heparin chromatography is an efficient purification step, decreasing the benefit of a poly-histidine tag, and antibodies to the native proteins are commercially available. Therefore, we will describe a method for purification of recombinant proteins that lack additional sequence. Nonetheless, it is worth noting that sFRPs with tags linked to their carboxyl termini retain functional activity *(8, 12)*. Considering the potential applications of these derivatives, we also include a protocol for their isolation by chelating chromatography. For recombinant expression of sFRPs, we made a serendipitous observation that MDCK cells released a surprisingly large amount of sFRP-1 and sFRP-2 into their culture fluid (yielding milligrams per liter quantities of purified material) (*see* **Note 1**). Moreover, MDCK cells are hardy and form extremely adherent monolayers, allowing for the harvesting of multiple rounds of conditioned serum-free media. The following description outlines the protocol we developed with the MDCK recombinant expression system (*see* **Note 2**).

3.1. Preparation of Conditioned Medium for Protein Purification

1. Grow sFRP-1/MDCK transfectants in T175 flasks until monolayers are confluent (*see* **Notes 3** and **4**). After washing with PBS, maintain the cells in 25 or 30 mL of serum-free DMEM, and collect conditioned media every 3 days for five to seven harvests. MDCK monolayers can be cycled from serum-free to serum-containing DMEM for 3-day intervals to prolong cell viability and protein synthesis (after aspirating serum-containing medium, rinse monolayers with PBS before adding serum-free medium for conditioning).
2. Clarify conditioned medium by centrifugation at 10,000×*g* for 10 min at 4°C and filtration (*see* **Note 5**).
3. Place samples on dry ice for rapid freezing, and store them at –20°C or lower temperature for subsequent purification.

3.2. Heparin Chromatography

1. Thaw conditioned medium and clarify by centrifugation at 10,000×*g* for 10 min at 4°C, followed by filtration.

2. Equilibrate heparin resin with 10 column volumes of equilibration solution (0.05 M phosphate buffer, pH 7.4, and 0.1 M NaCl). For samples derived from as much as 2 L of conditioned medium, use a 1-mL bed volume and a flow rate of 1 mL/min. Chromatography can be performed either in a cold room or at room temperature.
3. Load sample on resin and collect flow through (fraction that is not retained on resin) for subsequent immunoblotting (*see* **Note 6**).
4. Wash resin with equilibration solution for ~10 column volumes, or until optical density of effluent is near background of equilibration buffer (*see* **Note 7**).
5. Elute retained protein with solutions containing stepwise increases in NaCl concentration. For sFRP-1, elute sequentially with 0.05 M phosphate buffer, pH 7.4, solutions containing 0.3, 0.7, 1.0, and 1.2 M NaCl; 5–10 column volumes per step, and collect fractions corresponding to half or one column volume at the peak (0.5 or 1.0 mL). sFRP-1 is recovered with 1.2 M NaCl (**Fig. 3.1**). For sFRP-2, elute with a slightly modified

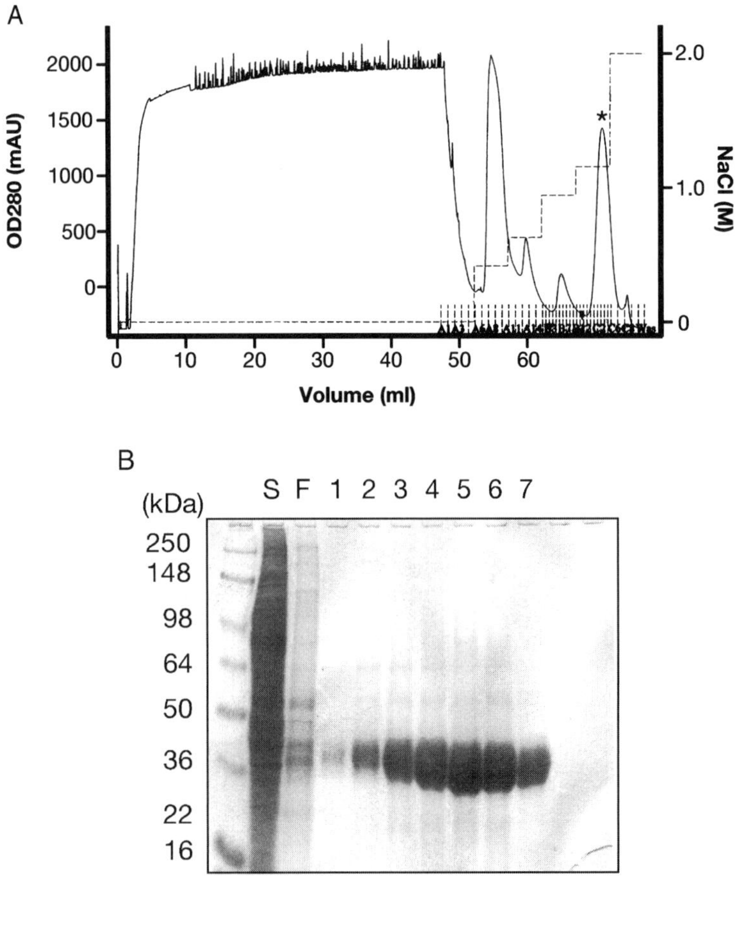

set of solutions, containing 0.3, 0.5, 0.8, and 1.0 M NaCl. sFRP-2 emerges with 0.8 M NaCl (**Fig. 3.2**) (*see* **Note 8**).

6. Remove aliquots of fractions expected to contain sFRP protein for analysis by immunoblotting and protein staining (*see* **below**). Snap-freeze the remainder of these fractions and store

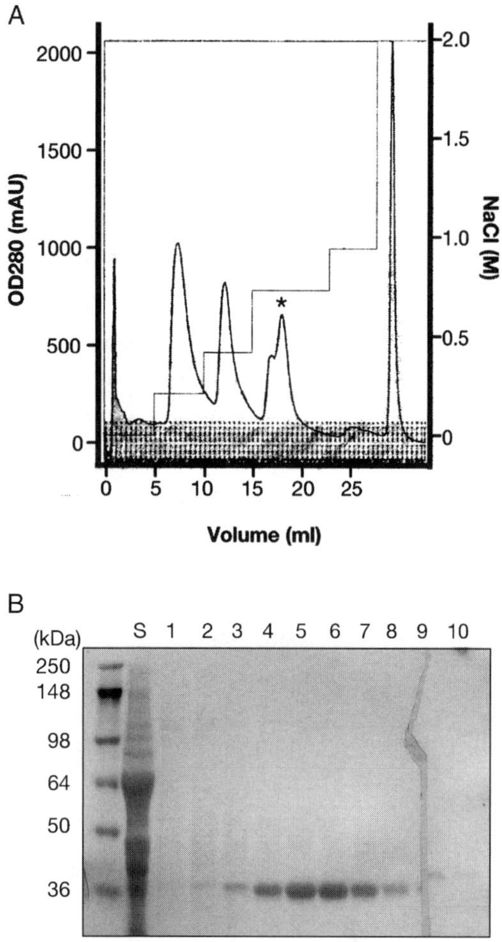

Fig. 3.2. Purification of recombinant sFRP-2. **A** Concentrated conditioned medium from sFRP-2/MDCK cells was applied to Hi-Trap heparin resin (bed volume, 1 mL). Samples were eluted with increasing NaCl concentration (*dashed line*) and protein content was determined by monitoring optical density at 280 nm (*solid line*). Fractions are indicated on the horizontal axis. The *asterisk* indicates the peak containing sFRP-2. **B** Equal volumes (10 μL) of selected fractions eluted with 0.8 M NaCl were analyzed by 12% SDS-PAGE, and visualized by staining with CBB. S: starting material.

Fig. 3.1. Purification of recombinant sFRP-1. **A** Concentrated conditioned medium from sFRP-1/MDCK cells was applied to Hi-Trap heparin resin (bed volume, 1 mL). Samples were eluted with increasing NaCl concentration (*dashed line*) and protein content was determined by monitoring optical density at 280 nm (*solid line*). Fractions are indicated on the horizontal axis. The *asterisk* indicates the peak containing sFRP-1. **B** Equal volumes (10 μL) of selected fractions eluted with 1.2 M NaCl were analyzed by 12% SDS-PAGE, and visualized by staining with CBB. S: starting material; F: flow through.

them in a freezer. Subsequently, peak fractions can be divided into aliquots and refrozen. Once the purity of peak fractions is confirmed, protein concentration can be estimated from optical density (sFRP-1 and sFRP-2 extinction coefficients at 280 nm are 0.968 and 0.871, respectively, for 1 mg/mL solutions at neutral pH) (*see* **Note 9**).

3.3. Chelating Affinity Chromatography

1. As with heparin chromatography, freshly prepared or thawed medium (concentrated or unconcentrated) should be clarified by centrifugation and filtration to minimize the likelihood of fouling the resin when the sample is applied.
2. Wash chelating resin (1.0 mL) in a column with 5.0 mL of water (flow rate of 1.0 mL/min).
3. Charge resin with 0.5 mL of 0.1 M $NiSO_4$.
4. Wash resin again with 5.0 mL distilled water.
5. Equilibrate resin with 5–10 mL of 0.05 M phosphate buffer with 0.01 M imidazole, pH 7.4.
6. Load sample on resin, then wash with 5–10 mL of equilibration buffer.
7. Elute protein with eluants containing stepwise increases in imidazole; typically use buffers that have 0.05 and 0.10 M imidazole, with sFRPs eluting at the higher concentration.
8. Analyze and store protein as indicated above for heparin chromatography.

3.4. SDS-PAGE, Followed by Protein Staining

1. Combine a standard-size aliquot (typically 5–10 μL) of eluted fractions with Laemmli sample buffer, and boil at 100°C for 10 min to denature proteins.
2. Load samples in adjacent lanes of an SDS polyacrylamide gel, along with a lane containing a set of molecular mass standards. Typically we use a pre-poured 12% or 4–20% polyacrylamide mini-gel.
3. After clamping the gel into the electrophoresis apparatus, add Tris–glycine–SDS running buffer into the upper and lower reservoirs.
4. Turn on the power supply, setting at constant voltage (140 V), and run for ~1.5 h at room temperature, until the dye front reaches the bottom of the gel.
5. After electrophoresis, transfer the gel to a clean glass tray and rinse the gel 3× 5 min with 100–200 mL of ultrapure water.
6. Replace water with CBB staining solution, and gently rock the tray (either manually or with a platform rocker). Maximal staining intensity usually is achieved within 15–30 min, although the gel can be stained overnight (*see* **Notes 10** and **11**).

7. After staining/destaining has been completed, it is advisable to dry the gel for scanning and/or long-term storage. After multiple washes with ultrapure water using the rocker, incubate the gel in gel drying solution for 15–20 min in a closed container on the rocker.
8. Prepare two square pieces of cellophane (each larger than the gel) and immerse them, one at a time, in gel-drying solution for 15–20 sec; ensure that both sides of each sheet are thoroughly wetted (wear gloves during this procedure).
9. Place the side of the gel-drying frame with holes at the corners onto a flat surface, and center one of the sheets of prewetted cellophane on top of it.
10. Lay the gel on the cellophane, so that the gel is positioned inside the boundaries of the frame. Make sure no bubbles are trapped between the cellophane and the gel.
11. Place the second piece of cellophane over the gel, again avoid any bubbles between gel and cellophane. Eliminate any wrinkles by smoothly rubbing the assembly with a gloved hand.
12. Align the other side of the gel-drying frame such that the corner pins fit into the holes in the bottom side, and the position plastic clamps at the four corners to secure the assembly.
13. Put the assembly into an upright, vertical position and let it remain standing until the gel is completely dry (12–36 h).
14. Remove the cellophane/gel sandwich from the frame, trim away the excess cellophane, and store the sandwich between the pages of a notebook under light pressure for ~2 days. The gel now should remain flat, suitable for scanning and long-term storage.

3.5. Immunoblotting

1. Remove the gel cassette from the electrophoresis unit and rinse it with distilled water.
2. Briefly submerge the gel in chilled transfer buffer.
3. Cut out two rectangular pieces of gel blot paper, each exceeding the length and width of the gel by ~1.5 cm; cut out a portion of polyvinylidene fluoride (PVDF) membrane that closely matches the size of the gel.
4. Wet the PVDF membrane in 100% methanol for 10 sec, then immerse the membrane in distilled water; soak the paper and two foam sheets in transfer buffer.
5. Assemble a multi-layered sandwich of components in the transfer cassette: wet foam sheet, paper, PVDF membrane, gel, paper, and second wet foam sheet; remove any bubbles from this assembly.
6. Place this cassette in the transfer apparatus, with the membrane between the gel and the anode (this orientation is

essential to ensure transfer of proteins onto membrane). Fill the tank with transfer buffer, and put the lid on the tank.

7. Turn on the power supply, and run transfer at constant current (0.4 A) for 1 h. Pre-stained molecular weight markers should be clearly visible on the membrane.
8. Incubate the membrane in blocking solution for 1 h at room temperature on a platform rocker.
9. Remove blocking solution, and incubate the membrane in primary antibody solution (use goat anti-sFRP-1 reagent at a 1:500 dilution) overnight (~16 h) at 4°C on a rocking platform.
10. Remove primary antibody solution, and wash membrane 4× 4 min with TTBS at room temperature.
11. After the last wash, incubate the membrane in solution containing HRP–secondary antibody conjugate (donkey anti-goat IgG-HRP at a 1:1,000 dilution) for 1 h at room temperature; follow with 4× 4 min washes in TTBS.
12. Combine peroxide buffer and enhancer solutions for electrochemiluminescence (ECL) detection just prior to use for maximal effectiveness. Pour the ECL mixture over the entire membrane, and rock it manually to ensure even coverage. After 1 min, place the membrane on a paper towel to remove excess solution and then transfer the membrane into a plastic sheet protector securely fastened in a film cassette.
13. In a dark room equipped with a safe light, place X-ray film on top of the membrane in the sheet protector and expose for varying intervals, typically ranging from several seconds to 5–10 min.

3.6. Bioassays of sFRP Effects on Wnt Activity

1. The protocol will depend on the bioassay, but the key point is to pre-incubate the Wnt sample with the sFRP preparation. Typically, incubate Wnt and sFRP for 30 min at 37°C prior to addition to cells. A pilot dose–response titration of sFRP concentration is advisable, as the effective dose will vary depending on multiple experimental parameters (such as cell type, Wnt family member, and extracellular matrix).
2. Examples of sFRP inhibition of Wnt-3a-dependent, β-catenin stabilization and Dvl phosphorylation/mobility shift in SDS-PAGE are illustrated in **Figs. 3.3** and **3.4**, respectively. In these experiments, confluent monolayers of the indicated cells were incubated in serum-free medium overnight, followed by addition of serum-free conditioned medium from parental or Wnt-3a-expressing L929 cells. After the specified time, cells were lysed and whole cell lysates immunoblotted with antibodies to β-catenin (1:1,000), Dvl-2 (1:500), or Dvl-3 (1:500). The secondary antibody reagent was diluted at 1:5,000. Other aspects of the immunoblotting protocol were as described above.

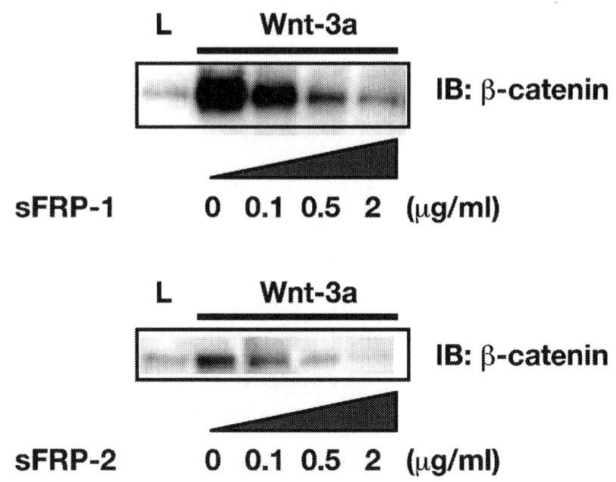

Fig. 3.3. Dose-dependent inhibition of Wnt-3a-dependent, β-catenin stabilization in L929 cells by purified sFRP-1 and sFRP-2. After growth in serum-containing medium, L cells were incubated overnight in serum-free medium and then treated for 6 h with serum-free conditioned medium from parental L cells (L) or Wnt-3a/L cell transfectants (Wnt-3a). The Wnt-3a medium had been pre-incubated with varying concentrations of sFRPs for 30 min at 37°C before addition to cells. Cell extracts were resolved by 10% SDS-PAGE and immunoblotted (IB) with primary antibody directed against β-catenin.

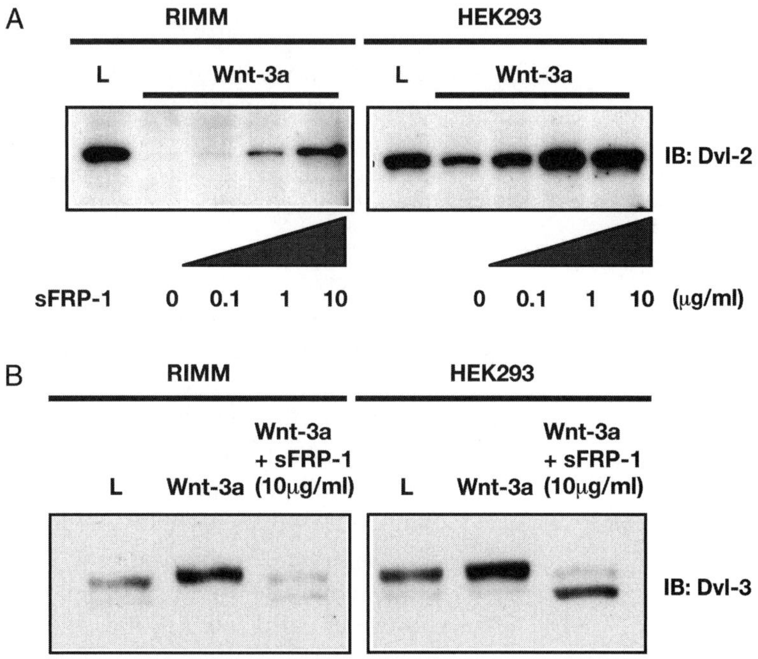

Fig. 3.4. sFRP-1 regulation of Wnt-dependent changes in Dvl-2 (**A**) and Dvl-3 (**B**) mobility/cross-reactivity in western blots. Serum-starved RIMM-18 and HEK293 cells were treated for 3 h either with conditioned medium from parental L cells (L) or Wnt-3a/L cell transfectants. Varying concentrations of sFRP-1 had been added to the Wnt-3a medium 30 min before incubation with cells. Cell extracts were processed for immunoblotting (IB) with a monoclonal antibody directed against Dvl-2 that exhibits reduced cross-reactivity following Wnt-dependent, Dvl-2 phosphorylation *(13)*, or an antibody that recognizes Dvl-3. sFRP-1 blocked the activity of added, and potentially endogenous Wnt.

4. Notes

1. To obtain clonal cell lines, we seeded the sFRP/MDCK mass cultures at a 1:50,000 dilution in collagen-coated wells and subsequently transferred colonies to culture dishes. Serum-free conditioned media were collected from several clones, and equivalent amounts of total protein were analyzed by immunoblot analysis for expression of sFRP. Depending on the level of expression and antibody sensitivity, the media might have to be concentrated with a device such as a Centricon-10 microconcentrator (Millipore) prior to immunoblotting.
2. Other commonly used mammalian expression systems have particular advantages. For instance, HEK293 can be adapted for growth in suspension, which facilitates large-scale production. Chinese hamster ovary (CHO) cells transfected with a vector encoding dihydrofolate reductase as well as the protein of interest will undergo methotrexate-dependent amplification of integrated plasmid copy number, increasing the synthesis of recombinant protein. The latter was used to produce sFRP-3 *(9)*. Both cell lines are amenable to clonal selection by plating cells at sparse density and/or use of cloning cylinder.
3. T-flasks and culture plates are convenient containers for cell growth, suitable for small or intermediate production. For larger scale operations, other devices such as cell factories (multilevel surfaces) or bioreactors for cells adapted for growth in suspension or on microcarriers (beads) would be better options.
4. Transfected cells were selected and routinely passaged in medium containing Geneticin at 500 μg/mL (prior to transfection, one should titrate parental cells with varying concentrations of the selective agent to determine a concentration that will efficiently kill cells). Omit antibiotic from cell cultures to be used for collection of conditioned medium; selection is not necessary at this point, and the antibiotic might interfere with bioassays of crude medium.
5. For intermediate preparations (1–2 L), concentrate clarified medium ~40-fold by ultrafiltration in a stirred chamber apparatus (Amicon M2000) under nitrogen gas pressure, using a YM membrane with either a 10- or 3-kDa molecular mass cutoff. For larger preparations (several liters or more), conditioned medium can be concentrated by tangential flow ultrafiltration with a Pellicon system (Millipore). Like unconcentrated media, concentrated samples can be snap-frozen and stored for future purification.

6. sFRP should not be present in the flow-through fraction. If detected there, that would indicate the capacity of the resin had been exceeded or there were a problem in resin equilibration or sample preparation. For additional capacity, link Hi-Trap columns in series, or use a 5-mL column.
7. Ideally, chromatography would be performed with a system that includes a pump, on-line spectrophotometer operating in the ultraviolet range, and a fraction collector. However, a sophisticated chromatography system is not required. Access to an ultraviolet spectrophotometer for periodic reading of optical density is desirable to optimize the timing of changes in elution buffers. However, once a routine is established, a standard elution protocol can be used without concomitant monitoring of optical density.
8. For both preparations, the resin should be washed with 0.05 M phosphate buffer, pH 7.4, and 2.0 M NaCl to remove residual protein. The resin can be re-equilibrated with several column volumes of equilibration solution and stored in a refrigerator for future use. Inclusion of 0.02% (w/v) sodium azide is prudent to prevent microbial growth.
9. Both sFRP-3 and sFRP-4 elute from heparin resin at lower NaCl concentrations (~0.7 and 0.5 M NaCl, respectively) where more contaminants are present (lack of purity was exacerbated by a lower level of recombinant expression by sFRP-3/MDCK and sFRP-4/MDCK transfectants). Therefore, additional chromatography is required to obtain homogeneous preparations.
10. If CBB is too intense, the gel can be destained to improve clarity. Decant the staining solution, replace it with destaining solution, and let the gel stir on a rocker until the background is transparent and the bands well visualized. Then wash the gel 3–5 times with water for 1–2 h prior to drying the gel.
11. Silver stain is more sensitive than CBB, and may detect trace contaminants not observed with the latter. However, the relative intensity of protein staining is more variable with silver stain than CBB.

Acknowledgments

This research was supported by the Intramural Research Program of the National Institutes of Health, National Cancer Institute.

References

1. Kawano, Y. and Kypta, R. (2003) Secreted antagonists of the Wnt signalling pathway. *J. Cell Sci.* **116**, 2627–2634.
2. Rubin, J.S., Barshishat-Kupper, M., Feroze-Merzoug, F., and Xi, Z.F. (2006) Secreted WNT antagonists as tumor suppressors: pro and con. *Front. Biosci.* **11**, 2093–2105.
3. Lee, H.X., Ambrosio, A.L., Reversade, B., and De Robertis, E.M. (2006) Embryonic dorsal-ventral signaling: secreted frizzled-related proteins as inhibitors of tolloid proteinases. *Cell* **124**, 147–159.
4. Rodriguez, J., Esteve, P., Weinl, C., Ruiz, J.M., Fermin, Y., Trousse, F., et al. (2005) SFRP1 regulates the growth of retinal ganglion cell axons through the Fz2 receptor. *Nat. Neurosci.* **8**, 1301–1309.
5. Hoang, B., Moos, M., Vukicevic, S., and Luyten, F.P. (1996) Primary structure and tissue distribution of Frzb, a novel protein related to Drosophila frizzled, suggest a role in human skeletal morphogenesis. *J. Biol. Chem.* **271**, 26131–26137.
6. Finch, P.W., He., X., Kelley, M.J., Uren, A., Schaudies, R.P., Popescu, N.C., et al. (1997) Purification and molecular cloning of a secreted, Frizzled-related antagonist of Wnt action. *Proc. Natl. Acad. Sci. U.S.A.* **94**, 6770–6775.
7. Roszmusz, E., Patthy, A., Trexler, M., and Patthy, L. (2001) Localization of disulfide bonds in the Frizzled module of Ror1 receptor tyrosine kinase. *J. Biol. Chem.* **276**, 18485–18490.
8. Üren, A., Reichsman, F., Anest, V., Taylor, W.G., Muraiso, K., Bottaro, D.P., et al. (2000) Secreted Frizzled-related protein-1 binds directly to Wingless and is a biphasic modulator of Wnt signaling. *J. Biol. Chem.* **275**, 4374–4382.
9. Dann, C.E., Hsieh, J.-C., Rattner, A., Sharma, D., Nathans, J., and Leahy, D.J. (2001) Insights into Wnt binding and signalling from the structures of two Frizzled cysteine-rich domains. *Nature* **412**, 86–90.
10. Levashova, Z.B., Plisov, S.Y., and Perantoni, A.O. (2003) Conditionally immortalized cell line of inducible metanephric mesenchyme. *Kidney Int.* **63**, 2075–2087.
11. Endo, Y., Wolf, V., Muraiso, K., Kamijo, K., Soon, L., Üren, A., et al. (2005) Wnt-3a-dependent cell motility involves RhoA activation and is specifically regulated by Dishevelled-2. *J. Biol. Chem.* **280**, 777–786.
12. Bafico, A., Gazit, A., Pramila, T., Finch, P.W., Yaniv, A., and Aaronson, S.A. (1999) Interaction of Frizzled-related protein (FRP) with Wnt ligands and the Frizzled receptor suggests alternative mechanisms for FRP inhibition of Wnt signaling. *J. Biol. Chem.* **274**, 16180–16187.
13. Gonzales-Sancho, J.M., Brennan, K.R., Castelo-Soccio, L.A., and Brown, A.M.C. (2004) Wnt proteins induce Dishevelled phosphorylation via an LRP5/6-independent mechanism, irrespective of their ability to stabilize b-catenin. *Mol. Cell. Biol.* **24**, 4757–4768.

Chapter 4

Measuring GSK3 Expression and Activity in Cells

Adam R. Cole and Calum Sutherland

Abstract

Glycogen synthase kinase (GSK)-3 is a key signalling intermediate in the action of Wnts. This protein kinase is ubiquitously expressed and has high inherent activity but is inhibited by activation of Wnt signalling *or* activation of growth factor receptor tyrosine kinases (e.g. insulin, nerve growth factor [NGF], platelet-derived growth factor [PDGF], etc.). The degree of inhibition of GSK3 in cells treated with such reagents is dependent on the cell type and the stimulus used. Therefore, the ability to accurately measure GSK3 activity in cells is an important aspect of GSK3 research. The activity of GSK3 is reduced by posttranslational modification (phosphorylation) and this can be measured by immunoblot with specific reagents (indirect), or by immunoprecipitation and assay (direct), as long as the modification is protected during these procedures. However, inhibition by phosphorylation is specific to cellular activation by growth factors and nutrients. Wnt inhibition of GSK3 does not involve phosphorylation of these residues on GSK3 and therefore it cannot be measured using this modification. Currently, the simplest way to assess Wnt inhibition of GSK3 is to monitor phosphorylation of specific GSK3 substrates in cells (e.g. β-catenin). Alternatively, Wnt inhibition of GSK3 can be measured by partial purification of cellular GSK3 by ion exchange chromatography and assay of fractions or possibly by immunoprecipitation and assay. In this chapter, we demonstrate the use of the different approaches to measure GSK3 activity in SH-SY5Y cells, describe the best antibodies currently available, and discuss the potential drawbacks of each method.

Key words: GSK3, Wnt, Assay, Phosphorylation, Growth factor.

1. Introduction

Glycogen synthase kinase (GSK)-3 is an evolutionarily conserved Ser/Thr kinase that is ubiquitously expressed in all mammalian tissues and in all major subcellular organelles. Two isoforms are encoded by separate genes; GSK3β (chromosome 19q13.2) encodes a protein of 51 kDa, while GSK3β (chromosome 3q13.3) encodes a protein of 47 kDa *(1–3)*. The difference in size is due to a glycine-rich N-terminal insertion in GSK3α. The bulk of each isoform is comprised of a central kinase domain (~350

Elizabeth Vincan (ed.), *Wnt Signaling, Volume I: Pathway Methods and Mammalian Models, vol. 468*

Book doi: 10.1007/978-1-59745-249-6

amino acids) with 98% sequence homology between isoforms and is closely related to the cyclin-dependent kinase family. A splice variant of GSK3β that contains a 13-amino acid insert in the catalytic domain exhibits enhanced expression in the brain *(4)*.

GSK3 was first identified as the third kinase known to phosphorylate glycogen synthase (hence its name) as part of the insulin-mediated regulation of glucose metabolism *(5)*. Since then, it has been shown to participate in many additional cellular functions, such as regulation of gene transcription, development, apoptosis, cell cycle, cytoskeletal dynamics, and molecular transport. GSK3 activity is reported to be deregulated in diabetes and Alzheimer's disease (AD), and inhibitors are being assessed for treatment of both disorders *(6)*. To date, almost 50 substrates of GSK3 have been reported, with the majority being transcription factors, metabolic proteins, and cytoskeleton-associated proteins *(7)*.

GSK3 is one of the most unusual kinases of the 500 or so encoded in the human genome. Firstly, most (if not all) substrates require prior phosphorylation by another kinase before they can be efficiently phosphorylated by GSK3. This process is known as "priming" and occurs 4 or 5 residues C-terminal to the site phosphorylated by GSK3. Secondly, GSK3 is constitutively active in cells under basal conditions. This is partly due to constitutive phosphorylation of a conserved tyrosine residue on the activation loop of the kinase domain (Tyr279 in GSK3α, and Tyr216 in GSK3β), which is absolutely required for kinase activity. Thirdly, phosphorylation of GSK3 at a conserved N-terminal serine residue inhibits its kinase activity. This phosphoserine acts as a pseudo-substrate and binds to the phosphate-binding pocket on GSK3, preventing interaction with primed substrates *(8)*. Phosphorylation at this site is mediated by members of the AGC family of kinases (e.g. PKB, PKC, p70S6K, and p90RSK) and commonly occurs downstream of growth factor and phosphatidylinositol 3 kinase (PI3K) signalling. Activation of the canonical Wnt pathway also inhibits GSK3 activity, although this is not mediated by N-terminal phosphorylation but by protein–protein interactions *(9–11)* (see earlier chapters).

Interestingly, most evidence suggests that growth factors and Wnts regulate distinct pools of GSK3 in cells and thus each has a distinct set of downstream targets (**Figs. 4.1** and **4.2**). Part of this

→

Fig. 4.1. Wnt3a regulates GSK3-mediated phosphorylation of β-catenin, but not CRMP2. **A** Lysates of SH-SY5Y cells incubated with conditioned medium from control (–) or Wnt3a-expressing (+) L cells (1 h) were subjected to SDS-PAGE and transferred to nitrocellulose membranes. Membranes were probed with antibodies that recognise GSK3α/β when phosphorylated at Ser21/9, Tyr279/216, or total GSK3α/β. The band intensities were quantitated and the ratios of phosphorylated GSK3α/β to total GSK3α/β are shown (control lysates normalised to 100%). Error bars indicate the range of duplicate experiments. **B** Same as (**A**), except membranes were probed with antibodies that recognise phosphorylated β-catenin (Thr41/Ser37/Ser33), total β-catenin, phosphorylated CRMP2 (Thr514/509), or total CRMP2. *n.s.*, non-significant; *$p<0.05$ (Student's *t* test).

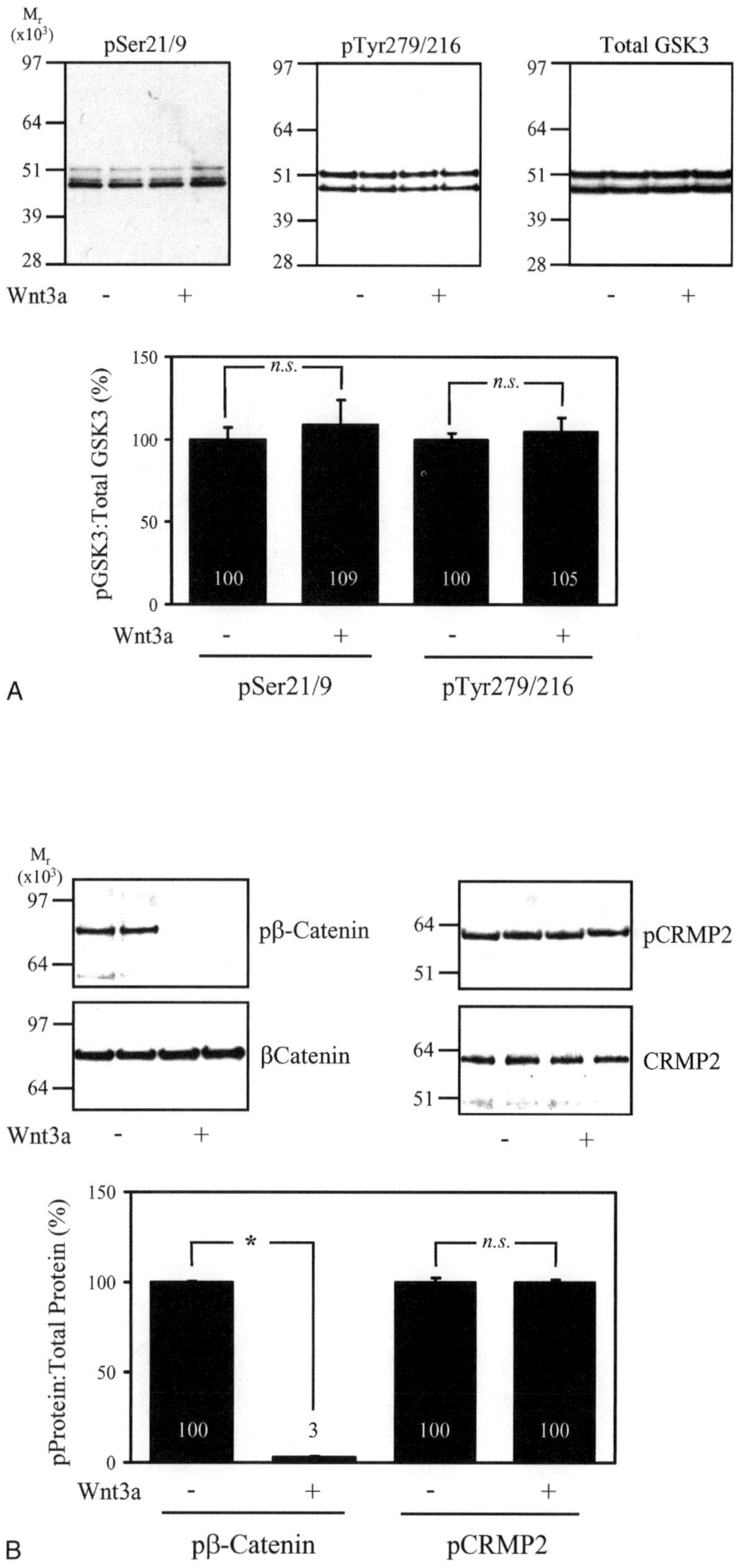

Mr (x10³)
pSer21/9
pTyr279/216
Total GSK3
97
64
51
39
28
Wnt3a - +
150
100
50
0
pGSK3:Total GSK3 (%)
n.s.
100
109
105
pSer21/9
pTyr279/216
A
Mr (x10³)
pβ-Catenin
βCatenin
pCRMP2
CRMP2
pProtein:Total Protein (%)
*
3
B

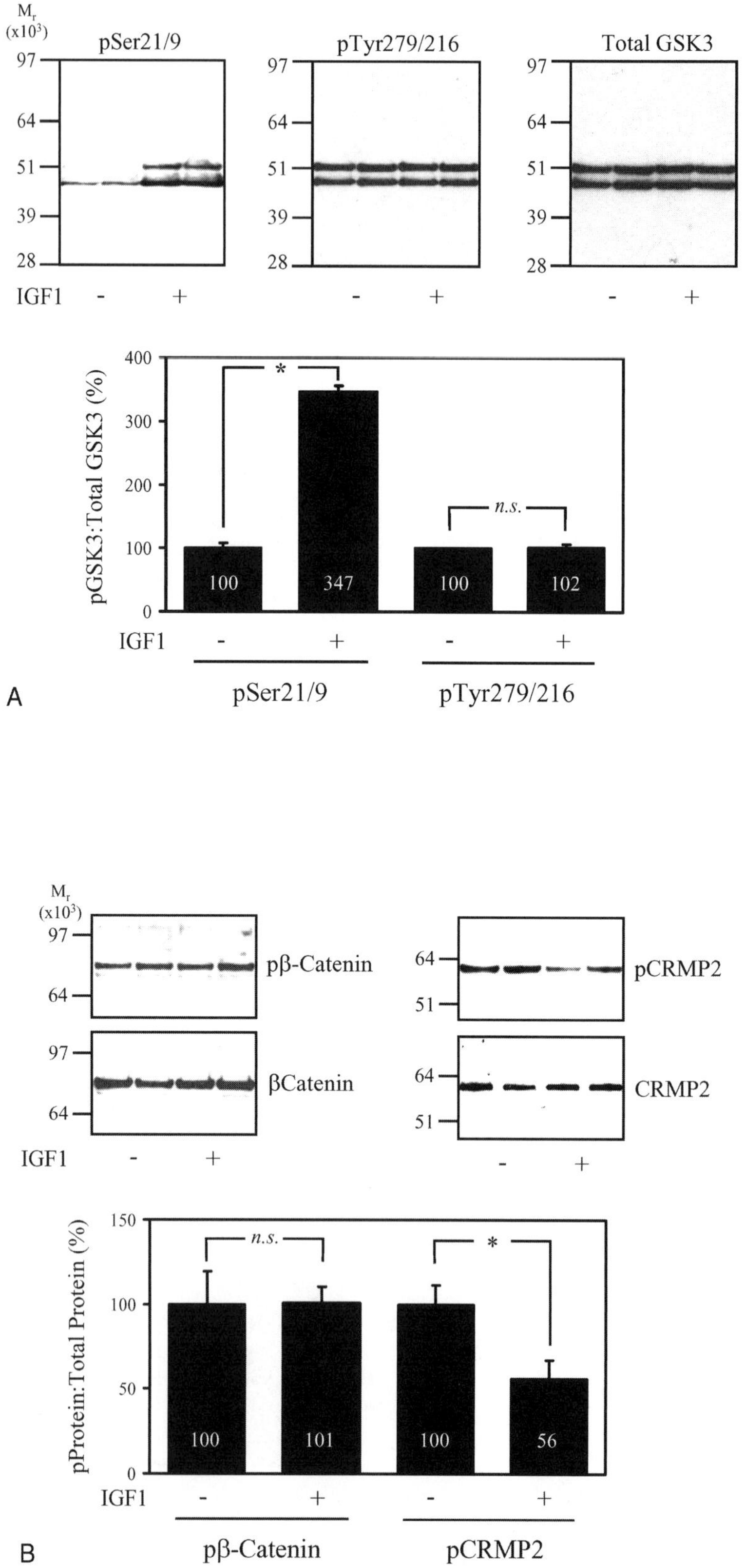
Mr (x10³)
pSer21/9
pTyr279/216
Total GSK3
97
64
51
39
28
IGF1 - +
pGSK3:Total GSK3 (%)
400
300
200
100
0
*
n.s.
100
347
100
102
IGF1 - + - +
A
pSer21/9
pTyr279/216
Mr (x10³)
pβ-Catenin
βCatenin
pCRMP2
CRMP2
IGF1 - +
pProtein:Total Protein (%)
150
100
50
0
n.s.
*
100
101
100
56
IGF1 - + - +
B
pβ-Catenin
pCRMP2

difference may be due to growth factor inhibition being mediated through N-terminal phosphorylation of GSK3, resulting in inhibition of primed substrates, and Wnt inhibition focussing on a specific GSK3-containing complex. In addition, growth factors regulate both GSK3α and GSK3β, while Wnts have so far only been shown to regulate GSK3β *(12)*. Therefore the measurement of GSK3 activity in cells must take these differences into account.

There are several approaches available to measure GSK3 activity in cells, however the approach chosen depends on the signalling pathway being analysed. For example, inhibition by phosphorylation (following growth factor stimulation of cells) can be monitored by:

1. Immunoprecipitation of total cellular GSK3 and assay (*see* **Sections 3.1** and **3.6**). This will provide a "specific activity" measurement relative to total protein in the cell, and is the most accurate measure of total GSK3 activity, or

2. Immunoblot using antibodies to phospho-Ser21 (GSK3α) or phospho-Ser9 (GSK3β) (*see* **Sections 3.1–3.4**). This is a comparative analysis showing relative changes in the inhibitory phosphorylation of GSK3 *(13)*. It should be noted that it does not give an absolute measurement of GSK3 activity, since the stoichiometry of phosphorylation cannot be assessed by this method. Immunoblot using antibodies to phospho-Tyr279/216 can assess alterations in activity independent of the N-terminal phosphorylation, although phosphorylation at this site is most likely to be constitutive *(14)*.

3. Immunoblot using antibodies to known substrates of GSK3, for example collapsin response mediator protein (CRMP)-2 (**Fig. 4.2** and ref. *(15)*).

On the other hand, assessment of Wnt regulation of GSK3 is more complex. Four methods have been used previously:

1. Immunoprecipitation of total cellular GSK3 and assay as above, although this may be a very transient effect as we (**Fig. 4.3**) and others *(10, 12)* do not see any change in GSK3 specific activity beyond 30 min exposure of cells to Wnt.

2. Comparative analysis of phosphorylation of the Wnt target protein β-catenin by immunoblot (*see* **Sections 3.1–3.5**); or

Fig. 4.2. IGF1 regulates GSK3-mediated phosphorylation of CRMP2, but not β-catenin. **A** Lysates of SH-SY5Y cells incubated without (–) or with (+) IGF1 (50 ng/mL, 1 h) were subjected to SDS-PAGE and transferred to nitrocellulose membranes. Membranes were probed with antibodies that recognise GSK3α/β when phosphorylated at Ser21/9, Tyr279/216, or total GSK3α/β. The band intensities were quantitated and the ratios of phosphorylated GSK3α/β to total GSK3α/β are shown as a graph (control lysates normalised to 100%). Error bars indicate the range. **B** Same as (**A**), except membranes were probed with antibodies that recognise phosphorylated β-catenin (Thr41/Ser37/Ser33), total β-catenin, phosphorylated CRMP2 (Thr514/509), or total CRMP2. *n.s.*, non-significant; $*p < 0.05$ (Student's *t* test).

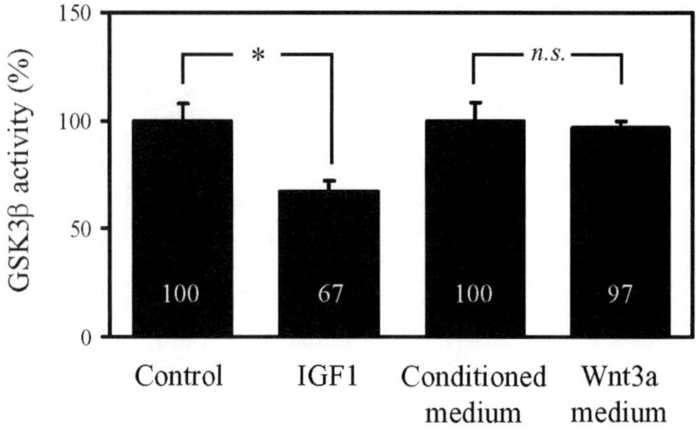

Fig. 4.3. IGF1, but not Wnt3a, reduces cellular GSK3β activity. SH-SY5Y cells were stimulated with IGF1 or Wnt3a (1 h), then harvested in cell lysis buffer. GSK3β was immunoprecipitated from lysates and its activity was measured in an in vitro kinase assay using pGS2 peptide. GSK3β activity is presented in a graph as a percentage of untreated cells (normalised; 100% activity from control cells equates to 52 pmol/min/mg). *n.s.*, non-significant; *, $p < 0.05$ (Student's *t* test).

3. Partial purification of cellular GSK3 by cation exchange chromatography *(12)*, followed by assay (*see* **Section 3.6**). This approach assumes that the inhibitory protein complex is maintained during the purification procedure and that a substantial proportion of cellular GSK3 is inhibited (hence, this is subject to the cell system used and variations in protein purification that may disrupt the adenomatous polyposis coli [APC] complex). As with the immunoprecipitation assay above, this measurement may be very transient; or
4. Immunoprecipitation of the Axin–APC complex rather than total GSK3 *(16, 17)* followed by assay. However, this approach assumes that this is the only GSK3 complex regulated by Wnt and that any inhibitory interactions are maintained during the isolation procedure. Interestingly, this method predicted relatively poor inhibition of GSK3 by Wnt compared with the other approaches.

In this chapter, we provide technical details of the immunoprecipitation and assay of GSK3 as well as immunoblot analysis of GSK3 and its substrates.

2. Materials

2.1. Cell Culture and Lysis

1. SH-SY5Y cells (ATCC #CRL-2266) were maintained in Dulbecco's Modified Eagle's Medium (DMEM; Cambrex, East Rutherford, NJ) supplemented with 10% (v/v) foetal bovine serum

(FBS; Gibco; Invitrogen, Carlsbad, CA), penicillin/streptomycin (Gibco) and 10 μM retinoic acid to induce neuronal differentiation (*see* **Note 1**).

2. Insulin-like growth factor (IGF)-1 (Invitrogen) was dissolved in phosphate-buffered saline (PBS) containing 0.1% (w/v) bovine serum albumin (BSA) at a concentration of 50 μg/mL and stored in aliquots at –20°C (*see* **Note 2**).
3. Mouse L cells that express and secrete Wnt3a (ATCC, Manassas, VA; #CRL-2647) and the control L cells that do not secrete Wnt3a (ATCC #CRL-2648) are grown in DMEM containing 10% (v/v) FBS and penicillin/streptomycin. Media is collected from these cells and stored at 4°C for up to 1 month or at -20°C for longer periods (*see* **Note 3**).
4. Cell lysis buffer: 50 mM Tris-HCl, pH 7.5, 1% (v/v) Triton X-100, 0.27 M sucrose, 1 mM EDTA, 0.1 mM EGTA, 1 mM sodium orthovanadate, 50 mM sodium fluoride, 5 mM sodium pyrophosphate, 0.1% (v/v) β-mercaptoethanol, and complete protease inhibitor cocktail (Roche, Basel, Switzerland).
5. Bradford reagent for determining lysate protein concentrations is from Sigma-Aldrich (St. Louis, MO; #B6916).

2.2. Sodium Dodecyl Sulphate (SDS)-Polyacrylamide Gel Electrophoresis (PAGE) and Electrotransfer

1. Pre-cast NuPAGE Bis-Tris 4–12% polyacrylamide gels from Invitrogen (see www.invitrogen.com) provide high-quality and reproducible separation of proteins, but are expensive, are not suitable for integral membrane proteins, and require specialised tanks and buffers for best performance. Standard SDS-PAGE is an alternative to this system *(18)*.
2. Stock solutions of morpholinepropane sulphonic acid (MOPS) electrophoresis running buffer (20×), transfer buffer (20×), antioxidant solution (1,000×), SDS sample buffer (4×), and SeeBlue Plus2 molecular weight markers are available from Invitrogen and are optimised for the NuPAGE system. These reagents are based on standard SDS-PAGE reagents (*Running Buffer*, 0.1 M glycine, 25 mM Tris-HCl, pH 7.4, and 0.05% (w/v) SDS; *Transfer Buffer*, 25 mM Tris-HCl, pH 7.4, 192 mM glycine, and 20% (v/v) methanol; *Sample Buffer*, 25 mM Tris-HCl, pH 6.8, 2% (w/v) SDS, 10% (v/v) glycerol, and 0.1 μg/mL Bromophenol Blue).
3. Nitrocellulose membranes are available from Amersham (Hybond C-extra, #RPN303E; GE Healthcare, Piscataway, NJ).

2.3. Immunoblotting

1. Many different types of phosphorylation-dependent and phosphorylation-independent antibodies for GSK3 are commercially available and are listed in **Table 4.1.** For the experiments shown in this chapter, the following antibodies were used; GSK3α/β (pS21/9) (#9331; Cell Signaling Technology,

Table 4.1
List of commercially available GSK3 antibodies

Protein	Source	Company	Specificity	Immunogen
GSK3α	Rabbit polyclonal	Cell Signaling Technology (#9338)	H, M, R	Synthetic peptide of human GSK3α (region not stated)
	Chicken polyclonal	Chemicon, Millipore, Billerica, MA (#AB3441)	H	Synthetic peptide of human GSK3α (aa 89–100)
	Chicken polyclonal	GeneTex, San Antonio, TX (#GTX17705)	H, C	aa 89–100 of human GSK3α
	Rabbit polyclonal	GeneTex (#GTX29767) or AbCAM (#Ab9767)	H	RARTSSFAEPC, corresponding to amino acids 16–25 of human GSK3α
	Mouse monoclonal	R&D Systems (#MAB2986, clone 288901)	H	*Escherichia coli*-derived rhGSK3α (aa 90–483)
	Rabbit polyclonal	Upstate (#07-389)	H, R	aa 471–483 rat GSK3α (QAPDATPTLTNSS)
	Rabbit monoclonal	Upstate (#05-737, clone AW103)	H, M, R	aa 471–483 rat GSK3α (QAPDATPTLTNSS)
GSK3α (pS21)	Mouse monoclonal	Cell Signaling Technology (#9337, clone 46H12)	H, M, R, Mk, Z	Synthetic peptide of residues surrounding Ser21 of human GSK3α
	Rabbit polyclonal	GeneTex (#GTX29765), AbCAM (#9765)	H	RARTS$_p$SFAEPC, corresponding to aa 16–25 of Human GSK3α
	Sheep polyclonal	Upstate (#07-420)	H, R	aa 16–26 of human GSK3α (RARTSsFAEPG)
GSK3β	Mouse monoclonal	BD Transduction Laboratories (#610201)	H, M, R, C	Rat GSK3β aa 1–160

	Rabbit polyclonal	Calbiochem, Gibbstown, NJ (#361528)	H, M, R	aa 334–348 (CAHSFFDELRDPNVK) of rat GSK3β C-terminus
	Rabbit monoclonal	Cell Signaling Technology (#9315, clone 27C10)	H, M, R, Mk	Synthetic peptide of human GSK3β (region not stated)
	Mouse monoclonal	Chemicon (#MAB8689, clone 1H8)	H, R, X	Recombinant *Xenopus* GSK3β
	Rabbit polyclonal	Chemicon (#AB8687)	H, M, R, B	Synthetic peptide of rat GSK3β (aa 335–349)
	Chicken polyclonal	Chemicon (#AB3443)	H	Synthetic peptide of human GSK3β (aa 21–36)
	Mouse monoclonal	GeneTex (#GTX22602), AbCAM (AB2602, clone G8)	H, R, X	Recombinant *Xenopus* GSK3β
	Rabbit polyclonal	GeneTex (#GTX19940)	H, M, R, B	Synthetic peptide from kinase subdomain XI region of the rat GSK3β
	Mouse monoclonal	R&D Systems (#MAB2506, clone 272536)	H, M, R	*E. coli*-derived rhGSK3β
	Mouse monoclonal	Sigma (#G6414, clone GSK-4B)	H, M, R, Rb	Recombinant rabbit GSK3β
GSK3β (pS9)	Rabbit polyclonal	Biosource (#44600G)	H, R	Synthetic peptide of residues surrounding Ser9 of human GSK3β
	Mouse monoclonal	Calbiochem (#361527, clone 2D3)	H, M	aa 6–12 of human GSK3β (RTTsFAE)
	Rabbit polyclonal	Calbiochem (#PS1018)	H, R	Synthetic peptide of residues surrounding Ser9 of human GSK3β
	Rabbit polyclonal	Calbiochem (#PC242)	H, R, X	Synthetic peptide of residues surrounding Ser9 of human GSK3β

(continued)

Table 4.1 (continued)

Protein	Source	Company	Specificity	Immunogen
	Rabbit polyclonal	Cell Signaling Technology (#9336)	H, M, R, Mk, Z	Synthetic peptide of residues surrounding Ser9 of human GSK3β
	Rabbit polyclonal	Covance, Denver, PA (#PRB-534P)	H, M, R	Synthetic peptide of residues surrounding Ser9 of GSK3β
	Rabbit polyclonal	Upstate (#07-835)	H, M, R, X	Not stated
	Rabbit polyclonal	GeneTex (#GTX29769) or AbCAM (AB9769)	H	RRTT$_p$SFAE, corresponding to aa 4–12 of Human GSK3β
	Rabbit polyclonal	Sigma (#G6542)	H, M, R	Synthetic peptide of residues surrounding Ser9 of human GSK3β
GSK3α/β	Mouse monoclonal	AbCAM, Cambridge, MA (#Ab17059, clone 1H8)	H, M, R, X	Recombinant *Xenopus* GSK3β
	Mouse monoclonal	Biosource International (Invitrogen) (#44610)	H, M, R, X	Recombinant *Xenopus* GSK3β
	Mouse monoclonal	Calbiochem (#368662, clone 1H8)	H, M, R, B, O, P, Hm, Rb	Recombinant *Xenopus* GSK3β
	Rabbit polyclonal	GeneTex (#GTX15314) or AbCAM (#Ab15314)	H, M, R, C, M	Synthetic peptide (human) (C terminal)
	Rabbit polyclonal	R&D Systems (#AF2157)	H, M, R	*E. coli*-derived rhGSK3β
	Mouse monoclonal	Sigma (#G4414, clone GSK-3B)	H, M, R, Rb	Recombinant rabbit GSK3β

	Rabbit monoclonal	Upstate (#05-903)	H, M, R	Recombinant full-length human GSK3β
GSK3α/β (pS21/9)	Rabbit polyclonal	Cell Signaling Technology (#9331)	H, M, R, Z	Synthetic peptide of residues surrounding Ser21 of human GSK3
	Rabbit polyclonal	R&D Systems (#AF1590)	H, M, R	Synthetic peptide of residues surrounding Ser9 of GSK3β
	Mouse monoclonal	Upstate (#05-643, clone 2D3)	H, M, R	aa 6–12 of human GSK3β (RTTsFAE)
GSK3α/β (pY279/216)	Mouse monoclonal	AbCAM (#AB33021)	H, M, R, Dros	aa 203–219 of *Drosophila* GSK3 KQLLHGEPNVSY$_p$ICSRY
	Mouse monoclonal	BD Transduction Laboratories (#612312)	H, M, R	Synthetic peptide surrounding Tyr216 of rat GSK3β
	Rabbit polyclonal	Biosource (#44604G)	H, M, R	Synthetic peptide surrounding Tyr279/216 of human GSK3α/β
	Rabbit polyclonal	Calbiochem (#ST1013)	H, M, R	Synthetic peptide surrounding Tyr279/216 of human GSK3α/β
	Rabbit polyclonal	Covance (#PRB-535P)	H, M, R	Synthetic peptide surrounding Tyr216 of GSK3β
	Rabbit polyclonal	GeneTex (#GTX24797) or AbCAM (#Ab4797)	H, M, R	Synthetic peptide
	Rabbit polyclonal	GeneTex (#GTX24797), AbCAM (#Ab4797), Covance (#PRB-535P)	H, M, R	Synthetic peptide
	Rabbit polyclonal	Sigma (#G5791)	H, M, R	Synthetic peptide surrounding Tyr279/216 of human GSK3α/β

H, human; M, mouse; R, rat; C, chicken; Mk, monkey; Z, Zebrafish; X, xenopus; B, bovine; Rb, rabbit; O, ovine; P, porcine; Hm, hamster; Dras, Drosophilla; aa, amino acid; rh; recombinant human;

Danvers, MA), GSK3α/β (pY279/216) (#612312; BD Transduction Laboratories, San Jose, CA) and GSK3α/β (#05-903; Upstate; Millipore, Billerica, MA) (*see* **Note 4**).

2. Other antibodies used were β-catenin (phospho-Thr41/Ser37/Ser33) (#9561; Cell Signaling Technology), β-catenin (#9562; Cell Signaling Technology), CRMP2 (pT514/509) (#PB-043; Kinasource, Dundee, Scotland) and CRMP2 (#AB-042; Kinasource).
3. Blocking buffer consists of 5% (w/v) dried skim milk in Tris-buffered saline/Tween-20 (TBS-T) buffer (50 mM Tris-HCl, pH 7.5, 150 mM NaCl, and 0.05% (v/v) Tween-20).
4. All antibodies are diluted 1:1,000 in 1% (w/v) dried skim milk in TBS-T, except the CRMP2 antibody (1:250) (*see* **Note 5**).
5. Rabbit anti-mouse (#31450), sheep anti-rabbit (#31460) and donkey anti-sheep (#31480) horseradish peroxidase-conjugated secondary antibodies are available from Pierce (Rockford, IL).
6. Enhanced chemiluminescence (ECL) reagent (#RPN2106) and X-ray films (#RPN3103K) are available from Amersham.

2.4. Image Analysis and Quantitation

1. Images are digitised by scanning and densitometry using Aida quantitation software. Data analysis is performed using Microsoft Excel software.
2. Images are prepared using Adobe Photoshop and Microsoft Powerpoint software.

2.5. Immuno-precipitation Kinase Activity Assays

1. The anti-GSK3β monoclonal antibody is from BD Transduction Laboratories (#610201) and protein G-Sepharose is from Amersham (#17-0618-01).
2. Radioactive [γ-^{32}P]ATP is from Amersham (#PB10218), P81 paper from Whatman (#3698-915; Maidstone, UK) and ortho-phosphoric acid from BDH AnalaR (#10173CD; VWR, Lutterworth, UK). Scintillation fluid is from Perkin Elmer (#6013371; Waltham, MA) and a Beckman LS-6500 scintillation counter is used to detect radioactivity (Beckman Coulter, Fullerton, CA).
3. Kinase buffer consists of 50 mM Tris-HCl, pH 7.5, 150 mM NaCl, 0.03% (v/v) Brij-35, and 0.1% (v/v) β -mercaptoethanol.
4. The 10× [γ-^{32}P]-MgATP stock solution contains 100 mM $MgCl_2$, 1 mM un-labelled ATP, and approximately 30 MBq/mL [γ-^{32}P]ATP.
5. Phospho-glycogen synthase 2 (pGS2) peptide is available from Upstate (#12-241) and the pGS2 10× stock solution contains 200 μM pGS2 in kinase buffer.

3. Methods

3.1. Cell Stimulation and Lysis

1. SH-SY5Y cells are grown on 10-cm tissue culture plates in medium containing retinoic acid for 4 days to induce neuronal differentiation. On the fourth night, remove the medium, rinse the cells once in pre-warmed serum-free DMEM containing penicillin/streptomycin and retinoic acid, and incubate in this medium overnight to serum-starve the cells. This maximally induces GSK3 activity.
2. The following morning, stimulate the cells by adding 50 ng/mL IGF1 or carrier solution for the desired time(s) (here, 1 h at 37°C). Alternatively, add an equal volume of conditioned medium from control or Wnt3a-expressing L cells to the medium already in the plate for the desired time(s) (here, 1 h at 37°C).
3. Harvest the cells by placing the plates on ice, aspirate the medium, rinse the cells once in ice-cold PBS, and scrape the cells directly into 0.2 mL of ice-cold cell lysis buffer. Leave on ice for 5 min and then centrifuge the samples (14,000×*g*, 10 min) to pellet insoluble material. Immediately snap-freeze the supernatant (lysate) in liquid nitrogen and discard the pellet.

3.2. Preparation for SDS-PAGE

1. Determine the protein concentrations of each sample using the Bradford assay *(19)*. Briefly, add various amounts of lysate (usually 0.1–10 μL in a total volume of 100 μL of water) to 1 mL of Coomassie (Bradford) reagent, vortex, and incubate at room temperature for 10 min. Vortex the solutions again, then measure the absorbance using a spectrophotometer at λ_{max} 595 nm. Compare the absorbance of each sample solution to a standard curve of BSA (serial dilution: 0 to 10 μg) in order to establish the concentration of protein in each sample. All samples and standards should be assayed in triplicate.
2. As a general rule, it is best to load equal *volumes* of samples as well as *amounts* of total protein for each sample to their respective lanes on the gel. For example, equal amounts of total protein from each sample in **Fig. 4.1** (160 μg, with volumes ranging from 136 to 231 μL) are diluted with different amounts of cell lysis buffer to give the same final volume (240 μL). Then, 80 μL of 4× SDS sample loading buffer containing 40 mM dithiothreitol (DTT) is added to each sample, mixed, and heated to 70°C for 15 min. Every denatured sample therefore contains 0.5 mg/mL total protein, and every lane is loaded with the same protein *amount* and *volume*.

3.3. SDS-PAGE and Protein Transfer

1. Rinse the wells of the pre-cast polyacrylamide gels with water to remove unpolymerised polyacrylamide, then insert the gels into the electrophoresis tank and lock them in place.
2. Add running buffer (1×) to the inner and outer chambers of the tank.
3. Load molecular weight markers (e.g. SeeBlue Plus 2–5 μL) into lane 1 of the gel using gel-loading pipette tips, then load equal volumes of each sample into the other lanes (usually 4–40 μL containing 2–20 μg of total protein).
4. Connect the tank to a power supply and run the gels at 120 V for 20 min at room temperature, then at 180 V until the gel front reaches the exposed strip of polyacrylamide at the bottom of the gel.
5. While the gel is running, prepare the equipment and reagents for the transfer step. Immerse the sponges in transfer buffer and remove any air bubbles by tapping with a spatula. Cut squares of blotting paper and nitrocellulose membranes slightly smaller in size than the sponges, label the membranes with a small mark in a corner using a pencil, and immerse them all in transfer buffer.
6. Remove the gel from the electrophoresis tank and open the plastic casing of the gel using a spatula. Carefully remove the gel and submerge it in transfer buffer.
7. Assemble the transfer cassette in the following order: sponge, blotting paper, gel, membrane, blotting paper, and sponge. Place the cassette into the electrophoresis tank with the gel towards the negative terminal and the membrane towards the positive terminal. When a current is applied, the negatively charged SDS–protein complexes will move to the membrane. Add transfer buffer, connect to a power source, and transfer the proteins from the gel to the membrane at 35 V for 1–2 h.
8. When the transfer is complete, remove the membrane from the cassette and incubate it in blocking buffer for 1 h at room temperature or overnight at 4°C (enough volume to ensure that no part of the membrane becomes dry).

3.4. Immunoblotting for GSK3 and GSK3 Substrates

1. Discard the blocking buffer and wash the membranes 3× 10 min in TBS-T.
2. Incubate the membranes with the primary antibody diluted in 1% (w/v) milk in TBS-T (*see* **Note 5**) with constant shaking or rotating to ensure even coverage of the membrane with the antibody. This step is usually performed at 4°C overnight, but can also be performed for shorter periods at room temperature (at least 1 h).
3. Remove the membranes from the primary antibody solution and wash them 3× 10 min in TBS-T. Most primary antibody

solutions can be re-used three to five times without loss of sensitivity or selectivity, but should be stored at 4°C for up to 1 month (add 0.02% [w/v] sodium azide to prevent bacterial growth). Dilute the appropriate secondary antibody in a solution containing 1% (w/v) milk in TBS-T according to the manufacturer's instructions (usually 1:2,500–1:10,000). Incubate the membranes in this solution for 1 h at room temperature with shaking or rotation.

4. Discard the secondary antibody solution and wash the membranes 3× 10 min in TBS-T.
5. The ECL reagents A and B should be removed from the fridge to allow them to equilibrate to room temperature before use. In a dark room under a safety light, mix equal volumes of A and B (e.g. 1 mL + 1 mL for a single membrane) for no more than 1 min. Drain excess TBS-T from the membranes and place them on a flat, clean surface. Pipette enough of the mixed ECL reagent to cover the entire membrane and let it stand for 1 min.
6. Quickly drain the ECL reagent from the membrane and place it in inside Saran wrap in an X-ray film cassette.
7. Place an X-ray film onto the membrane (initially try 1 min) then develop the film (*see* **Note 6**). While developing the first film, a second and third exposure to film can be performed for a longer or shorter period of time to provide different exposures of the image. This is important in order to establish the linearity of the ECL reaction for each sample with each antibody.
8. Membranes can be re-probed with different antibodies (*see* **Note 7**).

3.5. Image Analysis and Quantitation of Immunoblots

1. Choose an exposure with an intensity in the linear range of the ECL/immunoblotting technique. As a general rule, bands should be solid and black (light and grey suggests bands are under-exposed) and should contain sharp edges (blurred/fuzzy edges suggests bands are overexposed). Over-sized bands ("ballooned") also suggests that bands are overexposed.
2. Digitise the X-ray films containing images of the immunoblots by scanning. Suitable software (such as Aida) can be used to quantify band intensity (*see* **Note 8**).
3. To determine changes in phosphorylation of a protein, calculate the ratio of phosphorylated protein to total protein for each sample (e.g. divide the densitometric value of pSer21/9-GSK3 by the value for the corresponding total GSK3α/β). The membranes for phospho and total GSK3 analyses should be generated using the same denatured lysate solution and performed at the same time.

4. **Figure 4.1a** shows analysis of phosphorylation of GSK3 isoforms at Ser21/9 and Tyr279/216 in SH-SY5Y cells following stimulation with Wnt3a. Representative blots and quantified ratios of phosphorylation relative to total protein are provided. **Figure 4.1b** shows analysis of the same cell lysates using phospho-β-catenin and p-CRMP2, two established targets of GSK3. Note that only the phosphorylation of β-catenin is reduced following Wnt stimulation, therefore phosphorylation of GSK3 or CRMP2 cannot be used to monitor Wnt regulation of GSK3 activity in cells. In contrast, IGF1 stimulation of these cells (**Fig. 4.2a**) induces phosphorylation of Ser21/9 but does not affect Tyr279/216 phosphorylation, and reduces phosphorylation of CRMP2 but does not affect phosphorylation of β-catenin (**Fig. 4.2b**). Therefore, IGF1 regulation of GSK3 can be visualised by changes in GSK3 and/or CRMP2 phosphorylation.

3.6. Immunoprecipitation Kinase Activity Assays (See Note 9*)*

1. Pre-incubate anti-GSK3β (5 μg) monoclonal antibody with protein G-Sepharose (400 μL of a 50% slurry) in kinase buffer at 4°C for 1 h to conjugate the antibody to the protein G-Sepharose beads.
2. Centrifuge the beads at 6,000×*g* for 1 min, discard the supernatant, and wash the beads two times in ice-cold PBS to remove unbound antibody. Resuspend the beads in 250 μL of ice-cold PBS.
3. Incubate 100 μL of beads with 400 μg of lysate in a total volume of 1 mL of cell lysis buffer (4°C, 1 h) with shaking or rotating.
4. Centrifuge at 6,000×*g* for 1 min, wash once in cell lysis buffer containing 0.5 M NaCl, once in cell lysis buffer, then resuspended in 190 μL of kinase buffer (all performed on ice).
5. Separate the beads into four aliquots of 40 μL and add 5 μL of pGS2 peptide solution while on ice.
   NOTE: All steps from this point involve the use of radioactive [γ-^{32}P]-MgATP, so appropriate safety precautions and disposal of radioactive waste should be strictly employed.
6. To start the kinase reaction, add 5 μL of 10× [γ-^{32}P]-MgATP solution and incubate on an immunoprecipitation shaker at 30°C for 15 min.
7. To terminate the reaction, spot 40 μL onto 1.5×1.5-cm squares of P81 paper (marked numerically in pencil) and immediately immerse in 75 mM orthophosphoric acid.
8. Wash the P81 papers 3× 10 min in 75 mM orthophosphoric acid. Use a magnetic stirrer to continually agitate the acid to ensure efficient washing of the papers.

9. Discard the final wash and briefly immerse the papers in acetone. Discard the acetone and air-dry the papers (this can be expedited using a hair dryer).
10. Place each paper in individually labelled tubes. Add 2 mL of scintillant to each tube and measure the radioactivity. Alternatively, the radioactivity can be measured by Cerenkov counting (if an appropriate detector is available), which does not require the addition of scintillation fluid.
11. Calculation of GSK3 specific activity requires the measurement of the specific activity of the ATP. 1 μL of 10× [γ-^{32}P]-MgATP contains 1 nmole of ATP, therefore the counts per minute (cpm) in 1 μL of this solution (counted in 2 mL of scintillant as above) provides the counts per minute per nanomole for the substrate (usually around 200,000 cpm/nmole). This must be counted simultaneously with the papers. Therefore, if GSK3 precipitated from 100 μg of cell protein produces 10,000 cpm in 15 min, then the specific activity of the GSK3 in that cell lysate is 10,000/200,000 cpm/nmole phosphate transferred per 15 min/0.1 mg cell protein under the assay conditions used; or 0.0333 nmoles/min/mg or 33 pmoles/min/mg.

   Use of this assay allows direct comparison of GSK3 activities between experiments and cell systems from different days and in different locations.
12. **Figure 4.3** demonstrates that 1 h exposure to IGF1 produces a 33% reduction in the specific activity of total GSK3 in SH-SY5Y cells, while there is no measurable effect of 1 h exposure of Wnt3a on total GSK3 activity (it should be noted that others have reported a 40–50% reduction upon 10 min exposure to Wnt proteins in other systems *(10, 12)*).

3.7. Comparison of Methodology for Analysis of GSK3 Activity

As described above, care must be taken in selection of the most appropriate method when analysing GSK3 activity. Upon IGF1 stimulation, GSK3 Ser21/9 phosphorylation increases threefold (**Fig. 4.2a**), and this equates to a drop in phosphorylation of the GSK3 substrate CRMP2 of 44% (**Fig. 4.2b**) and a reduction in GSK3 specific activity of 33% (**Fig. 4.3**). If all other regulation of GSK3 is constant (e.g. Tyr279/216 phosphorylation, **Fig. 4.2a**) then the threefold induction of Ser21/9 phosphorylation predicts an increase of stoichiometry at these sites from 0.1 to 0.3 mol/mol, since 67% of GSK3 remains active. Interestingly, the analysis of CRMP2 phosphorylation is very close to that of the specific activity measurement. This would suggest that the phosphorylation of Thr514/509 of CRMP2 by GSK3 is very sensitive to changes in cellular GSK3 activity. Each proposed substrate of GSK3 has to be analysed in combination with GSK3 activity in order to establish whether it is representative of changes in GSK3 activity.

In contrast, Wnt3a treatment of cells for 1 h does not alter the specific activity of GSK3 (**Fig. 4.3**). This is similar to experiments where GSK3β associated to axin was isolated and measured following challenge of cells with Xwnt8 *(17)*. However, partial purification of GSK3 on a MonoS column followed by in vitro assay suggested that Drosophila Wingless (Wg) could reduce total cellular GSK3 activity in fibroblasts by as much as 50% *(12)*. It is likely that the discrepancy is due to a transient inhibition of GSK3 by Wnt *(10, 12)*. However, the phosphorylation of the known Wnt/GSK3 pathway target β-catenin is almost completely lost and this reduction is sustained even after 1 h exposure to Wnt (**Fig. 4.1b**). This strongly suggests that either only a small pool of Wnt-responsive GSK3 is present in SH-SY5Y cells or that the inhibition of GSK3 is transient but the dephosphorylation of β-catenin is sustained. Either way, β-catenin phosphorylation is purely a qualitative method for measuring regulation of GSK3 by Wnts in cells, and does not reflect either the degree or duration of regulation of GSK3 activity.

4. Notes

1. GSK3 is ubiquitously expressed. Therefore, the protocols described here could be adapted for any cell line/tissue.
2. GSK3 is inhibited by all growth factor receptor tyrosine kinases investigated (e.g. insulin, epidermal growth factor [EGF], NGF, PDGF, hepatocyte growth factor [HGF], brain-derived neurotrophic factor [BDNF], neurotrophin [NT]-3). Here, we have focussed our study on IGF1.
3. It is generally considered that any Wnt molecules that regulate β-catenin activity inhibit GSK3. Reduced cellular GSK3 activity has been demonstrated for Drosophila *Wingless (12)*, Xenopus Wnt8 *(17)*, and mammalian Wnt1 *(10)*. We have used mammalian Wnt3a in our studies.
4. Many commercial antibodies that recognise phosphorylated and non-phosphorylated epitopes on GSK3 are listed in Table 4.1. We have not systematically tested all of the antibodies listed. However, of the antibodies that we have used, we recommend the antibodies listed in **Section 2.3.1**.
5. Instead of milk in the blocking and antibody solutions, some antibodies work best in 1% BSA, especially antibodies that detect phospho-tyrosine epitopes. Also, some companies recommend a higher concentration of milk than 1%. For example, Cell Signaling Technology recommends the use of the pSer21/9-GSK3 antibody in 5% milk. However, our experience

shows that its sensitivity is improved in 1% milk without loss of specificity.

6. The luminescence emitted by the ECL/horseradish peroxidase (HRP)-conjugated secondary antibodies decreases with time, so it is best to expose the membranes to film immediately after addition of the ECL reagent, especially when using antibodies with low sensitivity.
7. Stripping then re-probing the nitrocellulose membranes with another antibody is an option when samples are in short supply, however in general it is not recommended. The stripping procedure is often inconsistent across the surface of the membrane, resulting in unreliable band intensities across the lanes, as well as higher-than-normal background. In addition, stripped membranes cannot be re-probed with phospho-specific antibodies, since the stripping solution also removes phosphate moieties from proteins on the membrane. As an alternative, membranes should be washed to remove the ECL reagent, re-blocked, and incubated with an antibody that recognises an antigen of a molecular weight that is sufficiently different to that of the first antibody/antigen. The resultant film will contain bands for both the first and second antibodies used on this membrane, but the large difference in molecular weight will easily distinguish the two antigens.
8. The immunoblotting approach described here is semi-quantitative. Provided the intensities of each band are in the linear range of detection, this will provide comparative data to bands within the same analysis (membrane). If absolute quantification is required, a standard curve of known amounts of purified antigen must be analysed, preferably on the same gel but at least developed at the same time.
9. The immunoprecipitation assay of GSK3 is described. However there are variations on this procedure that have proved useful. For example Cook et al. partially purified GSK3 on a MonoS FPLC column prior to in vitro assay of the fractions against the pGS2 peptide *(12)*. This could be used to measure either the growth factor or Wnt regulation of GSK3. Secondly, Cross et al. developed a procedure to measure GSK3 in a crude cell lysate by performing the in vitro GSK3 assay (*see* **Section 3.6**) with the dephosphorylated (unprimed) GS (glycogen synthase) peptide substrate as well as the pGS2 (phospho-glycogen synthase peptide 2). peptide *(20)*. The difference in the two measurements should be GSK3 activity. However, it should be noted that the peptide has several phosphorylatable residues and therefore it may be targeted by other kinases in the lysate (reducing the signal-to-noise ratio or interfering with GSK3 phosphorylation). In addition, the large number of kinases likely to be active (and potentially affected by growth factor stimulation) can deplete

the ATP and reduce the linearity of the assay. Finally, the other kinases in the assay may indirectly regulate GSK3 activity during the in vitro assay, hence this assay may not truly represent the activity state of GSK3 in cells. A third modification of the immunoprecipitation assay involves pre-incubating the GSK3 with or without the phosphatase PP2A *(21)*. This treatment removes the inhibitory phosphorylation at Ser21/9 thereby fully activating the enzyme. Comparison of GSK3 activity with or without pre-treatment with PP2A provides an estimate of the percentage of active versus inactive (Ser21/9-phosphorylated) GSK3 in the cells under each experimental condition.

Acknowledgments

This work was supported by the Alzheimer's Research Trust and Diabetes UK.

References

1. Woodgett, J. R. (1990) Molecular cloning and expression of glycogen synthase kinase-3/factor A. *EMBO J.* 9, 2431–2438.
2. Hansen, L., Arden, K. C., Rasmussen, S. B., Viars, C. S., Vestergaard, H., Hansen, T., et al. (1997) Chromosomal mapping and mutational analysis of the coding region of the glycogen synthase kinase-3alpha and beta isoforms in patients with NIDDM. *Diabetologia* 40, 940–946.
3. Shaw, P. C., Davies, A. F., Lau, K. F., Garcia-Barcelo, M., Waye, M. M., Lovestone, S., et al. (1998) Isolation and chromosomal mapping of human glycogen synthase kinase-3 alpha and -3 beta encoding genes. *Genome* 41, 720–727.
4. Mukai, F., Ishiguro, K., Sano, Y., and Fujita, S. C. (2002) Alternative splicing isoform of tau protein kinase I/glycogen synthase kinase 3beta. *J. Neurochem.* 81, 1073–1083.
5. Embi, N., Rylatt, D. B., and Cohen, P. (1980) Glycogen synthase kinase-3 from rabbit skeletal muscle. Separation from cyclic-AMP-dependent protein kinase and phosphorylase kinase. *Eur. J. Biochem.* 107, 519–527.
6. Cohen, P. and Goedert, M. (2004) GSK3 inhibitors: development and therapeutic potential. *Nat. Rev. Drug Discov.* 3, 479–487.
7. Frame, S. and Cohen, P. (2001) GSK3 takes centre stage more than 20 years after its discovery. *Biochem. J.* 359, 1–16.
8. Frame, S., Cohen, P., and Biondi, R. M. (2001) A common phosphate binding site explains the unique substrate specificity of GSK3 and its inactivation by phosphorylation. *Mol. Cell* 7, 1321–1327.
9. Thomas, G. M., Frame, S., Goedert, M., Nathke, I., Polakis, P., and Cohen, P. (1999) A GSK3-binding peptide from FRAT1 selectively inhibits the GSK3-catalysed phosphorylation of axin and beta-catenin. *FEBS Lett.* 458, 247–251.
10. Ding, V. W., Chen, R. H., and McCormick, F. (2000) Differential regulation of glycogen synthase kinase 3beta by insulin and Wnt signaling. *J. Biol. Chem.* 275, 32475–32481.
11. Liu, X., Rubin, J. S., and Kimmel, A. R. (2005) Rapid, Wnt-induced changes in GSK3beta associations that regulate beta-catenin stabilization are mediated by Galpha proteins. *Curr. Biol.* 15, 1989–1997.
12. Cook, D., Fry, M. J., Hughes, K., Sumathipala, R., Woodgett, J. R., and Dale, T. C. (1996) Wingless inactivates glycogen synthase kinase-3 via an intracellular signalling pathway which involves a protein kinase C. *EMBO J.* 15, 4526–4536.
13. Sutherland, C., Leighton, I. A., and Cohen, P. (1993) Inactivation of glycogen synthase kinase-3 beta by phosphorylation: new kinase connections in insulin and growth-factor signalling. *Biochem. J.* 296 (Pt 1), 15–19.

14. Cole, A., Frame, S., and Cohen, P. (2004) Further evidence that the tyrosine phosphorylation of glycogen synthase kinase-3 (GSK3) in mammalian cells is an autophosphorylation event. *Biochem. J.* 377, 249–255.
15. Cole, A. R., Knebel, A., Morrice, N. A., Robertson, L. A., Irving, A. J., Connolly, C. N. et al. (2004) GSK-3 phosphorylation of the Alzheimer epitope within collapsin response mediator proteins regulates axon elongation in primary neurons. *J. Biol. Chem.* 279, 50176–50180.
16. Hart, M. J., de los, S. R., Albert, I. N., Rubinfeld, B., and Polakis, P. (1998) Downregulation of beta-catenin by human Axin and its association with the APC tumor suppressor, beta-catenin and GSK3 beta. *Curr. Biol.* 8, 573–581.
17. Itoh, K., Krupnik, V. E., and Sokol, S. Y. (1998) Axis determination in Xenopus involves biochemical interactions of axin, glycogen synthase kinase 3 and beta-catenin. *Curr. Biol.* 8, 591–594.
18. Laemmli, U. K. (1970) Cleavage of structural proteins during the assembly of the head of bacteriophage T4. *Nature* 227, 680–685.
19. Bradford, M. M. (1976) A rapid and sensitive method for the quantitation of microgram quantities of protein utilizing the principle of protein-dye binding. *Anal. Biochem.* 72, 248–254.
20. Cross, D. A., Culbert, A. A., Chalmers, K. A., Facci, L., Skaper, S. D., and Reith, A. D. (2001) Selective small-molecule inhibitors of glycogen synthase kinase-3 activity protect primary neurones from death. *J. Neurochem.* 77, 94–102.
21. Cross, D. A., Alessi, D. R., Vandenheede, J. R., McDowell, H. E., Hundal, H. S., and Cohen, P. (1994) The inhibition of glycogen synthase kinase-3 by insulin or insulin-like growth factor 1 in the rat skeletal muscle cell line L6 is blocked by wortmannin, but not by rapamycin: evidence that wortmannin blocks activation of the mitogen-activated protein kinase pathway in L6 cells between Ras and Raf. *Biochem. J.* 303 (Pt 1), 21–26.

Chapter 5

Inhibition of Glycogen Synthase Kinase-3

Andrei V. Ougolkov and Daniel D. Billadeau

Abstract

There are two homologous forms of glycogen synthase kinase (GSK)-3, GSK-3α and GSK-3β, which play overlapping roles in the regulation of Wnt, Hedgehog, and insulin pathways, as well as the activation of nuclear factor (NF)-κB-mediated gene transcription. These signaling pathways regulate gene transcription, cell cycle, apoptosis, inflammation, glucose metabolism, stem-cell renewal, and differentiation. More than 50 GSK-3 inhibitors representing a wide range of chemical structures have already been identified, and their utility in the treatment of type II diabetes mellitus, Alzheimer's disease, bipolar diseases, cancer, and other human pathologies is currently being investigated. Here, we discuss two methods of GSK-3 inhibition, which can be used to determine the involvement of GSK-3 in a cellular process of interest.

Key words: Glycogen synthase kinase-3, β-catenin, glycogen synthase.

1. Introduction

Glycogen synthase kinase (GSK)-3 is a serine/threonine protein kinase which was first described as a component of insulin-mediated signaling pathway *(1)*. There are two homologous mammalian isoforms, encoded by different genes, GSK-3α and GSK-3β, sharing 97% sequence similarity within their kinase catalytic domains *(2)*. In contrast to most kinases, GSK-3 is inactivated by phosphorylation *(2)*. GSK-3 is active in nonstimulated cells, whereas other kinases require cell stimulation for activation *(3)*. GSK-3 has been shown to phosphorylate numerous physiological substrates, including metabolic enzymes and transcription factors *(2)*. Recently, GSK-3 has emerged as an attractive therapeutic target for the treatment of neurodegenerative diseases, noninsulin-dependent diabetes mellitus, inflammation, and cancer *(4–7)*.

Elizabeth Vincan (ed.), *Wnt Signaling, Volume I: Pathway Methods and Mammalian Models, vol. 468*

Book doi: 10.1007/978-1-59745-249-6

There are different pools of GSK-3 in the cell. The pool of GSK-3 participating in Wnt signaling is part of a multiprotein complex known as the β-catenin destruction complex, which includes β-catenin, axin, and the tumor suppressor, adenomatous polyposis coli (APC) protein *(3)*. In the absence of Wnt signaling, GSK-3 targets β-catenin for ubiquitin-mediated proteosomal degradation by phosphorylation of β-catenin, axin, and APC *(3)*. Phosphorylation of APC by GSK-3 facilitates the interaction of β-catenin and APC *(8)*, whereas phosphorylation of axin by GSK-3 stabilizes this protein *(9)*. Exposure of cells to Wnt ligands leads to β-catenin accumulation, most likely through disruption of the interaction of axin with GSK-3 *(3)*, although the exact mechanism still needs to be clarified.

Of note, the mechanism of GSK-3 inhibition by insulin signaling is different from that of Wnt signaling *(10)*. Insulin receptor signaling leads to the activation of the serine/threonine kinase Akt, which will phosphorylate GSK-3 at specific serine residues (Ser21 in GSK-3α and Ser9 in GSK-3β), leading to the inactivation of GSK-3 kinase activity *(10)*. However, this phosphorylation and inactivation of GSK-3 by Akt does not result in β-catenin nuclear accumulation *(10)*. These data indicate that the insulin/Akt signaling pathway does not affect the β-catenin destruction complex, but only affects a pool of GSK-3 that is involved in the regulation of glycogen synthase *(4)*. Accordingly, Wnt signaling does not lead to GSK-3 phosphorylation at Ser21/Ser9 *(10)*.

Although GSK-3β has been the primary focus of most research, it is clear that GSK-3α and GSK-3β have some overlapping substrates. In fact, it has been demonstrated that tissues from GSK-3β-deficient mice do not show accumulation of β-catenin *(11)*. In GSK-3β-deficient mouse cells, GSK-3α was also found precipitated with axin *(11)*. These data indicate that GSK-3α and GSK-3β can substitute for each other in the regulation of β-catenin. Thus, both GSK-3α and GSK-3β need to be inhibited to decrease β-catenin phosphorylation **(Fig. 5.1a)** and induce β-catenin nuclear accumulation and transcriptional activity.

1.1. GSK-3 Inhibitors

Currently, more than 50 inhibitors of GSK-3 have been identified and this list is growing. Chemical structure, potency, and other features of GSK-3 inhibitors have been extensively reviewed *(12–14)*. Due to the high amino acid sequence homology of GSK-3α and GSK-3β kinase domains, no pharmacological inhibitors of GSK-3 have been identified that can differentiate between the two isoforms. Thus, care should be taken when using GSK-3 inhibitors to assign a role for GSK-3 isoforms in specific cellular functions. In fact, it is advisable to use GSK-3 inhibitors along with genetic models that disrupt GSK-3 signaling (e.g., gene-targeted cells such as mouse embryonic fibroblasts [MEFs] lacking GSK-3α or GSK-3β, or RNA interference techniques)

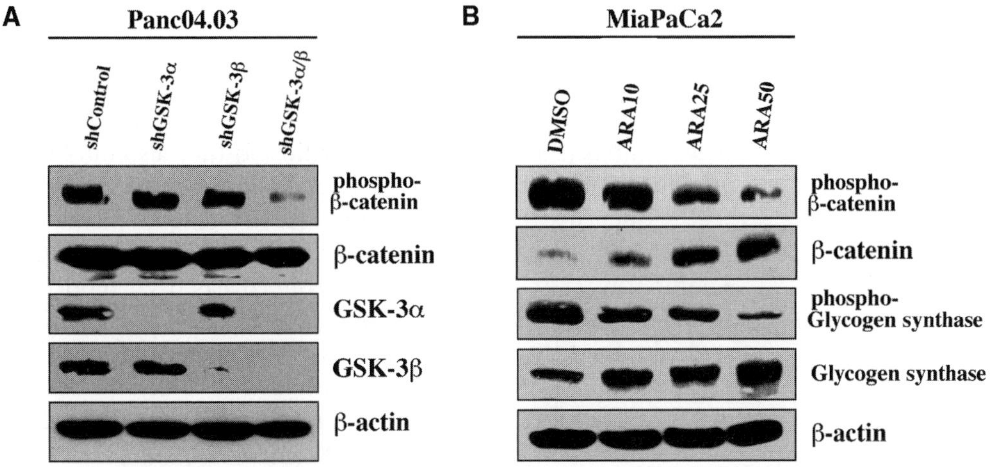

Fig. 5.1. Inhibition of GSK-3 activity. **A** Depletion of both GSK-3α and GSK-3β is required for a decrease of β-catenin phosphorylation at Ser33/37. Panc04.03 cells were transfected with shControl, shGSK-3α, and shGSK-3β vectors, as described in the Section 3, Methods; 48 h post-transfection, cell lysates were obtained, separated by SDS-PAGE, transferred to PVDF membrane, and probed with antibodies to indicated proteins. Because β-catenin is highly expressed in Panc04.03 cells, changes in total level of β-catenin are not detectable by Western analysis in transfected cells, although the nuclear pool of β-catenin might be increased in GSK-3α/β-depleted cells. To analyze activation of β-catenin/Tcf signaling, co-transfection with TOP/FOP luciferase reporter plasmids (TCF reporter plasmid kit; Upstate) is needed. **B** Pharmacological inhibition of GSK-3 activity. MiaPaCa2 pancreatic cancer cells do not express E-cadherin and there is free (free from binding to E-cadherin) cytosolic β-catenin pool in these cells. Although β-catenin is expressed in MiaPaCa2 cells, it is hardly detectable by Western analysis, because the β-catenin free cytoplasmic pool is rapidly phosphorylated by active GSK-3 and degraded. Inhibition of GSK-3 stabilizes β-catenin and leads to an increase of β-catenin expression easily detectable by Western analysis. To find out the optimal concentration of GSK-3 inhibitor in cells of interest at certain time point, cells are treated with different concentration of drug (e.g., AR-A014418 as presented at 10, 25, and 50 μM). At 24 h post-treatment, cell lysates were obtained; SDS-PAGE and Western blotting were performed as described in Section 3, Methods. AR-A014418 dose-dependent decreases of phospho-β-catenin (Ser33/37) and phospho-glycogen synthase (Ser645) expression, and increases in the total level of β-catenin and glycogen synthase protein expression were observed.

when examining the involvement of GSK-3 isoforms in the regulation of specific cellular processes.

1.1.1. Inorganic Elements as Inhibitors of GSK-3

Lithium was the first discovered GSK-3 inhibitor *(15)*. Despite the fact that millimolar concentrations (20–30 mM) of lithium are required to inhibit GSK-3 in cells, it has been used broadly as a pharmacological inhibitor of GSK-3 in vitro and in vivo *(13)*. Lithium inhibits GSK-3 through an unknown mechanism, although some hypotheses have been suggested. The first hypothesis proposes lithium as a magnesium (Mg^{2+})-competitive inhibition of GSK-3 *(16)*, whereas the second suggests lithium inhibitor by potassium deprivation *(17)*. Lithium is a nonspecific GSK-3 inhibitor and has alternative targets, e.g., inositol–phosphate phosphatases. Physiological concentrations of zinc ions have also been shown to directly inhibit GSK-3 in vitro in an ATP-noncompetitive manner *(18)*. It has been shown that treatment of HEK 293 cells with 20 μM $ZnCl_2$ for 2 h enhanced glycogen synthase activity twofold and increased cytoplasmic β-catenin levels *(18)*.

1.1.2. Small Molecule Inhibitors of GSK-3

Small molecule inhibitors of GSK-3 consist of two groups: ATP competitive and ATP noncompetitive. Among ATP-competitive inhibitors of GSK-3 are indirubins *(19)*, paullones *(20)*, anilinomaleimides *(21)*, and other small molecules *(22)*. However, these inhibitors are not GSK-3 specific and they also inhibit cyclin-dependent kinases (CDKs), due to the highly related amino acid sequence of the GSK-3 and CDK kinase domains *(19)*. Accordingly, numerous CDK inhibitors are very potent GSK-3 inhibitors *(19)*. Recently, AR-A014418 has been reported as an ATP-competitive, potent, and specific inhibitor of GSK-3 as determined by in vitro kinase assays toward cdk2, cdk5, and 25 other kinases *(23)*. Currently thiadiazolidinone (TDZD) is the only ATP-noncompetitive inhibitor of GSK-3 *(24)*. While the exact mechanism by which TDZD inhibits GSK-3 kinase activity is unclear, it has been suggested that the negative charge on the TDZD heterocycle recognizes the oxyanion-binding site of the enzyme *(24)*, thereby blocking its ability to recognize pre-phosphorylated substrates. The last and most specific method to inhibit cellular GSK-3 kinase activity is depletion of GSK-3 expression by RNA interference using small interfering RNAs (siRNAs) or short hairpin RNA (shRNA) targeting vectors specific for each isoform **(Fig. 5.1b)** *(25)*.

Although the inhibition of GSK-3 kinase activity by a certain inhibitor could be measured by in vitro kinase assay, we suggest Western blotting as a safer, faster, and easier way to measure the effect of an inhibitor on GSK-3 kinase activity. This can be achieved through the analysis of the phosphorylation state of GSK-3 substrates, e.g., glycogen synthase and β-catenin. Active GSK-3 phosphorylates β-catenin at Ser33/37 and glycogen synthase at Ser641/645. Commercial availability of antibodies that recognize phosphorylated forms of β-catenin (Ser33/37) and glycogen synthase (Ser641/645) provide straightforward assay protocols using immunoblotting.

2. Materials

2.1. Cell Culture and Lysis

1. MiaPaCa2 and Panc04.03 pancreatic cancer cells are maintained in DMEM and RPMI-1640 (Gibco/BRL, Bethesda, MD), respectively, supplemented with 10% fetal bovine serum (FBS; HyClone, Ogden, UT) and L-glutamine (Gibco/BRL).
2. Phosphate-buffered saline (PBS) containing 0.5 M ethylenediamine tetraacetic acid (EDTA).
3. Solution of trypsin (0.25%) containing 1 mM EDTA (Gibco/BRL).

4. Western lysis buffer for cell lysis: 10 mM Tris-HCl, pH 7.4, 50 mM NaCl, 5 mM EDTA, 50 mM NaF, 30 mM $Na_4P_2O_7$, and 1% (v/v) Triton-X100.
5. Protease inhibitors and phosphatase inhibitor are added fresh each time with the following final concentrations: 10 μg/mL aprotinin, 10 μg/mL leupeptin, 1 mM phenylmethanesulfonyl fluoride (PMSF), and 1 mM sodium orthovanadate.
6. Teflon cell scrapers (Fisher, Pittsburgh, PA).
7. 2× SDS sample buffer for electrophoresis: 0.0375 M Tris-HCl, pH 6.5, 8% (w/v) SDS, 10% (v/v) glycerol, 5% (v/v) β-mercaptoethanol, and 0.003% (w/v) bromophenol blue.

2.2. Sodium Dodecyl Sulfate (SDS) Polyacrylamide Gel Electrophoresis (PAGE)

1. Separating buffer (4×): 1.5 M Tris-HCl, pH 8.7, and 0.4% (w/v) SDS. Store at room temperature (RT).
2. Stacking buffer (4×): 0.5 M Tris-HCl, pH 6.8, and 0.4% (w/v) SDS. Store at RT.
3. 30% acrylamide/bis solution (37.5:1 with 2.6% C) and *N,N,N,N'*-tetramethyl-ethylenediamine (TEMED; Bio-Rad, Hercules, CA).
4. Ammonium persulfate: prepare 10% (w/v) solution in water and freeze aliquots at –20°C.
5. Water-saturated isobutanol. Store at RT.
6. Running buffer (10×): 250 mM Tris, 1.92 mM glycine, and 1% (w/v) SDS. Store at RT.
7. Prestained molecular weight markers: Rainbow marker (GE Healthcare, Buckinghamshire, UK).

2.3. Immunoblotting

1. Transfer buffer (10×): 250 mM Tris-Base and 1.92 M glycine. Store at RT.
2. Polyvinylidene difluoride (PVDF) membrane from Millipore (Bedford, MA) and gel blot paper from ISCBioExpress (Kaysville and UT).
3. Tris-buffered saline (TBS) with Tween (TBS-T): Prepare 10× stock with 1.37 M NaCl, 27 mM KCl, 250 mM Tris-HCl, pH 7.4, and 1% (v/v) Tween-20.
4. Blocking buffer: 4% (w/v) bovine serum albumin (BSA) and 0.1% (w/v) sodium azide in TBS.
5. Primary antibody dilution buffer: TBS-T supplemented with 2% (w/v) BSA.
6. Antibodies: Phospho-β-catenin Ser33/Ser37/Thr41, phospho-glycogen synthase Ser641, and glycogen synthase (Cell Signaling Technologies, Beverly, MA); β-catenin and GSK-3β (BD PharMingen, San Diego, CA); GSK-3α (Upstate, Lake Placid, NY); and β-actin (Sigma, St. Louis, MO).

7. Secondary antibodies: anti-mouse and anti-rabbit IgG conjugated to horseradish peroxidase (Santa Cruz Biotechnology, Santa Cruz, CA).
8. Enhanced chemiluminescent (ECL) reagents from Pierce (Rockford, IL).

2.4. Stripping Blots

1. Stripping buffer: 7 M Guanidine HCl in water. Store at RT.
2. Wash buffer: TBS-T.

2.5. Transfection of Cells with Silencing shGSK-3α/β Vectors

1. GSK-3α-and GSK-3β-specific targeting shRNA vectors were generated as previously described *(26)* using the target sequences 5′-TTGTGAGGCTGAGATACTT-3′ and 5′-GATTATACCTCTAGTATAG-3′, respectively.
2. Electro Square Porator ECM830 (BTX; Genetronics Inc., San Diego, CA).
3. Electroporation cuvettes with 4-mm gap (VWR Scientific Products, West Chester, PA).

2.6. GSK-3 Inhibitor AR-A014418

1. Prepare 50 mM stock solution of AR-A014418 *(23)* (Calbiochem, Gibbstown, NJ) in dimethylsulfoxide (DMSO). Store at 4°C.

3. Methods

3.1. Preparation of Samples for Assay of GSK-3 Activity by Western Blotting

1. Cells are treated with AR-A014418 or other GSK-3 inhibitors according to the protocol and the medium is then removed by aspiration. Place dishes on ice and wash with 2 mL of ice-cold PBS.
2. Add 0.5–1 mL of Western lysis buffer to each 10-cm plate, scrape cells, and transfer to a 1.5-mL microcentrifuge tube. Incubate tubes on ice for 15 min.
3. Clear cell lysate by centrifuging samples at 18,000×*g* in a refrigerated tabletop microcentrifuge for 15 min at 4°C. Transfer supernatant to a fresh microcentrifuge tube.
4. Measure proteins concentration by the Bradford method.
5. Mix 50 μg of proteins with 30 μL of 2× SDS-gel sample buffer and then boil for 5 min.

3.2. Sds-page

1. These instructions assume the use of a Hoeffer SE-600 gel system.
2. Prepare a 1.5-mm thick, 10% gel by mixing 8.0 mL of 4× separating buffer with 13.0 mL of water, 10.5 mL of acrylamide/bis solution, 100 μL of ammonium persulfate

solution, and 20 μL of TEMED. Pour the separating gel, and overlay with water-saturated isobutanol.

3. Pour off the isobutanol and rinse the top of the gel twice with water.
4. Prepare the stacking gel by mixing 1.3 mL of 4× stacking buffer with 3.0 mL of water, 0.8 mL of acrylamide/bis solution, 30 μL of ammonium persulfate solution, and 10 μL of TEMED. Pour the stack and insert the comb.
5. Add the running buffer to the upper and lower chambers of the gel unit and load the samples. Complete the assembly of the gel unit and run gel overnight at 8 mA.

3.3. Western Blotting

1. Once gel running is complete, disassemble the gel unit. Place the separating gel on top of the PVDF membrane submerged on top of the blot paper. Another sheet of paper is wetted in the transfer buffer and laid on top of the gel. Place the cassette into the transfer tank.
2. Transfer is accomplished at 90 V for 45 min in a transfer tank with a connection to a refrigerated/circulating water bath maintaining a temperature of 10–15°C.
3. After transfer, incubate the membrane in 50 mL blocking buffer for 1 h at RT on a rocking platform.
4. Next, incubate the membrane with 30 mL of primary antibody dilution buffer containing one of the following antibodies and incubate as indicated: phospho-β-catenin (dilution 1:1,000; overnight), phospho-glycogen synthase (dilution 1:1,000; 1 h at RT), GSK-3α (dilution 1:1,000; 1 h at RT), GSK-3β (dilution 1:1,000; 1 h at RT), or β-actin (dilution 1:10,000; 1 h at RT).
5. Wash the membrane three times for 5 min each in TBS-T.
6. Incubate the membrane for 1 h at RT with 30 mL of TBS-T containing anti-mouse or anti-rabbit IgG conjugated to horseradish peroxidase at a dilution of 1:5,000.
7. Wash the membrane three times for 5 min each in TBS-T.
8. Incubate the membrane with 5 mL of each enhanced chemiluminescent reagent for 5 min at RT, wrap in plastic wrap, and visualize on X-ray film. An example result is shown in **Fig. 5.1.**

3.4. Stripping and Reprobing Blots for Total β-Catenin and Glycogen Synthase

1. Incubate the membrane for 30 min at RT with 30 mL of stripping buffer. The PVDF membrane will turn opaque.
2. Rinse the membrane in water.
3. Wash the membrane three times for 10 min each in TBS-T.
4. Reprobe the membrane with β-catenin (dilution 1:1,000; 1 h at RT) or glycogen synthase (dilution 1:1,000; 1 h at RT) antibodies according to protocol described above. An example result is shown in Fig. 5.1b.

3.5. Transfection of Cells with Silencing shGSK-3α/β Vectors

1. Collect Panc04.03 cells from 10-cm culture dishes in the mid- to late-logarithmic phase of growth. First, wash the plate with 2 mL of PBS-EDTA solution and remove by aspiration. Next, add 1 mL of trypsin and incubate at 37°C for 2–3 min.
2. To the trypsinized cells, add 10 mL of complete RPMI medium, transfer to a 15-mL snapcap polypropylene tube, and centrifuge at 500×*g* at 4°C for 5 min.
3. Resuspend the cell pellet in the RPMI medium and measure the cell number using a hemocytometer.
4. Collect the cells by centrifugation, as described in step 3, and resuspend them in serum-free RPMI medium at RT at a concentration of 5×10^6 cells/mL.
5. Transfer a 400-μL aliquot of the cell suspension into a sterile 1.5-mL Eppendorf tube.
6. Add 30 μg of plasmid DNA into the 1.5-mL Eppendorf tube. Gently mix the cells and DNA by pipetting the solution up and down. Keep mixture for 10 min at RT.
7. Transfer the cell–DNA mixture to an electroporation cuvette and place in the holder for the electroporator. Discharge the device at 375 V for 10 ms. Remove the cuvette and keep at RT for 10 min (*see* **Note 1**).
8. Transfer the electroporated cells to a 10-cm tissue culture dish. Transfer the dish to a humidified incubator at 37°C with an atmosphere of 5–7% CO_2.
9. The next day, observe the plates and change growth medium.
10. Prepare cell lysates 24–72 h post-transfection.

4. Notes

1. Transient transfection can be conducted by three methods: lipofection, electroporation, and polybrene plus DMSO shock. We chose electroporation because this method is the most efficient procedure for Panc04.03 pancreatic cancer cells, with a transfection efficiency 90%.

Acknowledgments

This work was supported in part by the Mayo Foundation, and a Specialized Program of Research Excellence (SPORE) grant in pancreatic cancer (P50 CA102701) to DDB.

References

1. Plyte SE, Hughes K, Nikolakaki E, Pulverer BJ, Woodgett JR. (1992) Glycogen synthase kinase-3: functions in oncogenesis and development. *Biochim Biophys Acta*. **1114**, 147–162.
2. Frame S, Cohen P. (2001) GSK3 takes centre stage more than 20 years after its discovery. *Biochem J*. **359**, 1–16.
3. Doble BW, Woodgett JR. (2003) GSK-3: tricks of the trade for a multi-tasking kinase. *J Cell Sci*. **116**, 1175–1186.
4. Cohen P, Frame S. (2001) The renaissance of GSK3. *Nat Rev Mol Cell Biol*. **2**, 769–776.
5. Cohen P, Goedert M. (2004) GSK3 inhibitors: development and therapeutic potential. *Nat Rev Drug Discov*. **3**, 479–487.
6. Ougolkov AV, Billadeau DD. (2006) Targeting GSK-3: a promising approach for cancer therapy? *Future Oncol*. **2**, 91–100.
7. Dugo L, Collin M, Thiemermann C. (2007) Glycogen synthase kinase 3beta as a target for the therapy of shock and inflammation. *Shock*. **27**, 113–123.
8. Rubinfeld B, Albert I, Porfiri E, Fiol C, Munemitsu S, Polakis P. (1996) Binding of GSK3beta to the APC-beta-catenin complex and regulation of complex assembly. *Science*. **272**, 1023–1026.
9. Yamamoto H, Kishida S, Kishida M, Ikeda S, Takada S, Kikuchi A. (1999) Phosphorylation of axin, a Wnt signal negative regulator, by glycogen synthase kinase-3beta regulates its stability. *J Biol Chem*. **274**, 10681–10684.
10. Ding VW, Chen RH, McCormick F. (2000) Differential regulation of glycogen synthase kinase 3beta by insulin and Wnt signaling. *J Biol Chem*. **275**, 32475–32481.
11. Hoeflich KP, Luo J, Rubie EA, Tsao MS, Jin O, Woodgett JR. (2000) Requirement for glycogen synthase kinase-3beta in cell survival and NF-kappaB activation. *Nature*. **406**, 86–90.
12. Meijer L, Flajolet M, Greengard P. (2004) Pharmacological inhibitors of glycogen synthase kinase 3. *Trends Pharmacol Sci*. **25**, 471–480.
13. Martinez A, Castro A, Dorronsoro I, Alonso M. (2002) Glycogen synthase kinase 3 (GSK-3) inhibitors as new promising drugs for diabetes, neurodegeneration, cancer, and inflammation. *Med Res Rev*. **22**, 373–384.
14. Alonso M, Martinez A. (2004) GSK-3 inhibitors: discoveries and developments. *Curr Med Chem*. **11**, 755–763.
15. Stambolic V, Ruel L, Woodgett JR. (1996) Lithium inhibits glycogen synthase kinase-3 activity and mimics wingless signalling in intact cells. *Curr Biol*. **6**, 1664–1668.
16. Ryves WJ, Harwood AJ. (2001) Lithium inhibits glycogen synthase kinase-3 by competition for magnesium. *Biochem Biophys Res Commun*. **280**, 720–725.
17. Mora A, Sabio G, Gonzalez-Polo RA, et al. (2001) Lithium inhibits caspase 3 activation and dephosphorylation of PKB and GSK3 induced by K+ deprivation in cerebellar granule cells. *J Neurochem*. **78**, 199–206.
18. Ilouz R, Kaidanovich O, Gurwitz D, Eldar-Finkelman H. (2002) Inhibition of glycogen synthase kinase-3beta by bivalent zinc ions: insight into the insulin-mimetic action of zinc. *Biochem Biophys Res Commun*. **295**, 102–106.
19. Leclerc S, Garnier M, Hoessel R, et al. (2001) Indirubins inhibit glycogen synthase kinase-3 beta and CDK5/p25, two protein kinases involved in abnormal tau phosphorylation in Alzheimer's disease. A property common to most cyclin-dependent kinase inhibitors? *J Biol Chem*. **276**, 251–260.
20. Leost M, Schultz C, Link A, et al. (2000) Paullones are potent inhibitors of glycogen synthase kinase-3beta and cyclin-dependent kinase 5/p25. *Eur J Biochem*. **267**, 5983–5994.
21. Smith DG, Buffet M, Fenwick AE, et al. (2001) 3-Anilino-4-arylmaleimides: potent and selective inhibitors of glycogen synthase kinase-3 (GSK-3). *Bioorg Med Chem Lett*. **11**, 635–639.
22. Cross DA, Culbert AA, Chalmers KA, Facci L, Skaper SD, Reith AD. (2001) Selective small-molecule inhibitors of glycogen synthase kinase-3 activity protect primary neurones from death. *J Neurochem*. **77**, 94–102.
23. Bhat R, Xue Y, Berg S, et al. (2003) Structural insights and biological effects of glycogen synthase kinase 3-specific inhibitor AR-A014418. *J Biol Chem*. **278**, 45937–45945.
24. Martinez A, Alonso M, Castro A, Perez C, Moreno FJ. (2002) First non-ATP competitive glycogen synthase kinase 3 beta (GSK-3beta) inhibitors: thiadiazolidinones (TDZD) as potential drugs for the treatment of Alzheimer's disease. *J Med Chem*. **45**, 1292–1299.
25. Ougolkov AV, Fernandez-Zapico ME, Savoy DN, Urrutia RA, Billadeau DD. (2005) Glycogen synthase kinase-3beta participates in nuclear factor kappaB-mediated gene transcription and cell survival in pancreatic cancer cells. *Cancer Res*. **65**, 2076–2081.
26. Zakaria S, Gomez TS, Savoy DN, et al. (2004) Differential regulation of TCR-mediated gene transcription by Vav family members. *J Exp Med*. **199**, 429–434.

Chapter 6

Detection of Cytoplasmic and Nuclear Localization of Adenomatous Polyposis Coli (APC) Protein in Cells

Mariana Brocardo and Beric R. Henderson

Abstract

The adenomatous polyposis coli (APC) tumour suppressor gene is mutated in the majority of colon cancers. APC is a multi-domain protein whose distribution at different subcellular locations correlates with unique cellular processes. Our laboratory has focused on the link between APC subcellular location and function, and has characterized pathways for the trafficking of APC both into and out of the nucleus. Antibody specificity is an important factor in the determination of APC localization, and in this chapter we outline a strategy for the unambiguous detection of APC using a combination of biochemical and cell biology approaches.

Key words: APC, Western blotting, Subcellular localization, Colon cancer, Immunofluorescence microscopy, Antibodies, APC detection, Electrophoresis.

1. Introduction

Colorectal cancer is one of the most common malignancies in the world. The identification of mutations in the adenomatous polyposis coli (APC) gene in patients affected by familial adenomatous polyposis coli (FAP) was critical in understanding the molecular pathogenesis of colon cancer *(1–3)*. More important, APC mutations are detectable in >80% of all inherited and sporadic colorectal cancer patients and are directly linked with the early onset of colorectal cancer *(2)*. Most APC gene mutations are detected within the central mutation cluster region (MCR), and generate truncated APC peptides that lack a C-terminus. The truncated APC peptides display altered function, including

Elizabeth Vincan (ed.), *Wnt Signaling, Volume I: Pathway Methods and Mammalian Models, vol. 468*

Book doi: 10.1007/978-1-59745-249-6

a reduction in their ability to associate with microtubules (MT), facilitate correct chromosome segregation, and promote β-catenin degradation *(2, 3)*.

The APC gene encodes a large protein consisting of 2,843 amino acids with a predicted molecular weight of ~300 kDa, which interacts with proteins involved in the Wnt signalling pathway and in cytoskeletal organization *(3)*. APC can shuttle between the nucleus and cytoplasm and participates in a variety of cellular functions *(4)*. For instance, APC is known to regulate β-catenin localization and turnover *(1–3)*. In addition, APC is a cytoskeletal regulator and accumulates at the ends of MT bundles near the plasma membrane, where APC may contribute to cell migration *(5–7)*. Endogenous APC has been detected at several MT-associated locations, including membrane protrusions, kinetochores *(8, 9)*, the mitotic spindle *(10)*, and centrosomes *(11)* by immunofluorescence microscopy. The ability of APC to shuttle between nucleus and cytoplasm (reviewed in ref. *(4)*) implicates its movement into the nucleus, however many recent studies examining nuclear APC have lacked suitable controls. In this regard, the specificity of antibodies used to detect APC has caused discrepancies in reports relating to APC accumulation in the nucleus. In most cases, APC has been detected by cell staining and microscopy, but not confirmed using biochemical methods. We have shown, using cell fractionation and Western blot analysis, that endogenous forms of APC are predominantly cytoplasmic, although a fraction of APC is localized to the nucleus, and that APC nuclear export activity is not abolished by truncating cancer mutations *(12)*. These techniques are outlined in this chapter.

2. Materials

2.1. Cell Culture

1. Human SW480 (APC truncated at amino acid 1,338), HT-29 (APC truncated at amino acids 853 and 1,555), and HCT 116 (full-length APC) colon carcinoma cells are cultured in Dulbecco's Modified Eagle's Medium (DMEM) supplemented with 10% fetal bovine serum under standard tissue culture conditions.
2. A solution of trypsin (0.25%) and ethylenediamine tetra acetic acid (EDTA) (1 mM) is prepared, aliquoted, and can be stored at 4°C for 1 month and –20°C for longer periods.
3. Leptomycin B (LMB): working solution is diluted in ethanol at 2 μg/μL stored in aliquots at –20°C. Treatment of cells is for 5 h at a dose of 8 ng/mL.

2.2. RNA Interference to Silence the Expression of APC

1. LipofectAMINE 2000 from Life Technologies, Inc. (Invitrogen, Carlsbad, CA; Cat. No 11668-019) is supplied in liquid form at a concentration of 1 mg/mL. Store at 4°C. Do not freeze.
2. Double-stranded 21-mer RNA oligonucleotides homologous to sequences in human APC are purchased as purified duplexes (e.g. Qiagen-Xeragon Inc., Valencia, CA). The DNA target sequence is 5,867–AGGGGCAGCAACTGATGAAAA.

2.3. Cell Fractionation

1. Cytosolic and nuclear extracts are prepared using a fractionation NE-PER kit from Pierce biotechnology (Pierce, Rockford, IL; Cat. No 78833) according to the manufacturer's instructions.
2. Bio-Rad protein solution (5×): this is a commercial solution from Bio-Rad Laboratories (Hercules, CA; Cat. No 500-0006) based on the Bradford dye-binding procedure, which measures the color change of Coomassie Brilliant Blue G-250 dye when it binds to protein.

2.4. Sodium Dodecyl Sulfate (SDS)–Polyacrylamide Gel Electrophoresis (PAGE) for APC Truncated Forms

1. Laemmli Sample buffer (4×): 250 mM Tris-HCl, pH6.8, 8% (w/v) sodium dodecyl sulfate (SDS), 40% (w/v) glycerol, and 0.1% (w/v) bromophenol blue. Store at room temperature and add β-mercaptoethanol to 560 mM before using. The aliquot with β-mercaptoethanol should be discarded after use. Diluted concentrations of this buffer can be prepared when the volume of the samples is small.
2. Running buffer (5×): 0.125 M Tris base, pH to 8.3 with HCl, 0.96 M glycine, and 0.5% SDS. Store at room temperature.
3. Separating buffer (1×): 1.5 M Tris-HCl, pH 8.7, and 0.4% SDS. Store at room temperature.
4. Stacking buffer (1×): 0.5 M Tris-HCl pH 6.8, and 0.4% SDS. Store at room temperature.
5. Forty percent acrylamide/bis solution (29:1). This is a neurotoxin solution and so care should be taken to avoid exposure. Use gloves and a mask to prepare this solution. Store at 4°C.
6. *N,N,N,N'*-Tetramethyl-ethylenediamine (TEMED). Store at room temperature, but note that the quality declines after opening the bottle.
7. Ammonium persulfate (APS): prepare 10% solution in water and immediately freeze in single-use (500-μL) aliquots at –20°C. A working aliquot can be kept at 4°C for no more than 1 month (*see* **Note 1**).

8. Isopropanol. Store at room temperature.
9. Pre-stained commercial molecular weight markers.
10. Gel electrophoresis apparatus.

2.5. Western Blotting for APC Truncated Forms

1. Blotting buffer (5×): 0.125 M Tris-HCl pH 8.3, and 0.95 M glycine. Store at room temperature and cool before using. To prepare 1× solution, add 20% (v/v) methanol (*see* **Note 2**).
2. Nitrocellulose membrane from Millipore (Billerica, MA) and 3MM chromatography paper from Whatman (Maidstone, UK).
3. Red Ponceau solution (1×): 0.2% red Ponceau powder, 3% trichloroacetic acid (TCA) in water. Store at room temperature.

2.6. Agarose Gel Electrophoresis to Detect Full Length APC (~300 kDa)

1. Laemmli sample buffer (4×): 250 mM Tris-Cl, pH 6.8, 8% (w/v) SDS, 40% (w/v) glycerol, and 0.1% (w/v) bromophenol blue. Store at room temperature and add β-mercaptoethanol (to 560 mM) before using. The aliquot with β-mercaptoethanol should be discarded. Diluted concentrations of this buffer can be prepared when the volume of the samples are small.
2. Laemmli running buffer (5×): 125 mM Tris base, 960 mM glycine, and 0.5% (w/v) SDS.
3. Tris buffer EDTA (TBE) (1×): 89 mM Tris base, 89 mM boric acid, and 2 mM EDTA, pH 8.0. Store at room temperature.
4. 2.5% Agarose I in (TBE)/0.1% SDS.
5. Pre-stained commercial molecular weight markers.

2.7. Capillary Blotting for Full-Length APC

1. Tris-buffered saline (TBS) pH 7.6 (1×): 20 mM Tris-HCl, pH 7.6, and 137 mM NaCl. Store at room temperature.
2. Nitrocellulose membrane from Millipore and 3MM Chr chromatography paper (Whatman).

2.8. Detection of Full-Length and Truncated APC on Nitrocellulose Membrane

1. TBS pH 7.6 (10×): 200 mM Tris-HCl, pH 7.6, and 1.37 M NaCl. Store at room temperature.
2. TBS with Tween (TBS-T) (10×): prepare a TBS solution (X1) and add 0.1% (v/v) Tween-20.
3. 5% (w/v) *Non-fat* dry milk in TBS-T.
4. Primary antibody: APC (Ab-1) antibody from Oncogene Research.
5. Secondary antibody: anti-mouse IgG conjugated to horseradish peroxidase.
6. Enhanced chemiluminescent (ECL) reagents from Amersham (GE Healthcare, Piscataway, NJ).
7. Hyperfilm ECL: High performance chemiluminescence film (Amersham).

2.9. Immunofluorescence Microscopy for Detection of Full-Length and Truncated APC

1. Microscope cover slips (22×22 mm No. 1).
2. Phosphate-buffered saline (PBS) (10×): 1.37 M NaCl, 27 mM KCl, 100 mM Na_2HPO_4, and 18 mM KH_2PO_4 (adjust to pH 7.4 with HCl if necessary). Autoclave before storage at room temperature. To obtain a 1× solution, dilute 1 part with 9 parts of water.
3. Fixation solution: 3.7% (v/v) molecular biology-grade formalin. Prepare a solution in PBS for each experiment at room temperature.
4. Permeabilization solution: 0.2% (v/v) Triton X-100 in PBS.
5. Blocking solution: 3% (w/v) albumin from bovine serum (BSA).
6. The first, second, and third antibody solution are prepared in blocking solution.
7. Second antibody: anti-mouse or anti-rabbit conjugated to biotin from Dako Corp. (Glostrup, Denmark) diluted 1:500.
8. Third antibody: Texas Red conjugated to avidin from Vector Laboratories (Burlingame, CA) diluted 1:500.
9. Nuclear stain: 0.05 μg/mL of Hoechst 33258 is prepared in blocking solution.
10. Mounting medium: Vectashield (Vector Laboratories).

3. Methods

APC is a nuclear–cytoplasmic shuttling protein with diverse functions. However, the immunostaining of cells to detect nuclear localization of endogenous APC is prone to false positive results due to cross-reactivity of antibodies with other nuclear proteins. To obtain reliable results, it is important to compare APC localization using more than one approach, and if using fluorescence microscopy it is very important to complement those experiments by using cell fractionation and immunoblotting. The use of RNA interference (RNAi) provides a means to control for specificity of the cell staining of APC. In this section, we outline a procedure for the detection of full-length APC (using agarose gel electrophoresis) and truncated mutant APC (using SDS-PAGE) from colon tumor cell lines. A method for the staining and imaging of APC in fixed cells is also presented to complement the Western blotting, and some details are provided to incorporate the use of anti-APC small interfering RNA (siRNA) and of LMB (nuclear export inhibitor) to observe the specific silencing and nuclear sequestration of APC, respectively (illustrated later by appropriate data figures).

3.1. Cell Culture,Tranfection, and Preparation of Samples

1. The colon cancer cell lines are maintained in 75-cm^2 tissue flasks and upon reaching confluence are passaged using trypsin/EDTA. At 24 h before transfection, seed the experimental cultures into 25-cm^2 flasks for Western blot analysis, or seed the cells onto sterile coverslips for analysis by immunofluorescence microscopy. One 25-cm^2 flask or one coverslip is required for each experimental data point.
2. For RNAi, transfect the cells when at 50% of confluence (*see* **Note 3**) with 6 μg siRNA duplex per 25-cm^2 flask or 3 μg siRNA per standard coverslip in 2 mL or 1 mL of DMEM medium, respectively (without serum or antibiotics, *see* **Note 4**). Transfection of RNA duplexes is performed using LipofectAMINE reagent. Add the reagent to cell medium and incubate for 6 h. At the end of incubation, remove the medium and replace with complete DMEM. Continue incubation for 48 h.
3. To transfect cells with plasmid, seed the cells onto coverslips. Transfect the cells when at 75% of confluence (*see* **Note 3**) with 1–2 μg of DNA in 1 mL of DMEM (without serum and antibiotics, *see* **Note 4**) using LipofectAMINE reagent for 6 h. At this point, remove the medium and incubate the cells in complete DMEM for 48 h.

Treat the cells with 8 ng/mL LMB for 5 h and harvest the cells for cell fractionation.

4. To harvest and to obtain the cell fraction protein extracts, collect the medium of each flask in a 15-mL tube. Rinse the adherent cells with 2 mL of PBS and then collect them using 0.3 mL of trypsin/EDTA. Centrifuge the cellular suspension at 500×*g* for 7 min and remove the supernatant. Rinse the pellet with 500 μL of PBS and centrifuge at 500×*g* for 7 min. Again remove the supernatant and separate the cellular pellet into nuclear and cytoplasmic fractions using the NE-PER kit (as directed by manufacturer). Take an aliquot of each sample to determine the concentration of protein using Bradford solution.
5. When loading the gels, cytoplasmic and nuclear fractions are loaded in a way to reflect the proportional cellular protein contents. Equivalent amounts of each cell fraction (3:1 cytoplasmic to nuclear) are used; e.g. 60 μg of cytoplasm and 20 μg of nuclear extract.
6. Transfer the nuclear and cytoplasmic extracts to Eppendorf tubes. Add 4× Laemmli sample buffer to the samples to bring the buffer to 1×.
7. Close the tubes and heat them for 5 min at 95°C. After cooling on ice, pellet debris by centrifuging at 13,000×*g* for 5 min. The samples (supernatant) are now ready for separation on acrylamide or agarose.

3.2. SDS-PAGE for APC Truncated Forms

1. The instructions below are for the preparation of minigels using the BioRad Mini-Protean separation system. It is critical that the glass plates are cleaned with a detergent, rinsed extensively with distilled water, and dried without remaining residue.
2. Prepare 20 mL of a 1.5-mm thick, 8% gel by mixing: 3.8 mL of 40% acrylamide (29:1), 7.5 mL of separating buffer, 100 μL of 20% SDS, 100 μL of 10% APS, and 8.5 mL of water. Finally, apply 20 μL of TEMED, and pour the gel, leaving space for a stacking gel, and overlay with isopropanol. The gel should polymerize in about 20 min.
3. Pour off the isopropanol and rinse the top of the gel with water (*see* **Note 5**).
4. Prepare the stacking gel by mixing: 1.29 mL of 40% acrylamide (29:1), 1.25 mL stacking buffer, 50 μL of 20% SDS, 50 μL of 10% APS, and 7.4 mL water. Finally, apply 10 μL of TEMED. Use about 1 mL of this stacking gel solution to cover the top of the separating gel, and then insert the comb. The upper gel stack should polymerize within 20 min.
5. Prepare the "running" buffer by diluting 200 mL of the 5× running buffer up to final volume of 1 L with water in a measuring cylinder. Cover with Parafilm and mix by inverting.
6. Once the stacking gel has set, carefully remove the comb and, using a pipette or syringe with needle, flush the wells clean with water.
7. Assemble each gel unit to the electrical chamber and add the running buffer, ensuring that it covers the wells. Load 30 μL of sample into each well. Different volumes can be loaded depending on the sizes of the combs. Include one well for pre-stained molecular weight markers.
8. Complete the assembly of the gel unit and connect to a power supply. Initially set the voltage to 70 V to allow the samples to stack, and then, as they enter the separating gel, increase the voltage to 100 V (*see* **Note 6**). The blue dye fronts can be run off the gel if desired.

3.3. Western Blotting for APC Truncated Forms

1. The samples that have been separated by SDS-PAGE are then transferred to nitrocellulose membranes electrophoretically.
2. Prepare the blotting buffer by combining 200 mL of the 5× blotting buffer with 200 mL of methanol and water up to final volume of 1 L in a measuring cylinder. Cover with Parafilm and invert to mix.
3. Prepare a tray with blotting buffer and submerge two pieces of support foam (a component of the Bio-Rad Western transfer assembly), four sheets of 3MM paper, and a sheet of the nitrocellulose cut just larger than the size of the separating gel.

4. Disconnect the BioRad gel unit (or the equivalent) from the power supply and disassemble. Remove the stacking gel (optional) and rinse the separating gel in blotting buffer for 5 min. Take a transfer cassette and place on the top side (that which connects to the negative current) (*see* **Note 7**): one piece of support foam, two pieces of 3MM paper, the separating gel, then the sheet of nitrocellulose on top of the gel and two sheets of 3MM. Ensure that no bubbles are trapped in the resulting sandwich. The second wet foam sheet is laid on top and the transfer cassette is closed.
5. The cassette is placed into the transfer tank with blotting buffer, the lid put on the tank, and the power supply activated. Transfers can be carried out at either 0.090 A overnight or at 0.300 A for 2 h at 4°C.
6. Transfer the nuclear and cytoplasmic extracts to Eppendorf tubes. Add 4x Laemmli sample buffer to the samples to bring the buffer to 1x
7. Once the transfer is complete, take the cassette out of the tank and carefully disassemble, with the top sponge, sheets of 3MM, and gel removed. Rinse the nitrocellulose membrane with TBS. The coloured molecular weight markers should be clearly visible on the membrane. Stain the membrane with red Ponceau solution to detect the proteins.

3.4. Agarose Gel Electrophoresis to Detect Full-Length APC

Due to the large size of wild-type APC (~300 kDa), it is much more effectively visualized after separation by agarose gel electrophoresis rather than SDS-PAGE. Samples are separated on a 3% agarose gel made in TBE/0.1% SDS. The agarose gel is prepared using a vertical polyacrylamide gel apparatus with 1-mm spacers. First, pour a 1-cm 15% acrylamide plug to prevent leakage. Then pour the molten agarose while it is still warm. *There is no stacking gel.* The running buffer is Laemmli buffer. Run the gel at 70 V until the proteins in the 40–60 kDa range have migrated off the bottom of the gel.

3.5. Capillary Blotting for Full-Length APC

Transfer the proteins overnight onto a nitrocellulose membrane by downward capillary transfer as outlined below using TBS/0.04% SDS as the transfer buffer:

1. On a 2.5-cm stack of paper towels, place three pieces of gel-sized Whatman 3MM chromatography paper with the top piece pre-soaked in transfer buffer.
2. Place a pre-soaked, gel-sized piece of nitrocellulose membrane on top of the Whatman papers.
3. Place the gel on top of the membrane and cover the exposed papers with plastic wrap.
4. Place two gel-sized pieces of pre-soaked Whatman 3MM paper on top of the gel.

5. On opposite sides of the blotting stack, place a tray containing transfer buffer.
6. Lay two pre-soaked, long pieces of Whatman paper across the top of the gel setup so that the ends are submerged in the trays containing transfer buffer.
7. Cover the entire setup with plastic wrap and allow the transfer to proceed overnight.
8. After overnight transfer, mark the position of protein standards on the nitrocellulose membrane with a pencil and rinse the membrane with TBS-T. The detection of proteins with red Ponceau solution is not possible here.

3.6. Detection of Full-Length and Truncated Forms of APC

1. Incubate the membrane for 1 h in 5% non-fat dry milk in TBS-T at room temperature on a rocking platform.
2. Incubate the membrane with 1 μg/mL of APC (Ab-1) in 5% non-fat dry milk in TBS-T for 1 h at room temperature or overnight at 4°C on a rocking platform.
3. Wash the membrane three times, for 15 min each, in TBS-T at room temperature on a rocking platform.
4. Incubate the membrane with a secondary antibody freshly prepared for each experiment at a 1:8,000-fold dilution in 5% non-fat dry milk in TBS-T, at room temperature for 1 h.
5. Wash the membrane three times, for 15 min each, in TBS-T at room temperature on a rocking platform.
6. Develop the membrane using ECL chemiluminescent detection regents according to manufacturer instructions. Briefly, during the final wash, equal aliquots of each portion of the ECL reagent are warmed separately to room temperature and the remaining steps are done in a darkened room under safe light conditions. Once the final wash is removed from the blot, the ECL reagents are mixed together and then immediately added to the blot, which is placed between two pieces of sealable plastic (opened along one edge and closed after the ECL reagents are added). Then, rotate by hand for 1 min to ensure even coverage of the blot.
7. Remove the blot from the ECL reagents and place it between the leaves of an acetate sheet protector that has been cut to the size of an X-ray film cassette. Expose the membrane to film for 5 min. Adjust subsequent exposure times (from 30 sec until 24 h) as necessary (*see* **Fig. 6.1** for example of completed experiment).
8. The membranes can be re-probed with other antibodies as a fractionation or loading control.

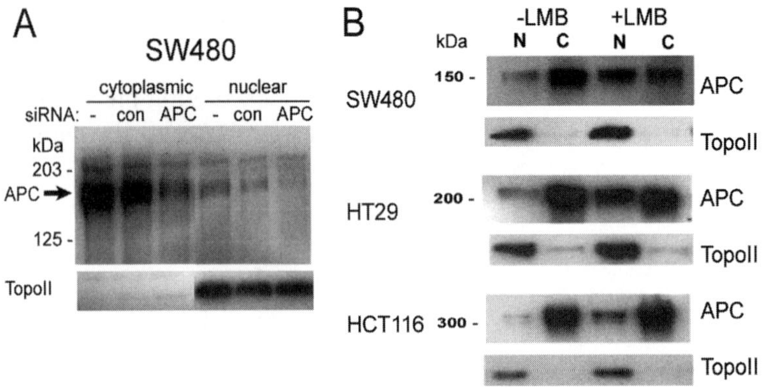

Fig. 6.1. Detection of APC by biochemical assay and its confirmation using RNAi. **A** Equivalent amounts of cytoplasmic or nuclear extracts from cells treated +/- siRNA (APC, or control BRCA1 siRNA [*con*]) were separated by SDS-PAGE and analysed by immunoblot. Truncated APC (150 kDa) was detected with antibody M-APC. The same membrane was reprobed for topoisomerase IIα (topoII; 170 kDa) as a fractionation control. APC siRNA effectively reduced cytoplasmic and nuclear APC by 80% relative to untransfected cells. **B** Nuclear (*N*) and cytoplasmic (*C*) cell extracts were prepared from SW480 ($APC^{mut/mut}$), HT29 ($APC^{mut/mut}$), and HCT116 ($APC^{wt/wt}$) colon cancer cell lines. The endogenous truncated forms of APC were separated by SDS-PAGE (Section 3.2), and full-length APC was separated by agarose gel electrophoresis (Section 3.4) prior to detection by Western blot using the M-APC antibody (1:4,000 dilution). (Reproduced from **ref.** *(12)* with permission from the Nature Publishing Group).

3.7. Immunofluorescence Microscopy of APC

The following is a general protocol for staining fixed cells and microscopic detection of APC. Cells can be transfected with anti-APC RNAi duplexes (*see* **Section 2** and **ref.** *(12)*) to confirm that nuclear or cytoplasmic staining observed is actually APC. Note that it is best to confirm siRNA efficacy first by Western blotting.

1. Remove the medium from cells (untransfected controls or 48 h post-transfection), and wash the coverslips three times with PBS.
2. Fix the cells with 3.7% formalin/PBS for 20 min at room temperature. The formalin is discarded (into a hazardous waste container) and the samples washed twice PBS.
3. Permeabilize the cells by incubation in PBS/Triton X-100 for 10 min at room temperature, and then rinse twice more with PBS.
4. Block the samples by incubating in 3% BSA solution for 1 h.
5. Remove the blocking solution without washing.
6. Apply 150 µL of the diluted primary antibody (e.g. Ali-28; Upstate) in blocking solution for 1 h at room temperature.
7. Wash three times with PBS.
8. Apply 150 µL of the diluted secondary antibody conjugated to biotin in blocking solution for 1 h at room temperature (*see* **Note 8**).

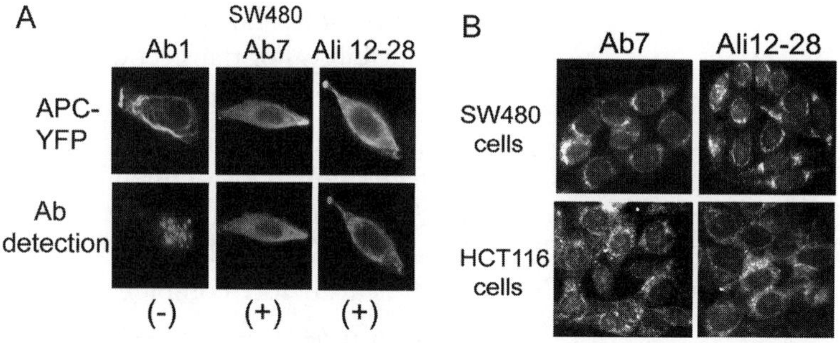

Fig. 6.2. Detection of endogenous APC in SW480 by immunofluorescence microscopy. **A** Full-length pAPC-YFP *(12)* was transfected into SW480 colon cancer cells. After 48 h, ectopic YFP localisation of APC-YFP was compared using different antibodies against APC protein by immunostaining. Positive co-localisation of each antibody with APC-YFP expression is indicated (+). **B** Cellular staining patterns observed with different APC antibodies (*Ab*). SW480 or HCT116 were fixed on glass coverslips in formalin and analyzed by immunofluorescence microscopy as outlined in Section 3.7. (Reproduced from **ref.** *(12)* with permission from the Nature Publishing Group).

9. Wash three times with PBS.
10. Apply 150 µL of avidin–Texas Red plus Hoechst solution in blocking solution for 40 min at room temperature.
11. Wash three times with PBS.
12. The samples are then ready to be mounted. The coverslip is carefully inverted onto a drop of mounting medium on a microscope slide. The slides are air-dried at room temperature and nail varnish is used to seal the sample. The sample can be viewed immediately after the varnish is dry, or can be stored in the dark at 4°C for long periods.
13. The slides are then examined with any standard fluorescence microscope system, and images collected using a CCD camera (*see* **Fig. 6.2**).

4. Notes

1. APS is not stable, and needs to be of good quality to polymerize. If the aliquot is not fresh, there will not be polymerization or the polymerization will be irregular.
2. The presence of methanol is necessary to fix the proteins to the membrane after transferring. Otherwise, the proteins will be lost from the gel into the buffer.
3. Transfect cells with siRNA when cells are at low density at the time of transfection to avoid cell overgrowth increasing the gene silencing. The optimal cell density should be determined empirically

for each cell line, and kept constant for all experiments to ensure reproducibility. For transfection of plasmid DNA, a higher confluence is recommended at the time of transfection.

4. The presence of serum may inhibit cationic lipid-mediated transfection. It is recommended that medium with reduced concentrations of serum such as Opti-MEM from Invitrogen (Cat. No. 31985-062) is used to dilute LipofectAMINE and nucleic acids before complexing.
5. The gel interface should be straight before applying the stacking gel. This is accomplished by using solvents of different density than the gel.
6. The different pH between the running and stacking gels enhances concentration of the samples at the stacking gel. The presence of SDS allows the samples to be separated based on their molecular weights (denatured gels).
7. It is vitally important to ensure this orientation or the proteins will be lost from the gel into the buffer rather than transferred to the nitrocellulose.
8. This step is included to amplify the signal but can also increase the background. It is possible to avoid this step by incubating with a second antibody anti-mouse or anti-rabbit conjugated to the fluorophore.

Acknowledgments

The authors thank the National Health and Medical Research Council of Australia and the Australian Research Council for funding support of this research.

References

1. Polakis, P. (2000) Wnt signaling and cancer. *Genes Dev* **14**, 1837–1851.
2. Fodde, R., Smits, R., and Clevers, H. (2001) APC, signal transduction and genetic instability in colorectal cancer. *Nat. Rev. Cancer* **1**, 55–67.
3. Lustig, B., and Behrens, J. (2003) The Wnt signaling pathway and its role in tumor development. (2003) *J. Cancer Res. Clin. Oncol.* **129**, 199–221.
4. Henderson, B.R. and Fagotto, F. (2002) The ins and out of APC and beta-catenin nuclear transport. *EMBO Reports* **3**, 834–839.
5. Näthke, I., Adams, C. L., Polakis, P., Sellin, J. H., and Nelson, W. J. (1996) The adenomatous polyposis coli tumor suppressor protein localizes to plasma membrane sites involved in active cell migration. *J. Cell Biol.* **134**, 165–179.
6. Mimori-Kiyosue, Y., Shiina, N., and Tsukita, S. (2000) Adenomatous polyposis coli (APC) protein moves along microtubules and concentrates at their growing ends in epithelial cells. *J. Cell Biol.* **148**, 505–518.
7. Jimbo, T., Kawasaki, Y., Koyama, R., Sato, R., Takada, S., Haraguchi, K., and Akiyama, T. (2002) Identification of a link between the tumour suppressor APC and the kinesin superfamily. *Nat. Cell Biol.* **4**, 323–327.

8. Kaplan, K. B., Burds, A. A., Swedlow, J. R., Bekir, S. S., Sorger, P. K., and Nathke, I. S. (2001) A role for the Adenomatous Polyposis Coli protein in chromosome segregation. *Nat. Cell Biol.* **3**, 429–432.
9. Fodde, R., Kuipers, J., Rosenberg, C., Smits, R., Kielman, M., Gaspar, C., van Es, J. H., Breukel, C., Wiegant, J., Giles, R. H., and Clevers, H. (2001) Mutations in the APC tumour suppressor gene cause chromosomal instability. *Nat. Cell Biol.* **3**, 433–438.
10. Dikovskaya, D., Newton, I. P., and Näthke, I. S. (2004) The adenomatous polyposis coli protein is required for the formation of robust spindles formed in CSF Xenopus extracts. *Mol. Biol. Cell* **15**, 2978–2991.
11. Louie, R. K., Bahmanyar, S., Siemers, K. A., Votin, V., Chang, P., Stearns, T., Nelson, W. J., and Barth, A. I. M., Adenomatous polyposis coli and EB1 localize in close proximity of the mother centriole and EB1 is a functional component of centrosomes. *J. Cell Sci.* **117**, 1117–1128.
12. Brocardo, M., Näthke, I. S., and Henderson, B. R. (2005) Redefining the subcellular location and transport of APC: new insights using a panel of antibodies. *EMBO Reports* **6**, 184–190.

Chapter 7

Detection of β-Catenin Localization by Immunohistochemistry

Nick Barker and Maaike van den Born

Abstract

β-catenin is a widely expressed 90-kDa protein with dual functions in cell adhesion and Wnt signalling. At the membrane, β-catenin forms complexes with E-cadherin to generate cell adhesion complexes responsible for maintaining the structural integrity of many epithelial tissues. On the other hand, accumulation of β-catenin in the nucleus in response to Wnt signalling facilitates complex formation with Tcf transcription factors, leading to activation of a genetic program influencing a range of cellular processes including cell growth, cell movement, and cell fate. Chronic activation of the Wnt signalling pathway as a result of mutations in key pathway components, including β-catenin itself, is a major cause of cancer. The associated increase in nuclear β-catenin protein is therefore considered to be a hallmark of Wnt-driven cancers and an invaluable tool to detect active Wnt signalling.

Key words: Immunohistochemistry: Antibody: Wnt signalling: β-catenin: Nuclear localization.

1. Introduction

β-catenin has long been known to have a crucial role in cell adhesion as a component of the E-cadherin complex at the cell membrane. Just over a decade ago, we discovered that β-catenin is not faithful to E-cadherin—it also forms specific complexes with Tcf transcription factors in the nucleus in response to activation of the canonical Wnt signalling cascade *(1–3)*.

In the absence of a Wnt signal, β-catenin is serially phosphorylated at a set of conserved serine/threonine residues at its N-terminus by the actions of the kinases, casein kinase (CK)-1 and glycogen synthase kinase (GSK)-3 (see Chapter 1 for an overview). This marks β-catenin for recognition by the β-transducin

Elizabeth Vincan (ed.), *Wnt Signaling, Volume I: Pathway Methods and Mammalian Models, vol. 468*

Book doi: 10.1007/978-1-59745-249-6

repeat-containing protein (β-TRCP) ubiquitin ligase and targets it for degradation by the proteasome. Activation of the canonical Wnt signalling cascade triggers the accumulation of non-phosphorylated β-catenin in the cytoplasm. β-Catenin subsequently enters the nucleus where it binds Tcf proteins to generate active transcription factor complexes capable of activating a set of Wnt target genes and eliciting a physiological response from the cell. This response is usually transient, with β-catenin being rapidly cleared from the nucleus following dissipation of the Wnt signal at the membrane. However, chronic accumulation of β-catenin in the nucleus sometimes occurs following mutation of key Wnt pathway components that prevent effective regulation of β-catenin levels in the cell. The resulting constitutive activation of the Wnt target gene program is considered to be a major contributing factor in many human disorders including cancers and oligodontia (tooth loss) *(4)*.

In simple terms, the presence of β-catenin in the nucleus can be viewed as evidence of Wnt signalling activity in a tissue. In tissue sections, this is most commonly visualized by immunohistochemistry using commercially available antibodies recognizing β-catenin.

2. Materials

2.1. Tissue Block Preparation

1. Formalin: 4% (w/v) formaldehyde in phosphate-buffered saline (PBS).
2. Tissue-dehydration solutions series: 70%, 96%, and 100% (v/v) ethanol.
3. Xylene.
4. Liquid paraffin (60°C).

2.2. Tissue Section Preparation

1. De-wax solvent: Xylene.
2. Tissue-rehydration solutions series: 100%, 96%, 80%, 70%, 60%, and 25% (v/v) ethanol.
3. Tissue cassettes (Klinipath, Duiven, The Netherlands).
4. Slide racks (Klinipath).
5. Cold plate (approximately –12°C).
6. Microtome.
7. Starfrost microscope slides.
8. Coverslips (Menzel-Gläser, Braunschweig, Germany).

2.3. Immunohistochemical Staining

1. Peroxidase (PO) blocking buffer: 0.040 M citric acid, 0.121 M disodium hydrogen phosphate, and 0.030 M sodium azide. Add 1 mL of H_2O_2 solution (30% [v/v] stock solution) per 10 mL PO blocking buffer before use (*see* **Note 1**).
2. Antigen retrieval solution: 10 mM Tris and 1 mM EDTA, pH 9.0. To prepare, dissolve 1.2 g Tris and 0.37 g EDTA in 1 L distilled water and dissolve by stirring. The resulting solution should already be at the required pH 9.0.
3. Steel pan for boiling slides (we use a high-sided asparagus pan).
4. Hot plate.
5. PBS.
6. Pre-blocking buffer: 0.05% (w/v) bovine serum albumin (BSA) in PBS.
7. 30% (w/v) Hydrogen peroxide (H_2O_2) stock solution.
8. Anti-β-catenin antibody diluted 1:100 (BD Transduction Laboratories, Franklin Lakes, NJ; #610154) in pre-blocking buffer.
9. Polymer horseradish peroxidase (HRP)-labelled anti-mouse Envision (Dako, Glostrup, Denmark) (*see* **Note 2**).
10. Diaminobenzidine (DAB) peroxidase substrate. To make up 10× DAB, dissolve 600 mg DAB (Sigma, St. Louis, MO) in 100 mL distilled water. Make aliquots of 1 mL and store at –20°C. Dilute the 10× DAB before use with 9 mL of phosphate–citrate buffer and then add 10 µL of 30% H_2O_2 stock. Protect from light. Diluted DAB/H_2O_2 solution is stable at room temperature for up to 6 h. Discard unused solution.
11. Mayer's haematoxylin. To prepare, dissolve 1 g haematoxylin-monohydrate (Merck), 0.2 g sodium iodate, 50 g aluminum potassium sulphate dodecahydrate (Alum), 50 g chloral hydrate, and 1 g citric acid in 1 L of distilled water by stirring well for at least 1 h.
12. Slide-mounting solutions series: 50%, 60%, 70%, 96%, and 100% (v/v) ethanol, and xylene.
13. Pertex mounting medium (Histolab, Gothenburg, Sweden).

3. Methods

Levels of β-catenin present in the adhesion complexes at the membrane are generally not influenced by Wnt signalling. This is in stark contrast to β-catenin levels in the cytoplasm and nucleus, which vary dramatically according to the status of the Wnt signalling pathway. If we take the intestine as an example, immunohistochemical staining using β-catenin antibodies readily detects β-catenin at the membrane, cytoplasm, and nucleus of cells located towards the base of the crypts of Lieberkuhn (*see* **Fig. 7.1a** and **b**). This reflects the presence of an active Wnt signalling pathway in this region of the tissue. In contrast, β-catenin is only visible at the membrane of epithelial cells on the villi, reflecting the absence of Wnt signalling in this tissue (*see* **Fig. 7.1a**). As we already discussed, several different human cancers are thought to be driven by inappropriate activation of the Wnt pathway, which leads to the uncontrolled accumulation of β-catenin in the cytoplasm and nucleus. This is particularly evident in intestinal polyps, which are the earliest stages of colon cancer (*see* **Fig. 7.1c**). Immunohistochemical staining using specific β-catenin antibodies is therefore a powerful tool for determining not only where active Wnt signalling is present within a tissue, but can also be used to detect early Wnt-driven cancers.

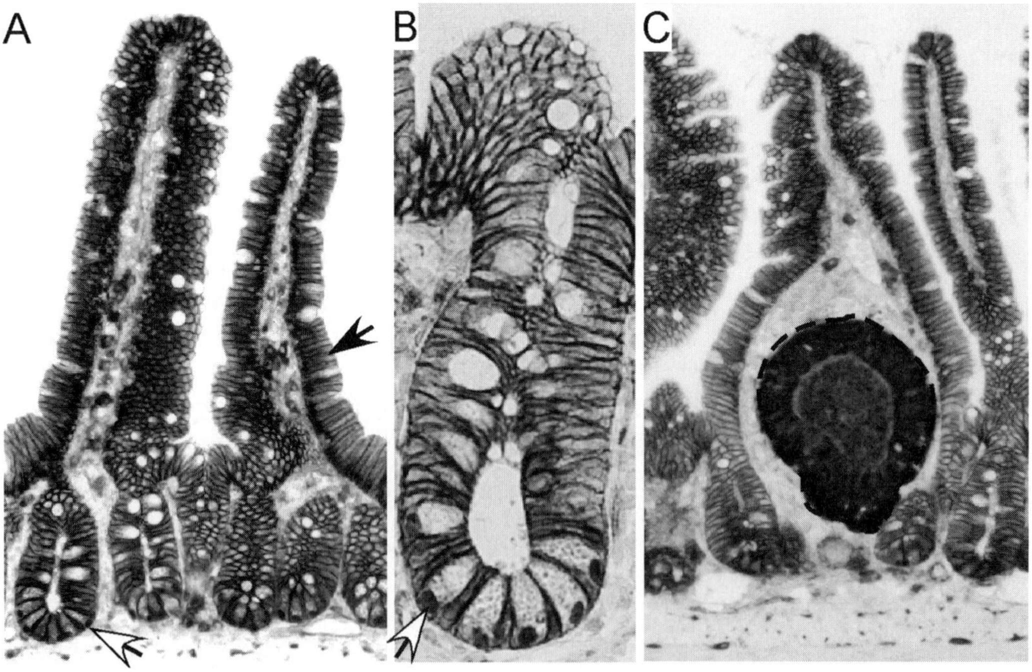

Fig. 7.1. Detection of β-catenin in tissue sections. Immunohistochemical staining of β-catenin in mouse small intestine using the BD Transduction Laboratories monoclonal antibody #610154. **A** β-Catenin is visible only at the membrane on the villus epithelium (*black arrow*). **A, B** β-Catenin accumulates in the nucleus of Paneth cells at the base of the intestinal crypts (*white arrows*), reflecting the presence of an activated Wnt signalling pathway in this region. **C** Massive accumulation of β-catenin in the cytoplasm and nucleus of an early intestinal adenoma as a result of constitutive activation of the Wnt signalling pathway (*circled*).

3.1. Preparation of Paraffin-Embedded Tissue Sections

1. Place the freshly isolated tissues immediately into a 50-mL Falcon tube containing at least a 10-fold volume of formalin (4% formaldehyde). Fix the tissues by constant mixing on a rolling platform overnight at room temperature.
2. Remove the formalin and wash the tissues twice for 20 min each in PBS at room temperature on a rolling platform.
3. Transfer the tissues into a Klinipath tissue cassette and label the front panel clearly using a pencil to record the tissue type, origin, and date.
4. Dehydrate the tissues by immersing the cassette in a 20-fold volume of 70% ethanol for 2 h at 4°C (*see* **Note 3**). The 70% ethanol is refreshed after 1 h. This procedure is repeated using 96% ethanol and then 100% ethanol.
5. Transfer the tissue cassettes to a 20-fold volume of xylene and incubate for 2 h at room temperature.
6. Remove the tissue cassettes from the xylene and dab onto tissues to remove any excess solvent. Then immerse the cassettes in liquid paraffin (60°C) overnight.
7. Remove the tissue cassettes from the liquid paraffin and transfer them to an embedding station, where they are embedded in paraffin using metal moulds. Carefully orient the tissues within the paraffin using heated forceps. Transfer the paraffin blocks to a cold-plate for 30 min to allow them to solidify.
8. Prepare 4-μm-thick sections using a microtome and transfer them to a clean water bath (at 40°C). The sections are allowed to spread out on the water and then floated onto the upper surface of coated microscope slides.
9. Transfer the slides into a slide rack and de-wax the slides by immersing in xylene (2× 5 min).
10. Hydrate the tissue sections by serial immersion in 100% ethanol (2× 1 min), 96% ethanol (2× 1 min), 70% ethanol (2× 1 min), and then distilled water (2× 1 min).

3.2. Immunohistochemical Staining Using Anti-β-catenin Antibodies

1. Rinse the slides three times for 15 sec each in distilled water.
2. Block endogenous peroxidase activity by immersing the slide rack containing the slides in a Klinipath container filled with PO-peroxidase blocking buffer for 15 min at room temperature (*see* **Note 4**).
3. Rinse the slide rack containing the slides three times for 15 sec each in distilled water.
4. Antigen retrieval is achieved by immersing the slides in a pan of boiling Tris-EDTA, pH 9.0, for 20 min (*see* **Note 5**).
5. Wash the slides three times for 2 min each by immersion in PBS.

6. Transfer the slides onto a staining platform of choice with the tissue surface facing upwards.
7. Add pre-blocking buffer drop-wise from a height of ~10 cm using a disposable Pasteur pipette (Sarstedt) so that the entire slide surface is covered. Then incubate the slides at room temperature for 30 min to facilitate blocking of non-specific binding of the antibody.
8. Drain the pre-blocking buffer off the slide onto tissue paper and dab the edges of the slides onto the tissues to ensure excess fluid is removed from the surface of the slide.
9. Add 125 μL of the primary β-catenin antibody (diluted 1:100 in pre-blocking buffer—*see* **Note 6**) drop-wise from a height of ~10 cm to cover the entire slide surface (*see* **Note 7**). Then incubate the slides in a humid atmosphere at room temperature for 2 h.
10. Transfer the slides to a slide rack and rinse them three times for 2 min each in PBS.
11. Return the slides to the staining platform and add 125 μL of anti-mouse Envision (polymer HRP-labelled) to the surface of the slide as described above. Incubate the slides in a humid atmosphere at room temperature for 1 h.
12. Transfer the slides to a slide rack and rinse them three times for 2 min each in PBS.
13. Return the slides to the staining platform and detect bound peroxidase activity by adding the DAB substrate to the slide surface as described above. Incubate the slides at room temperature for 10 min.
14. Transfer the slides to a slide rack and rinse them three times for 2 min each in PBS.
15. Counterstain the nuclei by immersing the slides in Mayer's haematoxylin for 2 min. Remove excess counterstain by rinsing the slides under running tap water for 5 min (*see* **Note 8**).
16. The samples are then ready to be mounted. Dehydrate the slides by serial immersion for 1 min each in 50% and 60% ethanol, followed by 2 min each in 70% and 96% ethanol, and finally three times 1 min each in fresh volumes of 100% ethanol. The slides are then immersed twice for 1 min each in xylene.
17. Add mounting medium (Pertex) and gently place a coverslip over the tissue sample. Remove air bubbles from under the coverslip by gently applying pressure from above using forceps.
18. The slides are then viewed under a phase-contrast microscope for evidence of β-catenin staining. Nuclei will be stained blue, whilst β-catenin will be readily visible as a brown stain.

Examples of β-catenin staining at the membrane and in the nucleus are shown in **Fig. 7.1** (note that the tissue in **Fig. 7.1** has only been lightly counterstained to allow grey-tone printing).

4. Notes

1. Unless otherwise stated, all solutions are prepared using distilled water.
2. In our experience, when detecting binding of mouse primary antibodies such as the BD Transduction Laboratories anti-β-catenin, the anti-mouse Envision secondary antibodies labelled with HRP consistently give the best signal-to-background ratio.
3. All dehydration steps using ethanol are performed at 4°C to ensure a gradual reduction in hydroxyl (water) bonds within the tissue, thereby reducing tissue damage and minimizing the possibility of disrupting antibody binding sites.
4. The PO-peroxidase blocking buffer should always be freshly prepared.
5. It is important to start timing 20 min only after the pan has again reached the boiling point following immersion of the slides. Formalin fixation generates protein cross-links that mask the antigenic sites in tissue specimens, thereby producing weak or false negative staining for immunohistochemical detection of some proteins. The Tris-EDTA-based antigen retrieval procedure is designed to break these protein cross-links, thereby unmasking the antigens and epitopes in formalin-fixed and paraffin-embedded tissue sections and enhancing the staining intensity of the β-catenin antibodies.
6. We have found the BD Transduction Laboratories monoclonal antibody to be the most consistent in detecting nuclear β-catenin on paraffin tissue sections. However, this procedure can, in principle, be adapted for other commercially available antibodies.
7. Great care should always be taken to ensure sufficient blocking. Also, the antibody solutions should be added to the slides in sufficient volumes to achieve complete submersion of the tissue section. The tissue sections should never be allowed to dry out during the staining procedure to avoid the appearance of non-specific background staining.
8. Mayer's haematoxylin is a commonly used counterstain that colors the nuclei blue.

References

1. Behrens, J., von Kries, J. P., Kuhl, M., Bruhn, L., Wedlich, D., Grosschedl, R., et al. (1996) Functional interaction of beta-catenin with the transcription factor LEF-1. *Nature* **382**, 638–642.
2. Molenaar, M., van de Wetering, M., Oosterwegel, M., Peterson-Maduro, J., Godsave, S., Korinek, V., et al. (1996) XTcf-3 transcription factor mediates beta-catenin-induced axis formation in Xenopus embryos. *Cell* **86**, 391–399.
3. van de Wetering, M., Cavallo, R., Dooijes, D., van Beest, M., van Es, J., Loureiro, J., et al. (1997) Armadillo coactivates transcription driven by the product of the Drosophila segment polarity gene dTCF. *Cell* **88**, 789–799.
4. Barker, N. and Clevers, H. (2006) Mining the Wnt pathway for cancer therapeutics. *Nat. Rev. Drug. Discov.* **5**, 997–1014.

Chapter 8

Assaying β-Catenin/TCF Transcription with β-Catenin/TCF Transcription-Based Reporter Constructs

Travis L. Biechele and Randall T. Moon

Abstract

Transcription-based reporters have been instrumental in characterizing the Wnt/β-catenin signaling pathway and will be essential in the search for therapeutics aimed at combating diseases linked to aberrant signaling. In this chapter, we introduce a new improved Wnt/β-catenin reporter system, β-catenin-activated reporter (BAR), and its accompanying control reporter system, found unresponsive BAR (fuBAR). Its enhanced sensitivity, increased dynamic range, and lentiviral platform provide a reporter system that will keep pace with the needs of scientists in the field.

Key words: Wnt, b-Catenin, Luciferase, Transcription, Reporter, BAR, TCF, LEF.

1. Introduction

The Wnt/β-catenin pathway is the best-studied Wnt pathway in part due to robust tools for measuring pathway activation both in vivo and in vitro. Among the earliest and still commonly used assays of Wnt/β-catenin signaling include phenotypic assays in *Drosophila (1)*, dorsal axis duplication in *Xenopus (2)*, and proliferation of C57MG mammary epithelial cells *(3)*. These assays were crucial for a substantial amount of the early characterization of the pathway. Much of the more recent characterization of the pathway has relied on the convenience of transcription-based reporter systems. The first transcription-based luciferase reporter of Wnt/β-catenin signaling, TOPFlash, was designed by Korinek et al. *(4)*. The TOPFlash reporter contains three TCF response elements (CCTTTGATC) upstream of a basal

Elizabeth Vincan (ed.), *Wnt Signaling, Volume I: Pathway Methods and Mammalian Models, vol. 468*

Book doi: 10.1007/978-1-59745-249-6

c-fos promoter while the control reporter, FOPFlash, contains three mutant TCF response elements (CCTTTGGCC). TOP-Flash was later modified by Upstate Biotechnology to contain three TCF response elements upstream of a minimal thymidine kinase (TK) promoter. The TOPFlash reporter system provided a reliable assay of pathway activation and was crucial for the identification and characterization of several pathway components.

As the characterization of the Wnt/β-catenin pathway turned to identifying modifiers of the core pathway components, the need for a more sensitive reporter developed. This niche was filled by Ajamete Kaykas in the Moon lab with the construction of the SuperTOPFlash reporter. SuperTOPFlash contains eight TCF response elements upstream of Clontech's minimal TA promoter *(5)*. This modification greatly enhances the sensitivity and dynamic range of the reporter (**Fig. 8.1a**) providing a better tool for characterizing modifiers of the pathway as well as the ability to identify new components in *Drosophila* genome-wide RNA interference (RNAi) screens *(6)*. This was significant as Super-TOPFlash was the first Wnt/β-catenin reporter responsive to wingless in *Drosophila* cells.

The necessity to monitor Wnt/β-catenin signaling in non-transfectable cells and achieve even greater sensitivity for high

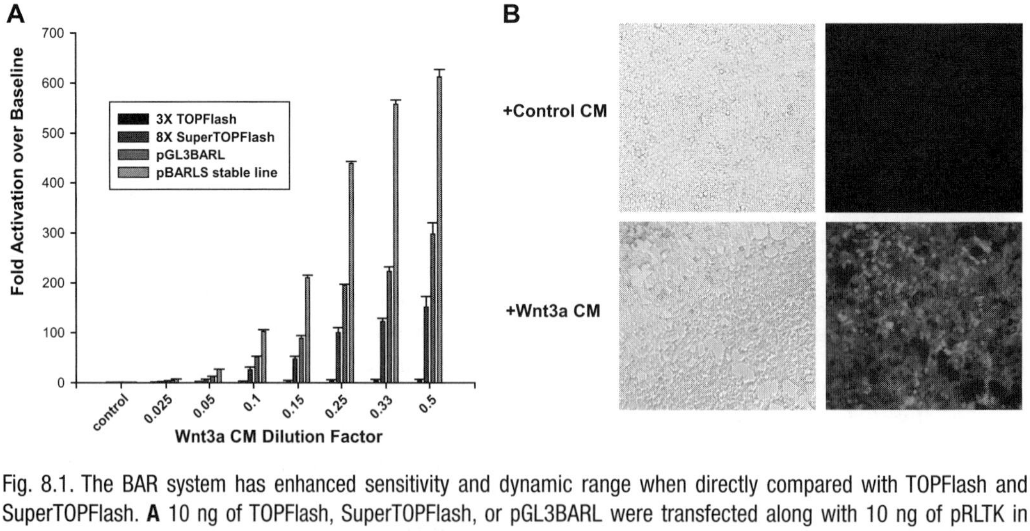

Fig. 8.1. The BAR system has enhanced sensitivity and dynamic range when directly compared with TOPFlash and SuperTOPFlash. **A** 10 ng of TOPFlash, SuperTOPFlash, or pGL3BARL were transfected along with 10 ng of pRLTK in HEK293T cells seeded in a 48-well plate. HEK293T cells stably expressing pBARLS were generated as described in Section 3.4. Cells were treated with specified doses of Wnt3a-conditioned media (*CM*) for 18 h. Luciferase activity was measured as described in Section 3.5.2 and data are presented as fold activation over control conditioned media-treated cells. **B** A monoclonal HEK293T cell line stably expressing pBARVS was generated as described in Section 3.4. Cells were treated with either control conditioned media or Wnt3a-conditioned media for 30 h.

throughput screening inspired the construction of the β-catenin activated reporter (BAR) system. The BAR system contains a concatemer of 12 TCF response elements separated by unique five-nucleotide linkers specifically designed to minimize recombination that can lead to loss of TCF binding sites. This series of TCF response elements is inserted upstream of Promega's minP minimal promoter, completing a functional promoter that drives the transcription of either Firefly luciferase (pBARL), renilla luciferase (pBARRen), or β-globin intron-linked Venus (pBARV) (Venus is a variant of EYFP *(7)*). These reporters were inserted between the long terminal repeats (LTRs) of a lentiviral-transducing plasmid. The result is a highly sensitive luciferase reporter with an unmatched dynamic range and Venus reporter that allows a spatial report of pathway activation (**Fig. 8.1**).

Control reporters, found unresponsive BAR (fuBAR), were constructed using the same strategy. They are identical to their respective parent reporter with the exception that each TCF DNA binding element contains a two-base substitution conferring a non-functional element (pfuBARL and pfuBARV). The essentially identical nature of the control reporters provides the most optimal experimental control, as well as allowing for identical lentiviral titer production when generated side by side with the responsive reporter.

A second version of the reporter constructs containing a PGK promoter driving a puromycin- or hygromycin-resistance gene was constructed for antibiotic selection in mammalian cells (pBARLS, pfuBARLS, pBARLHyg, pfuBARLHyg, pBARVS, pfuBARVS, pBARVHyg, and pfuBARLHyg). A third version containing a PGK or EF1α promoter driving dsRed (pBARVR and pfuBARVR) was constructed for visual detection of cells containing the reporter independent of reporter activation (**Fig. 8.2**).

It should be noted that although the Wnt/β-catenin reporters appear to be very specific readouts of signaling, they are in fact artificial promoters that may not faithfully reflect the activity of endogenous TCF/LEF response elements *(8)*. Therefore it is important to complement reporter data with a measure of the transcription profile of known Wnt/β-catenin target genes (see http://www.stanford.edu/~rnusse/pathways/targets.html and **Note 1** for target genes). It is also important to note that TCF/LEF-independent β-catenin-mediated transcriptional activation will not be detected with these reporter systems *(9)*.

In the following sections, we outline protocols for using BAR transiently, generating stable BAR cell lines, and measuring BAR luciferase activity.

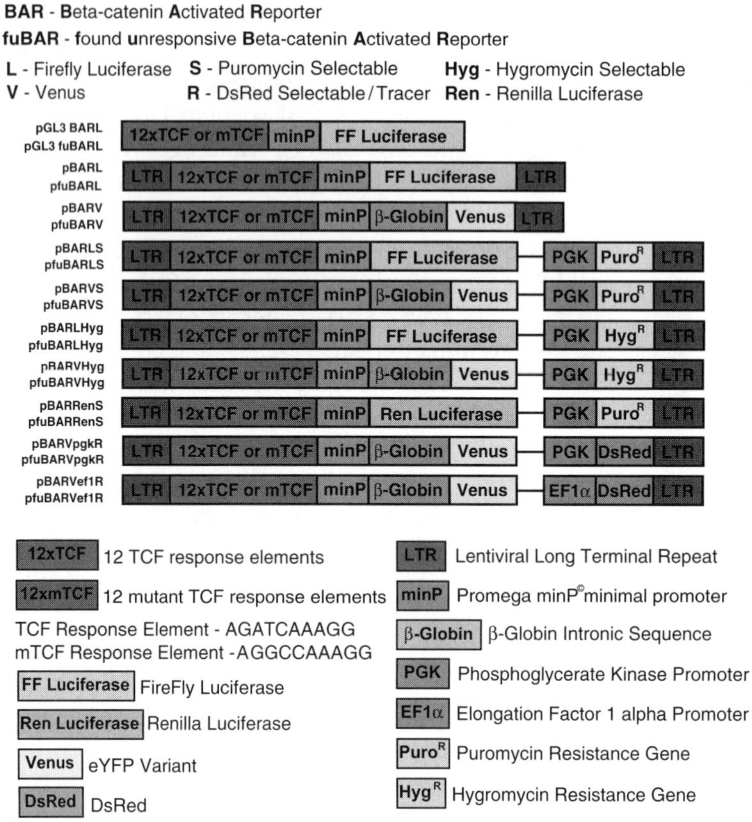

Fig. 8.2. Multiple platforms of the BAR system make it a versatile reporter system.

2. Materials

2.1. Transient Transfection of Reporter for Complementary DNA (cDNA) Overexpression or Small Interfering RNA (siRNA) Knockdown

1. 48-well cell culture plate.
2. HEK293T or other transfectable cells.
3. Lipofectamine 2000 (Invitrogen, Carlsbad, CA; cat. #11668-027) or transfection reagent of choice.
4. Optimem (Invitrogen; cat. #31985-088).
5. Plasmids (*see* **Notes 2** and **3**): pGL3BARL, pGL3fuBARL, pRLTK (Promega, Madison, WI; cat. #E2241), cDNA of interest.
6. siRNA, shRNA, and carrier plasmid (backbone of cDNA expression plasmid or empty vector).
7. L-cell control and Wnt3a-conditioned media or purified Wnt3a (*ref. (10)*; *see* **Chapter 2**; ATCC, Manassas, VA; CRL-2648 and CRL-2647) or (R&D Systems, Minneapolis, MN; cat. #1324-WN-002).

2.2. Lentivirus Production

1. HEK293T cells.
2. Media: DMEM supplemented with 10% (v/v) fetal bovine serum (FBS) and 1% (v/v) penicillin/streptomycin.
3. 2× HEPES-buffered saline (HBS) pH 7.1 (0.22-μm filtered). To prepare 40 mL of 2× HBS, pH 7.1, add 5.6 mL of 2 M NaCl, 4 mL of 0.5 M HEPES, pH 7, 60 μL of 1 M Na_2HPO_4, and 30.4 mL of dH_2O.
4. 2.5 M $CaCl_2$ (0.22-μm filtered).
5. Sterile water (0.22-μm filtered).
6. Plasmids (*see* **Note 3**)—pSL3, pSL4, pSL5, pSL9/rLuc, BAR, and fuBAR in lentiviral platform.

2.3. Lentivirus Concentration

1. 150 mL Millipore Stericup-GP PES filters (Millipore, Billerica, MA; cat. #SCGP U01 RE).
2. Beckman ultracentrifuge tubes (Beckman Coulter, Fullerton, CA; cat. #344058).
3. Beckman SW-28 swinging bucket rotor.
4. Pasteur pipettes.
5. 1× Tris-buffered saline (TBS): 50 mM Tris-HCl, pH 7.5, and 150 mM NaCl. To prepare, dissolve 6.05 g Tris and 8.76 g NaCl in 800 mL of ddH_2O. Adjust pH to 7.5 with 1 M HCl and make volume up to 1 L with ddH_2O. TBS is stable at 4°C for 3 months.

2.4. Generating Stable Reporter Cell Lines

2.4.1. Stable Luciferase Reporter Cell Line

1. pBARLS and pfuBARLS or pBARLHyg and pfuBARLHyg virus.
2. pSL9/rLuc virus.
3. Puromycin or hygromycin.
4. 6-well and 48-well cell culture plates.
5. L-cell control and Wnt3a-conditioned media or purified Wnt3a (ref. *(10)*; see **Chapter 2**; ATCC, CRL-2648 and CRL-2647) or (R&D Systems; cat. #1324-WN-002).

2.4.2. Stable Venus Reporter Cell Line

1. pBARVS and pfuBARVS.
2. Puromycin.
3. 6-well and 100-mm cell culture plates.
4. L-cell control and Wnt3a-conditioned media or purified Wnt3a (ref. *(10)*; see Chapter 2; ATCC, CRL-2648 and CRL-2647) or (R&D Systems; cat. #1324-WN-002).

2.5. Luciferase Assay

2.5.1. Low-Throughput Assay

1. 1× Passive lysis buffer.
2. Firefly luciferase reagent.
3. Stop & Glo® reagent (Renilla luciferase substrate and Firefly luciferase antagonist; Promega).
4. 96-well plate with white wells.

2.5.2. High-Throughput Assay

1. Dual-Glo^{TM} Firefly luciferase reagent (Promega).
2. Dual-Glo^{TM} Stop & Glo® reagent (Renilla luciferase reagent and Firefly luciferase antagonist).

3. Methods

3.1. Transient Transfection of Reporter for cDNA Overexpression or siRNA Knockdown

Prior to the BAR system, the majority of Wnt/β-catenin luciferase reporter assays were performed by transiently transfecting cells with the Firefly luciferase reporter, a Renilla luciferase normalization plasmid, and cDNAs/siRNA/shRNA to be analyzed. Although the transient reporter assay has a decreased dynamic range compared with the stably integrated reporter, it is still very robust and alleviates the production of lentivirus. The following method is based on a 48-well plate format and can be modified for other plate formats by scaling based on the surface area of the well. Each experimental condition is performed in triplicate. Specific transfection details for the transfection reagent used should be followed according to manufacturer's specifications (Lipofectamine 2000 protocol: http://www.invitrogen.com/content/sfs/manuals/lipofectamine2000_man.pdf).

3.1.1. Transiently Transfecting Reporter with cDNA Expression Plasmids

1. Day 1: Plate cells at a density such that they will be 80% confluent the following day.
2. Day 2: Transfect cells with 10 ng pGL3BARL or 10 ng pGL3fuBARL, 10 ng pRLTK, your construct(s) of interest, and the appropriate amount of carrier plasmid using the manufacturer's protocol.
3. Day 3: If the cells will not be treated with a source of Wnt3a or other modulators, then proceed to **Section 3.5** to read luciferase activity. Otherwise, treat the cells with Wnt3a or other modulators (*see* **Note 4**).
4. Day 4: Proceed to **Section 3.5** for measuring luciferase activity.

3.1.2. Transiently Transfecting Reporter with shRNA or siRNA

1. Day 1: Plate cells at a density such that they will be 40% confluent the following day.
2. Day 2: Transfect cells with siRNA or shRNA using the manufacturer's protocol for the transfection reagent used (Lipofectamine 2000 protocol: http//www.invitrogen.com/content/sfs/manuals/lipofectamine2000_man.pdf).
3. Day 3: Transfect cells with 10 ng pGL3BARL or 10 ng pGL3fuBARL, 10 ng pRLTK, and the appropriate amount of carrier plasmid using the manufacturer's protocol.
4. Day 5: Treat cells with Wnt or other modulator if necessary.
5. Day 6: Proceed to section **Section 3.5** for measuring luciferase activity.

3.2. Production of Lentivirus-Containing BAR and Stable Cell Line Production

The transduction plasmid backbone used for the lentiviral-compatible BAR constructs and the lentiviral helper plasmids were provided by the Naldini lab, Vita-Salute San Raffaele University, Milan, Italy. This lentivirus is replication incompetent and does not carry oncogenic cDNAs, which makes it Biosafety Level 2. The virus is, however, competent for human infection, requiring the use of personal protective guidelines, including double gloving, the use of barrier tips, and collection of all liquids in a non-aspirating system for inactivation with 10% bleach. The following protocol will yield high-titer virus that can be used to generate stable reporter cell lines, assay Wnt/β-catenin signaling in cells that are difficult to transfect such as primary cultures, or assay signaling in vivo. BAR and fuBAR virus is made at the same time to ensure equal titer. The pSL9/rLuc plasmid can be used to generate lentivirus containing a constitutive EF1α promoter driving Renilla luciferase for reporter assay normalization.

1. Day 1: Seed a 100-mm dish with HEK 293T cells such that they will be 70–80% confluent the next day. If very high titer virus is needed, scale up production to several 150-mm dishes and adjust the transfection suggested guidelines based on dish surface area.
2. Day 2: Prepare DNA cocktails for transfection as in **Table 8.1.** Add 500 μL (1,250 μL for 150-mm dish) of 2× HBS drop-wise to the above cocktail and bubble with 10 strokes of your pipette. Add drop-wise to you cells, gently mix, and return to incubator.
3. Day 3: Remove media and dispose of media following proper procedures for inactivation in 10% bleach. Replace with fresh media.
4. Day 4: Collect media and centrifuge for 5 min at 3,000×*g* to remove cellular debris. This media may now be used to infect cells or can be concentrated to achieve higher viral titer.

Table 8.1
DNA cocktails for transfection

	100-mm Dish	150-mm Dish
ddH_2O	450 μL – volume of DNA	1,125 μL – volume of DNA
2.5 M $CaCl_2$	50 μL	125 μL
Transducing vector (e.g., pBARLS)	4 μg	10 μg
Packaging vector (pSL4)	8 μg	20 μg
Envelope (pSL3)	2 μg	5 μg
Rev (pSL5)	4 μg	10 μg

3.3. Lentivirus Concentration

There are two methods for concentrating virus. Concentrating virus with 30-kDa molecular weight cut-off centrifugation filters (Millipore Amicon Ultra cat. #UFC903024) is a simple approach to yield a 50× concentration. A limitation to this approach is concurrent concentration of other components in the media including serum and this may have deleterious affects on the cell line to be infected. A second approach involves pelleting the virus by ultracentrifugation. This technique is slightly more labor intensive but allows you to completely exchange the media and yield a 500× concentration.

1. Aliquot 30–35 mL of viral containing media into Beckman ultracentrifuge tubes, match tubes by weight (use fresh media to balance the tubes), and spin at 50,000×*g* for 2 h at 4°C in the SW28 swing bucket rotor.
2. Carefully decant the supernatant and invert the tube on a paper towel for 5 min (will have ~50 µL supernatant plus virus left in the tube).
3. Add 50 µL (or desired volume) of 1× TBS or 1× PBS to each tube, seal with paraffin, and leave at 4°C overnight with no shaking.
4. Pipette up and down three to five times and combine the resuspended virus from each tube. Filter pooled virus with a 0.45-µm filter.
5. Aliquot, snap-freeze in liquid nitrogen, and store at –80°C. Virus should only be freeze–thawed once, dictating the size of the aliquots.

3.4. Generating Stable Reporter Cell Lines

For assays that do not require stable cell lines or the DsRed tracer, the reporters without a selectable marker are recommended, as they will produce higher titer virus. In this section, we cover methods for generating stable luciferase reporter cell lines as well as stable Venus reporter cell lines. The volume of virus used will vary depending on the cell line and viral titer.

3.4.1. Stable Luciferase Reporter Cell Line

The following method describes the production of a polyclonal reporter line. We want to stress that this is a general protocol and variable factors such as a cell line's responsiveness to Wnt and the sensitivity of your luminometer will determine the amount of virus needed to generate a perfect reporter line. Infecting the reporter line with pSL9/rLuc virus provides constitutive expression of Renilla luciferase providing normalization for siRNA experiments or assays that do not involve transfection. To date, we have generated over 40 stable reporter lines in vastly different cell types. Although rare, we have found cell line exceptions in which the reporter is not responsive to pathway activation.

1. Day 1: Seed a 6-well plate such that the cells will be 50% confluent the following day.

2. Add three different doses of reporter virus and matching doses of control reporter virus to the 6-wells. We typically start with 200 μL, 50 μL, and 10 μL of virus that has been concentrated 50×.
3. Day 2: Replace the media with fresh media. As always, inactivate viral containing media in 10% bleach.
4. Day 3: Transfer cells from each well to a 100-mm dish containing the appropriate concentration of puromycin or hygromycin for selection.
5. Allow several days for selection and repopulation of the cells.
6. Test each reporter line by seeding each line in several wells of a 48-well cell culture plate. Treat each line with several doses of L-cell control or Wnt3a-conditioned media and measure luciferase activity the following day (*see* **Section 3.5**).
7. Choose the best reporter line and corresponding control reporter line based on the dynamic range, expand the cells, and freeze back several vials, as reporter activity has been found in some cases to diminish over several passages.
8. For constitutive Renilla luciferase expression, seed the reporter cells in a 6-well plate such that they will be 50% confluent the following day and treat the cells with different doses of pSL9/rLuc virus.
9. Repeat steps 6 and 7.

3.4.2. Stable Venus Reporter Cell Line

A stable polyclonal Venus reporter line can be generated using an identical approach as the stable luciferase reporter line. The only difference is that reporter activity is measured by fluorescence using a microscope or plate reader. The following protocol details the use of fluorescence-activated cell sorting (FACS) to refine the heterogeneity of the line. Briefly, a stable pBARVS virus infected cell line is generated. A monoclonal or polyclonal line with the highest possible dynamic range is generated with two rounds of FACS. In the first round, cells are stimulated with an EC50 dose of Wnt3a-conditioned and a population of high Venus-expressing cells are collected. The population is cultured for several days without Wnt3a-conditioned medium and then resorted for cells that are not expressing Venus. This protocol yields a reporter line with very low basal activity and robust response to pathway activation (**Fig. 8.1b**).

1. Perform steps 1–5 from **Section 3.4.1**.
2. Test each reporter line by seeding each individual line in several wells of a 48-well cell culture plate. Treat each line with several doses of L cell control or Wnt3a-conditioned media and visualize or measure Venus fluorescence the following day. Choose the best cell line based on dynamic range and determine the EC_{50} dose of the Wnt3a-conditioned media.

3. Seed a 100-mm culture dish with reporter cells such that they will be 70% confluent the next day.
4. The following day, treat the cells with the EC_{50} dose of conditioned media.
5. 18–24 h following treatment, sort the cells by FACS with a narrow gate of the highest Venus-expressing cells.
6. Replate the sorted cells in standard growth media for at least 4 days to allow Venus expression to return to baseline. Before proceeding to step 7, you can restimulate a fraction of the cells with the EC_{50} dose of Wnt3a-conditioned media to check the integrity of the FACS.
7. Repeat the FACS and collect a narrow window of the lowest expressing cells. The entire sorted population can be collected as a single population or plated individually in a 96-well plate to create a monoclonal line (**Fig. 8.1b**).
8. Expand the line(s) and freeze back several vials of cells.

3.5. Luciferase Assay

The sensitivity and robustness of the BAR reporter allows for measuring luciferase activity in a broad range of luminometers and plate formats. BAR activity has been measured in luminometers ranging from single-tube luminometers to high-throughput plate readers. A greater than 1,000-fold dynamic range was achieved in a 384-well plate format and it is foreseeable that this can be achieved in a 1,536-well format as well. The luciferase assay reagent to be used depends on throughput of the assay. The standard low-throughput reagent is Promega's Dual-Luciferase® reporter assay system (cat. #E1910). For high-throughput assays, Promega's Dual-Glo™ (cat. #E2940) is recommended. The robustness of the BAR reporter allows you to use a fraction of Promega's suggested volume of reagent. The following low-throughput method has been optimized for use on a Berthold Mitras LB940 luminometer and the high-throughput method has been optimized on the Perkin Elmer Envision plate reader.

3.5.1. Low-Throughput Assay

In the following method, the cells were plated and treated in a 48-well plate and the luciferase activity was measured in a 96-well plate.

1. Aspirate cell culture media from each well.
2. Add 50 µL of 1× passive lysis buffer and moderately rotate for 20 min at room temperature.
3. Transfer 5 µL of each sample in duplicate to a 96-well white well plate.
4. Program your luminometer with the following settings:
   (a) Inject 10 µL of Firefly luciferase reagent.
   (b) Read total luminescence.

(c) Inject 10 μL of Stop & Glo® reagent.

(d) Read total luminescence.

5. Express data as a ratio of Firefly relative light units to Renilla relative light units.

3.5.2. High-Throughput Assay

In the following method, the cells were plated, treated, and the luciferase activity measured in a 384-well plate. The volume of culture media in each well prior to measuring luciferase activity is 40 μL.

1. Add 10 μL of Dual-Glo™ Firefly luciferase reagent using a liquid dispenser and incubate for 10 min at room temperature (if using clear bottom 384-well plates, bare nuclei will can be visualized if lysis is complete) (*see* **Note 5**).
2. Read total luminescence.
3. Add 10 μL of Dual-Glo™ Stop & Glo® luciferase reagent using a liquid dispenser and incubate for 10 min at room temperature (*see* **Note 5**).
4. Read total luminescence.
5. Express data as a ratio of Firefly relative light units to Renilla relative light units.

4. Notes

1. Two common Wnt target genes are axin2 and lef1. Real-time PCR primers for analyzing the human transcripts are as follows:

   Axin2 forward: CTCCCCACCTTGAATGAAGA.
   Axin2 reverse: TGGCTGGTGCAAAGACATAG.
   Lef1 forward: GACGAGATGATCCCCTTCAA.
   Lef1 reverse: AGGGCTCCTGAGAGGTTTGT.

2. The reporters in the lentiviral backbones cannot be used for transient reporter assays as the episomal form contains a constitutive promoter upstream of the TCF response elements that drives transcription independent of Wnt/β-catenin signaling.
3. All reporter plasmids and lentiviral helper plasmids can be obtained from the Moon lab by contacting either author.
4. The dose of Wnt3a-conditioned media and incubation time will vary based on the potency of the conditioned media. We have treated cells with Wnt3a-conditioned media for as little as 4 h and measured reporter activity above baseline. Common incubation times are 12–24 h.

5. The 10 μL of Dual-Glo™ luciferase and Stop & Glo® reagent used in **Section 3.5.2** may be reduced even further. The only concern is incomplete cell lysis, which may be overcome by supplementing the Firefly reagent with Promega's passive lysis buffer.

Acknowledgments

We thank the Howard Hughes Medical Institute for funding.

References

1. Nusslein-Volhard, C., and Wieschaus, E. (1980). Mutations affecting segment number and polarity in Drosophila. *Nature* **287**, 795–801.
2. McMahon, A.P. and Moon, R.T. (1989). Ectopic expression of the proto-oncogene int-1 in Xenopus embryos leads to duplication of the embryonic axis. *Cell* **58**, 1075–1084.
3. Jue, S.F., Bradley, R.S., Rudnicki, J.A., Varmus, H.E., and Brown, A.M. (1992). The mouse Wnt-1 gene can act via a paracrine mechanism in transformation of mammary epithelial cells. *Mol Cell Biol* **12**, 321–328.
4. Korinek, V., Barker, N., Morin, P.J., van Wichen, D., de Weger, R., Kinzler, K.W., Vogelstein, B., and Clevers, H. (1997). Constitutive transcriptional activation by a beta-catenin-Tcf complex in APC−/− colon carcinoma. *Science* **275**, 1784–1787.
5. Veeman, M.T., Slusarski, D.C., Kaykas, A., Louie, S.H., and Moon, R.T. (2003). Zebrafish prickle, a modulator of non-canonical Wnt/Fz signaling, regulates gastrulation movements. *Curr Biol* **13**, 680–685.
6. DasGupta, R., Kaykas, A., Moon, R.T., and Perrimon, N. (2005). Functional genomic analysis of the Wnt-wingless signaling pathway. *Science* **308**, 826–833.
7. Rekas, A., Alattia, J.R., Nagai, T., Miyawaki, A., and Ikura, M. (2002). Crystal structure of Venus, a yellow fluorescent protein with improved maturation and reduced environmental sensitivity. *J Biol Chem* **277**, 50573–50578.
8. Barolo, S., and Posakony, J.W. (2002). Three habits of highly effective signaling pathways: principles of transcriptional control by developmental cell signaling. *Genes Dev* **16**, 1167–1181.
9. Olson, L.E., Tollkuhn, J., Scafoglio, C., Krones, A., Zhang, J., Ohgi, K.A., Wu, W., Taketo, M.M., Kemler, R., Grosschedl, R., et al. (2006). Homeodomain-mediated beta-catenin-dependent switching events dictate cell-lineage determination. *Cell* **125**, 593–605.
10. Willert, K., Brown, J. D., Danenberg, E., Duncan, A. W., Weissman, I. L., Reya, T., et al. (2003) Wnt proteins are lipid-modified and can act as stem cell growth factors. *Nature*, **423**, 448–452.

Chapter 9

Native Promoter Reporters Validate Transcriptional Targets

Otto Schmalhofer, Simone Spaderna, and Thomas Brabletz

Abstract

The transcriptional activator β-catenin is the key mediator of the canonical Wnt signaling pathway. However, β-catenin does not itself bind DNA, but functions via interaction with T-cell factor (TCF)/lymphoid-enhancing factor (LEF) transcription factors. These proteins contain a high-mobility group (HMG) box that binds DNA in a sequence-specific manner. Thus, in the case of active Wnt signaling, β-catenin activates, in cooperation with proteins of the TCF/LEF family, the expression of a wide variety of genes. To date, the list of established Wnt targets is far from complete. The establishment of plasmids harbouring reporter genes under control of the native promoter sequences provides a tool to validate novel putative Wnt targets by directly quantifying the β-catenin-dependent activation of each specific gene. In this chapter, we describe how to generate such reporter plasmids using the MMP7 promoter as an example.

Key words: Wnt pathway, Native promoter regions, TCF recognition sites, Reporter plasmids, Bacmid, Firefly luciferase.

1. Introduction

β-Catenin, the main effector of the canonical Wnt signaling pathway, in cooperation with DNA-binding proteins of the TCF/LEF family, activates the expression of a wide variety of genes *(1)*. Numerous direct target genes are already described (a listing is available at: http://www.stanford.edu/~rnusse/pathways/targets.html). For a more profound understanding of the different effects of Wnt signaling, it is necessary to define additional β-catenin targets. A useful strategy to track down putative candidates is the screening of complementary DNA (cDNA) microarray data, for instance by comparing data sets of either "normal" cell lines with the respective cell lines exhibiting an activated Wnt pathway, or

Elizabeth Vincan (ed.), *Wnt Signaling, Volume I: Pathway Methods and Mammalian Models, vol. 468*

Book doi: 10.1007/978-1-59745-249-6

tumor cell lines containing stabilized β-catenin with the corresponding lines showing inhibited Wnt signaling *(2)*. Other possibilities are the comparison of tumors with and without *CTNNB1* mutations *(3)* or the analysis of tumors with different mutations and normal tissue *(4)* (a list of relevant published cDNA arrays is available at: http://www.stanford.edu/~rnusse/pathways/array.html). We use arrays of colon cancer cells transfected with small interfering RNA (siRNA) against β-catenin or green fluorescent protein (GFP) as a control, and arrays of colon carcinomas versus normal epithelium (unpublished).

First, confirmation of putative targets may be performed via detection of endogenous protein levels in the respective cell lines or tissue samples, e.g. by Western blot or immunohistochemistry, if appropriate antibodies are available. A method that does not rely on the availability of antibodies is to determine the messenger RNA (mRNA) levels using quantitative reverse transcriptase (RT)–polymerase chain reaction (PCR). The next step after verification of differential expression levels of the specified mRNA/protein is to validate that the regulation through β-catenin occurs at the level of transcription. One way to confirm this is to establish reporter constructs containing the native promoter region of the putative target gene. Usage of reporter genes, like firefly luciferase, allows for direct quantification of the promoter activity in response to β-catenin/TCF. This chapter mainly presents the steps of generating such native reporter constructs, emphasizing the in silico procedures and theoretical considerations that are necessary for cloning the reporter plasmids. The exact protocol for luciferase reporter assays is described in **Chapter 8, Volume 1**. To demonstrate that activation of reporter genes is directly mediated via binding of β-catenin/TCF to its recognition sites within the promoter region, additional experiments like electrophoretic mobility shift assays (EMSA) or chromatin immuno precipitations (ChIP) are required.

2. Materials

2.1. Preparation of Bacteria Artificial Chromosome (BAC) DNA

1. LB plates: 10 g Bacto-Tryptone, 5 g Bacto yeast extract, 10 g NaCl, 15 g Bacto-agar, and ddH_2O to 1 L. Autoclave to sterilize, cool to 55°C, add suitable antibiotic, and pour into sterile petri dishes.
2. LB medium: 10 g Bacto-Tryptone, 5 g Bacto-yeast extract, 10 g NaCl, and ddH_2O to 1 L, adjust pH to 7.0 and autoclave to sterilize, cool to room temperature (RT), and add appropriate antibiotics.

3. Antibiotics for supplementation of LB media (dependent on the resistance gene encoded by the plasmids/bacmids used): Ampicillin at 100 μg/mL medium, kanamycin at 25 μg/mL medium, and chloramphenicol at 12.5 μg/mL medium.
4. Buffer P1, P2, and P3 taken from an arbitrary Qiagen DNA Isolation Kit (Qiagen, Valencia, CA).
5. Isopropanol.
6. 70% (v/v) ethanol.

2.2. PCR

1. Genomic BAC clone (e.g. from imaGenes, Berlin, Germany).
2. Reporter vector: pGL3basic (Promega, Madison, WI).
3. Specific primers (described in **Section 3.2**).
4. 10 mM dNTP Mix (e.g. from Fermentas, Burlington, Ontario, Canada).
5. 2.5 U/μL Pfu DNA polymerase (e.g. from Fermentas), supplied with appropriate 10× reaction buffer.
6. 96% Tetramethylene sulfoxide (TMSO) (e.g. from Sigma, St. Louis, MO).
7. Qiaquick PCR Purification Kit (Qiagen) or similar kit from other companies.

2.3. Restriction Digest and Ligation

1. Restriction endonucleases (e.g. from Fermentas), the enzymes are supplied with appropriate 10× reaction buffers.
2. Calf intestine alkaline phosphatase (CIAP; e.g. from Fermentas).
3. 5 U/μL T4 DNA Ligase (e.g. from Fermentas), supplied with 10× reaction buffer.

2.4. Transformation

1. Competent bacteria: DH5α subcloning efficiency (e.g. from Invitrogen, Carlsbad, CA), store at –80°C.
2. SOC Medium: 1 g Bacto-yeast extract, 4 g Bacto-Tryptone, 400 μL of 5 M NaCl, 500 μL of 1 M KCl, and ddH_2O to 193 mL, adjust pH to ~7, autoclave to sterilize. All the following ingredients must be sterile filtered and added under a sterile laminar flow hood: 2 mL of 1 M $MgCl_2$, 2 mL of 1 M $MgSO_4$, and 2 mL of 2 M glucose, store these in suitable aliquots of about 5 mL.

2.5. Identification of Clones

1. 10 μM Screening primer forward: 5′-CTAGCAAAATAGGCTGTCCC-3′.
2. 10 μM Screening primer reverse: 5′-CTTTATGTTTTTGGCGTCTTCC-3′.
3. 10 mM dNTP Mix (e.g. from Fermentas).

4. 96% TMSO (e.g. from Sigma).
5. 5 U/μL Platinum Taq Polymerase (e.g. from Invitrogen), the enzyme is supplied with the suitable 10× reaction buffer and $MgCl_2$ solution (50 mM).
6. Qiaprep Spin Miniprep Kit (Qiagen).
7. Agarose (e.g. from SERVA Electrophoresis, Heidelberg, Germany).
8. 5× TBE-buffer: 54 g Tris base, 27.5 g boric acid, 3.72 g EDTA-Na_2-salt, and ddH_2O to 1 L *(5)*, for gel electrophoresis, prepare a 1:10 dilution resulting in 0.5× TBE-buffer.
9. 1-kb Ladder (e.g. from Fermentas).

3. Methods

For validation of putative transcriptional targets of the canonical Wnt pathway via reporter assays, it is necessary to identify the native promoter region of the respective genes and subsequently clone it into a suitable reporter plasmid. Naturally, the exact procedure differs depending on the specific targets to be investigated. Additionally, there are of course a variety of different reporter plasmids available, which may all be more or less useful, depending on the facilities of the specific laboratory. In order to exemplify the description of the procedure, we have chosen the published Wnt target MMP7 *(6, 7)* for the detailed protocol, including general considerations to be made for each specific target gene (*see* **Note 1**).

Concerning the reporter plasmid, we focus on pGL3basic (Promega) harboring the firefly luciferase reporter gene.

3.1. Identification of TCF Sites in Promoter Regions

1. Visit the National Center for Biotechnology Information (NCBI) website at http://www.ncbi.nlm.nih.gov/.
2. Choose the *Gene* database from the search box. Enter different aliases or the gene ID of your gene of interest (GOI), in our case MMP7 (gene ID: 4316) in the text box (**Fig. 9.1**).

Fig. 9.1. The NCBI homepage.

3. Choose the entry of your species of interest, e.g. homo sapiens (**Fig. 9.2**).
4. On the *entrez gene* site, click the nucleotide link in the *Genomic regions, transcripts, and products* text field and choose the *genbank* link (**Fig. 9.3**).
5. On the *entrez nucleotide* site, the sequence of the GOI, MMP7, is displayed. In the header its physical location,

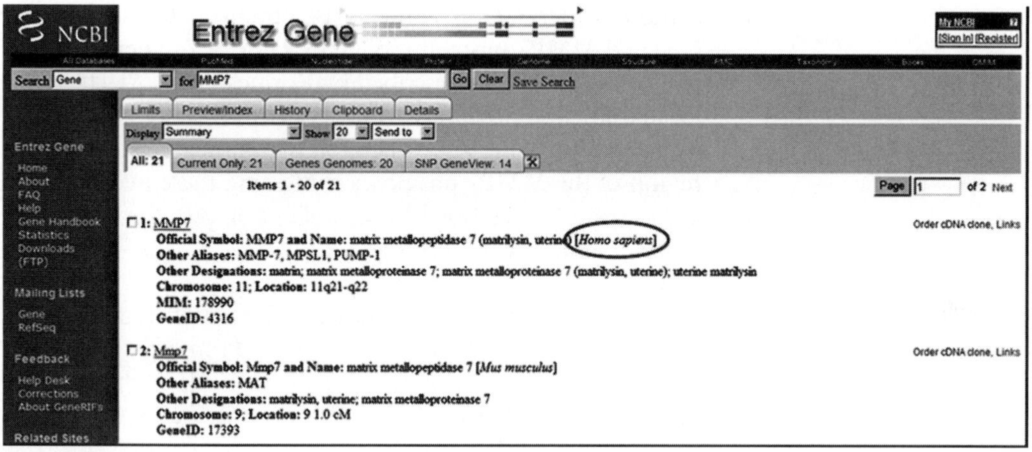

Fig. 9.2. The Entrez Gene site.

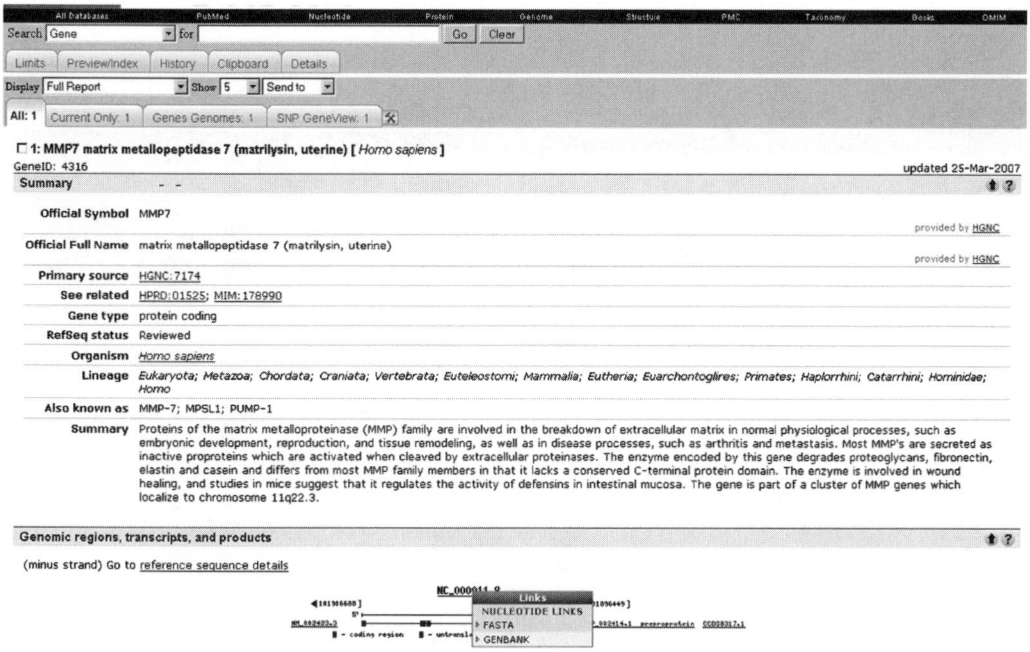

Fig. 9.3. MMP7 gene entry at the Entrez Gene site.

nucleotide 101896449 to 101906688 on chromosome 11, is depicted (**Fig. 9.4**).

6. To display the promoter region upstream of the transcription start of the MMP7 gene, add 5,000 bp in the *to* text box (*see* **Note 2**) and click the *refresh* button (**Fig. 9.5**). The transcription start (*see* **Note 3**) of the GOI now is at position 5,001 of the displayed sequence (**Fig. 9.6**).
7. Screen basepairs (bp) 1 to 5,000 of the sequence for the TCF binding sites listed in **Table 9.1**.
8. Although there are as many as 14 putative TCF sites in the screened MMP7 upstream sequence (**Fig. 9.7**), there are only two "classic" sites, as described by Korinek et al. *(8)* These are located in close proximity to the transcription start (indicated by an arrow in **Fig. 9.7**). Therefore we chose a relatively short region of the MMP7 promoter containing these two sites and a third one, as described by Ghiselli et al. *(9)*, for analysis in reporter assays (*see* **Notes 4** and **5**).

3.2. Selection of Primers

1. For cloning this fragment by means of a traditional strategy into the reporter plasmid pGL3-Basic from Promega (**Fig. 9.8**), choose restriction enzymes from the vector's multiple cloning site (MCS) that do not cut within the region of interest (*see* **Note 6**). Use one of the numerous online tools, e.g. the New England Biolabs (NEB) cutter at the NEB homepage at http://tools.neb.com/NEBcutter2/index.php.

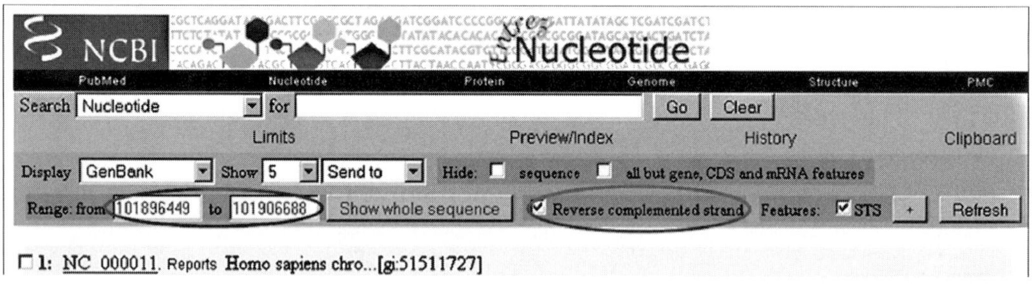

Fig. 9.4. MMP7 GenBank entry at the Entrez Nucleotide site.

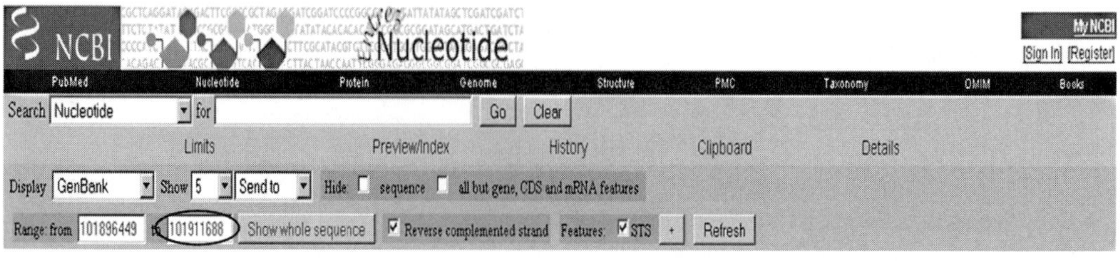

Fig. 9.5. Range of the displayed sequence.

2. Paste the sequence from bp –1,000 to +48 relative to the transcriptional start site of the MMP7 gene in the sequence window and submit (**Fig. 9.9**).
3. *XhoI* and *Hind*III from the the MCS of pGL3-Basic do not cut within this region, as you can learn from the 0-cutter list (**Fig. 9.10**).

```
mRNA            join(5001..5155,7765..7991,8076..8224,10684..10812,
                12347..12508,14944..15240)
                /gene="MMP7"
                /product="matrix metallopeptidase 7 (matrilysin, uterine)"
                /exception="unclassified transcription discrepancy"
                /note="Derived by automated computational analysis using
                gene prediction method: BestRefseq. Supporting evidence
                includes similarity to: 1 mRNA"
                /transcript_id="NM_002423.3"
                /db_xref="GI:75709180"
                /db_xref="GeneID:4316"
                /db_xref="HGNC:7174"
CDS             join(5048..5155,7765..7991,8076..8224,10684..10812,
                12347..12508,14944..14972)
                /gene="MMP7"
                /EC_number="3.4.24.23"
                /GO_component="extracellular matrix (sensu Metazoa);
                extracellular space [pmid 1497627]"
                /GO_function="calcium ion binding; matrilysin activity;
                zinc ion binding"
                /GO_process="collagen catabolism; peptidoglycan
                metabolism; proteolysis [pmid 1497627]"
                /note="matrin; matrix metalloproteinase 7 (matrilysin,
                uterine); uterine matrilysin"
                /codon_start=1
                /product="matrix metalloproteinase 7 preproprotein"
                /protein_id="NP_002414.1"
                /db_xref="GI:4505219"
                /db_xref="CCDS:CCDS8317.1"
                /db_xref="GeneID:4316"
                /db_xref="HGNC:7174"
```

Fig. 9.6. Position of mRNA in the displayed sequence.

Table 9.1
Pipetting instructions for PCR amplification of promoter sequences

TCF site	Publication
5′-WWCAAAG-3′	Korinek et al. Constitute transcriptional activation by a beta-catenin-Tcf complex in APC-/- colon carcinoma *(8)*
5′-ATCAAG-3′	Rockman et al. Id2 is a target of the β-catenin/T-cell factor pathway in colon carcinoma *(11)*
5′-TTCAAAC-3′	Gradl et al. The Wnt/Wg signal transducer β-catenin controls fibronectin expression *(12)*
5′-TTCAAAA-3′	Ghiselli et al. The cohesin SMC3 is a target the for β-catenin/TCF4 transactivation pathway *(9)*

```
3841 aaaagccatg gtgttctccc aagtaatgta taatataata aaagagacag acctattaca
3901 aaatgaatag gcagtgcagt gggatagaaa aagcactgag ctactatctg tgtgagctgg
3961 gagaagtata ataagttaaa ttatcatctt ggcctcactt tcatttttgg taagaatgga
4021 gtcattgggc tagaatctgt aggtaatgga ttcttactgc tatacgtaga ataaccattt
4081 tgtgtactaa ggaccaatgc agccctacct gtagctgggc agcagccaga gtcagagtga
4141 ttggaagaaa aaaattggat ctccaagttg aaggtctagc cttggaagaa tctgttacta
4201 taaaatgaga agcagaataa gtaagccaga tgaagagtta aacccgtgct tgtagatttt
4261 tgttggcttg gtattttttt gtgtttaatt caagaagaat tagaggcagt gttccccatt
4321 aagaaaagac ataaactgat taaaaggaga cccaaagaag ggaattatca ctgctctgct
4381 aagggacgtg gaaggtgagg ggacacagca cagttgtata tagagtggcc actaatccag
4441 ccacacagca gcatttccat ctgcctcctg ccatctttcc cctgtatgga gaaccacagg
4501 attttggtga tgttctattt cttgactttg gtgacggtta cagtatttgc ttaattatta
4561 tttcattatt tacatgtttt gtgcactttt ccaaaccttt taattgttaa gaatatatgg
4621 taccatgtat ttaagaatac atggtaccat ataataagaa tatatattta agaatatata
4681 ttattcttta agaatatatg gtaccataat gtcctgaatg atacctatga gagcagtcat
4741 ttgactttgg caaaaaaatg aggtttctca tggagtcaat ttatgcagca gacagaaaaa
4801 aaaatccttt gaaagacaaa tacattgtgt gcttcctgcc aataacgatg taatacttcc
4861 tcgttttagt taatgaaaaa taacacatac tttcaaagtt ctgtagactc taaaaagaaa
4921 gaaaacactc aaatgagtca cctatttcca cattcgaggc tgagaagcta tataaatttc
4981 tgcagtcact agcagaaaac accaaatcaa ccataggtcc aagaacaatt gtctctggac
5041 ggcagctatg cgactcaccg tgctgtgtgc tgtgtgcctg ctgcctggca gcctggccct
```

Fig. 9.7. Putative TCF binding sites in the MMP7 promoter region.

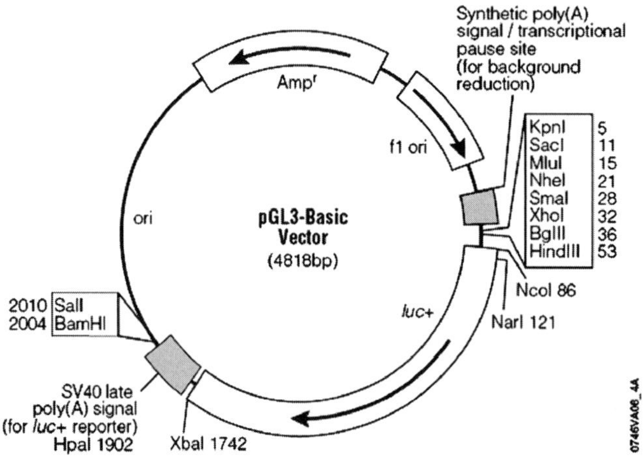

Fig. 9.8. pGL3 Basic vector from Promega.

4. For selection of primers, use one of the various tools available online, e.g. OligoPerfect™ Designer from Invitrogen at http://www.invitrogen.com/content.cfm?pageid=9716. Paste the sequence of interest into the *Target Sequence* text box, select *PCR:Cloning* from the application box, and submit (**Fig. 9.11**).

5. A suitable region of the MMP7 promoter to amplify is bp –844 to +35 (relative to the transcription start site), containing the three identified TCF sites. Add the recognition sequences of *Xho*I and *Hin*dIII and choose *Traditional Cloning* (**Fig. 9.12**).

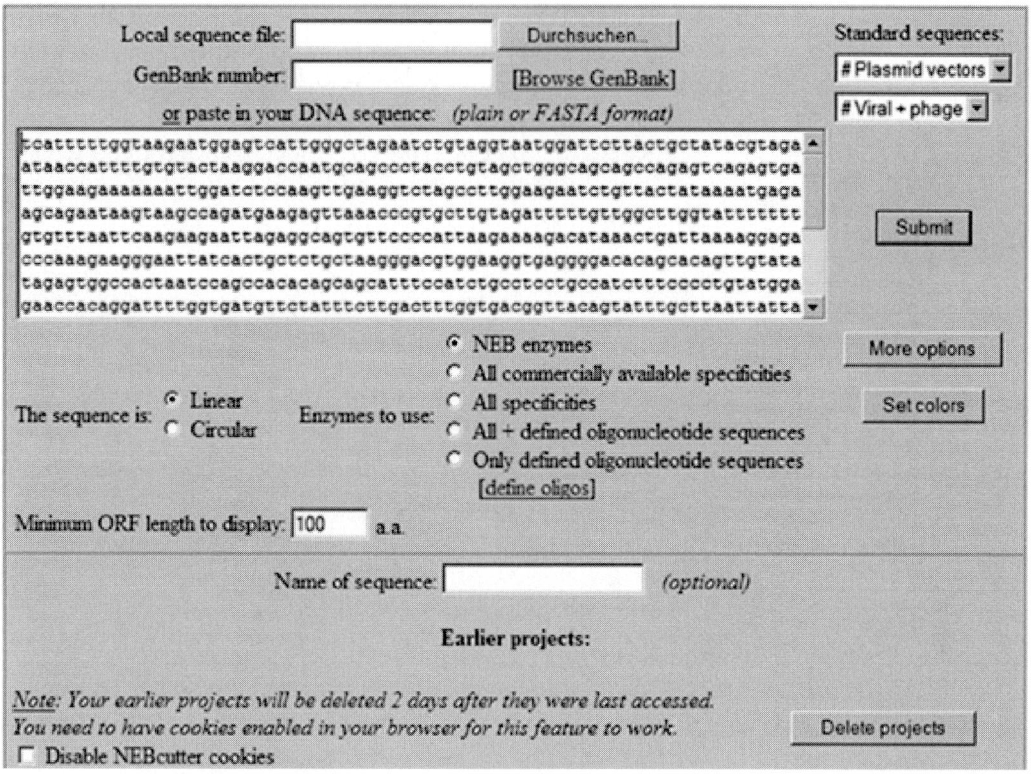

Fig. 9.9. Homepage of NEBcutter 2.0.

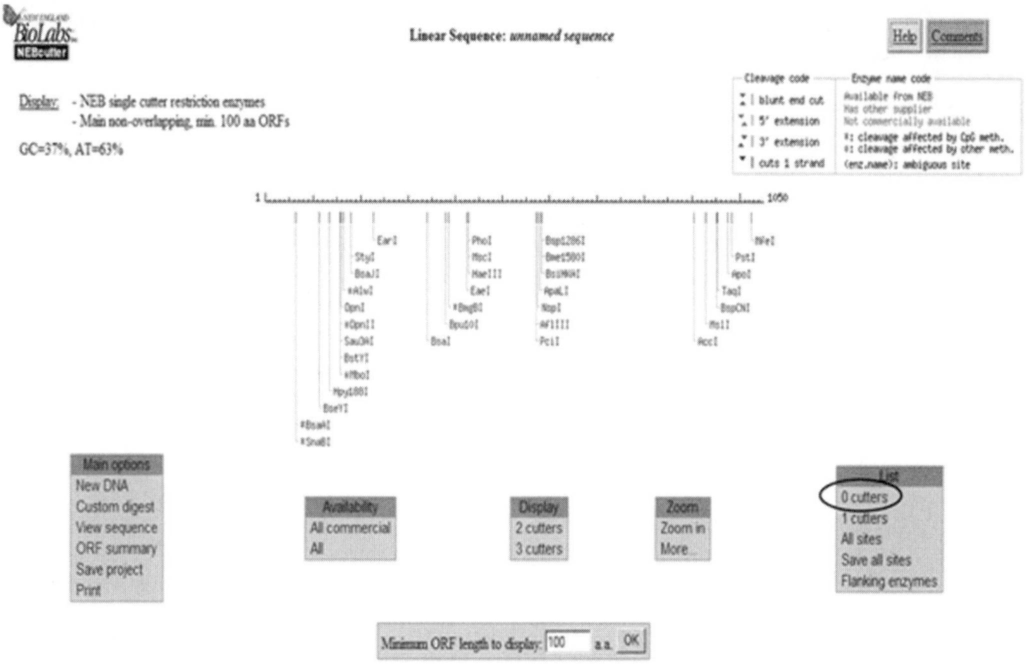

Fig. 9.10. Display of restriction analysis at the NEBcutter 2.0 site.

6. The Oligo Perfect™ Designer delivers different primer pairs (**Fig. 9.13**). For example, as forward primer 5′-acca<u>ctcgag</u>tggatctccaagttgaaggtc-3′ and as reverse primer 5′-atgc<u>aagctt</u>gagacaattgttcttggacctatg-3′. Underlined are the recognition sequences for *Xho*I or *Hin*dIII, respectively. Highlighted are additional oligonucleotides, added 5′ for improving restriction efficiency.

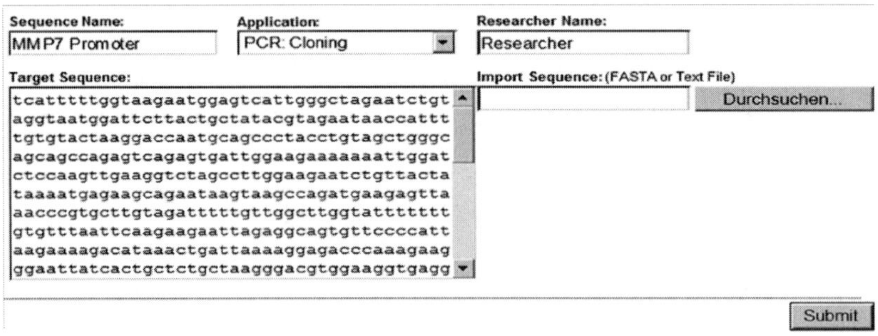

Fig. 9.11. OligoPerfect™ Designer at the Invitrogen homepage.

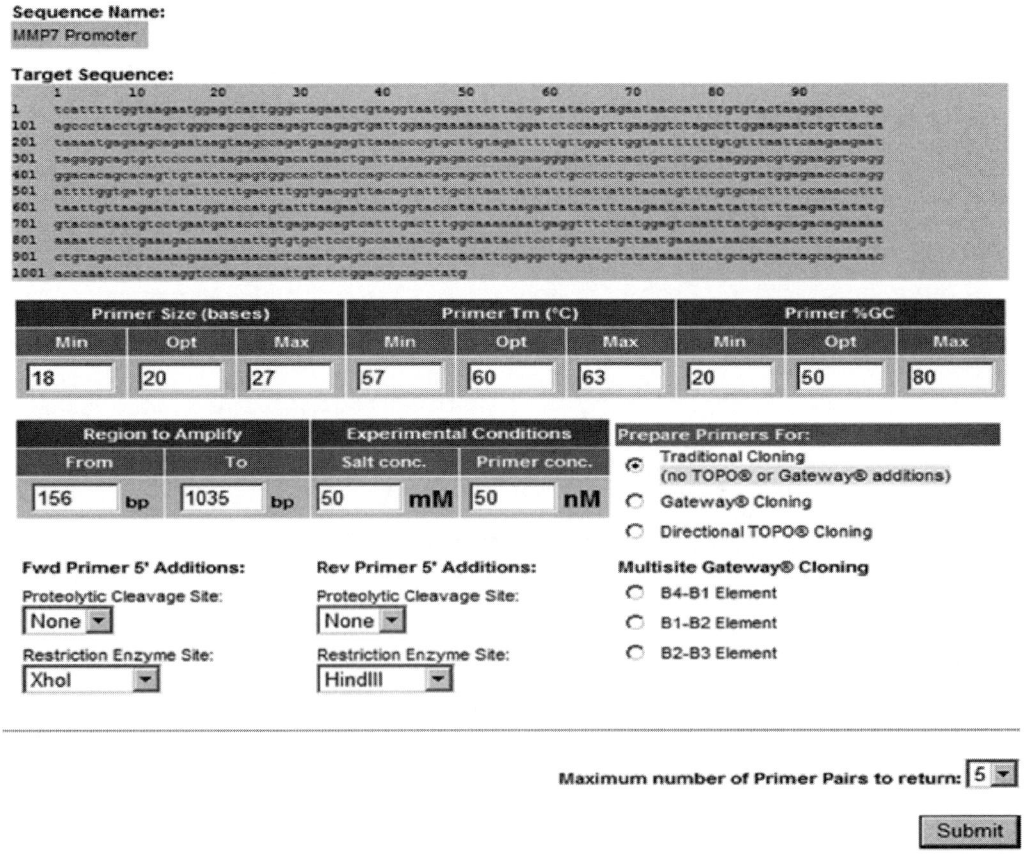

Fig. 9.12. Linear sequence display by Oligo Perfect™ Designer.

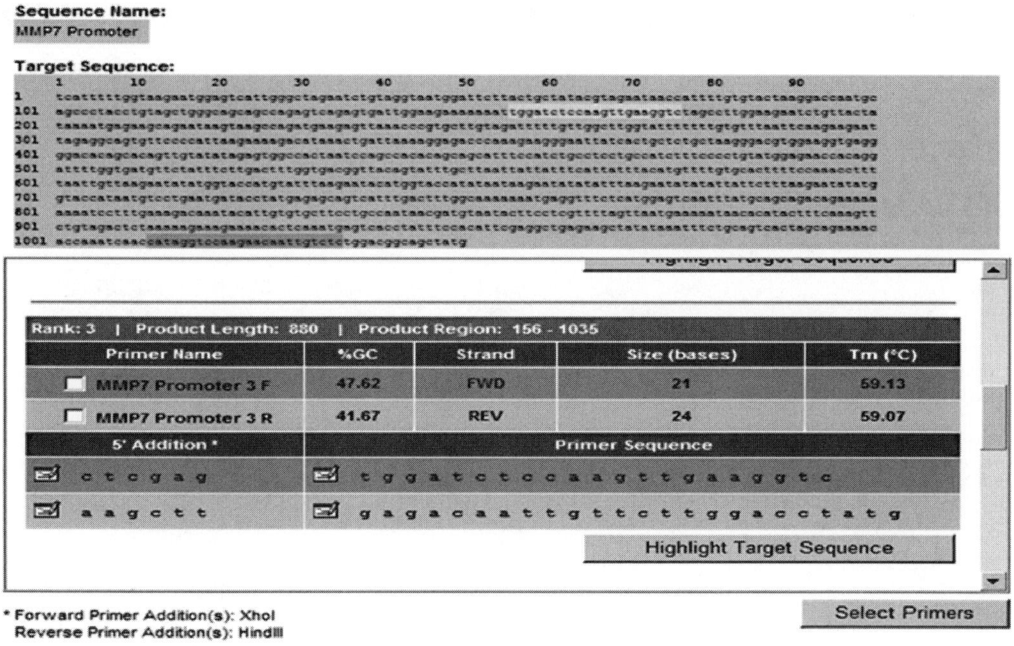

Fig. 9.13. Cloning primer search results for the MMP7 promoter by Oligo Perfect™ Designer.

Fig. 9.14. Search options of the GenomeCube.

3.3. Selection and Preparation of Bacmid

1. For identification of a BAC containing the genomic locus of human MMP7 for use as template in PCR (*see* **Note 7**), visit the website of imaGenes, formerly the German Resource Center for Genome Research, at http://www.imagenes-bio.de/ and choose *Product Search*.
2. Choose the species and the type of DNA clone of interest, which is homo sapiens and genomic clone, and search for MMP7 (**Fig. 9.14**).
3. Order the listed genomic DNA clone RZPDB737H032090D (**Fig. 9.15** and **Note 8**).
4. Preparation of Bacmid DNA is modified from Sambrook et al. *(5)* and closely follows instructions of the Qiagen mini-prep manuals. Buffers from Qiagen are used for isolation. The yielded DNA is of poor purity; however, the quality is sufficient for use in subsequent PCR.

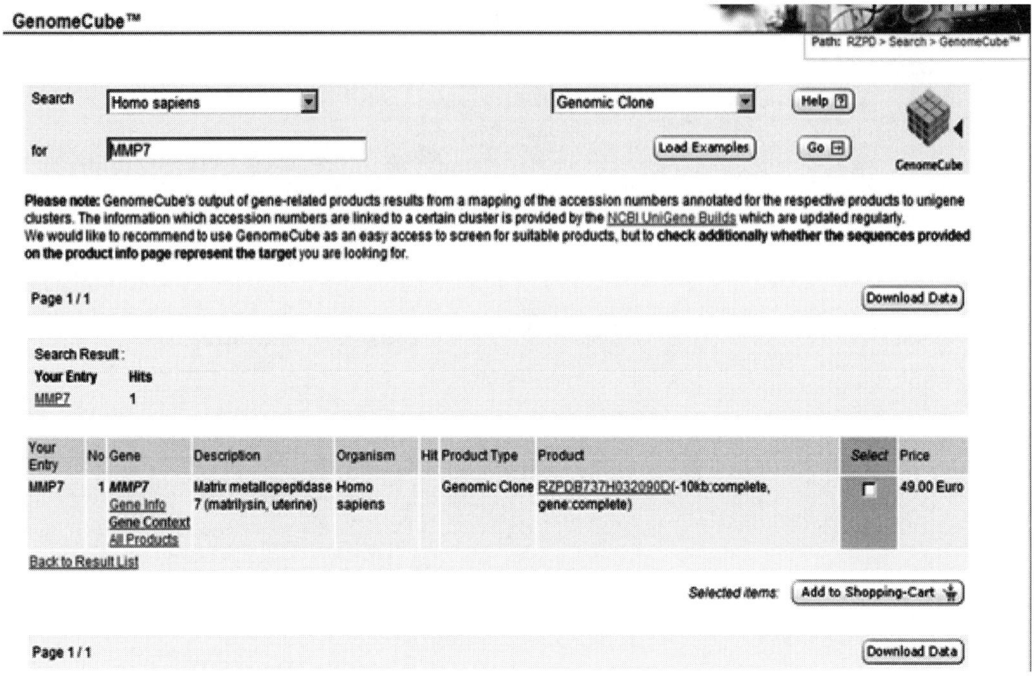

Fig. 9.15. MMP7 search results with the GenomeCube.

5. Pick a clone from a freshly streaked selective plate and incubate overnight at 37°C in 3 mL LB with the appropriate antibiotic on a shaking platform.
6. Harvest bacteria by centrifugation in a tabletop centrifuge for 5 min at 6,000 ×*g*.
7. Discard supernatant and resuspend pellet with 100 µL buffer P1.
8. Add 100 µL buffer P2, mix by gently inverting, and incubate for 5 min at RT.
9. Add 100 µL buffer P3, mix by gently inverting, and incubate for 10 min on ice.
10. Centrifuge with full speed in a microfuge (~17,900×*g*) for 10 min at RT.
11. Transfer the supernatant to a new microtube and add 200 µL isopropanol.
12. Centrifuge with full speed in a microfuge (~17,900×*g*) for 15 min at RT.
13. Discard the supernatant; add 200 µL of 70% ethanol and centrifuge for another 5 min.
14. Discard the supernatant and air-dry the pellet for 5–10 min (*see* **Note 9**).
15. Resuspend the DNA pellet by adding 50 µL H_2O. Determine and adjust the concentration to 100 ng/µL.

3.4. PCR

1. For amplification of the selected promoter fragment, prepare the PCR reaction mix according to **Table 9.2** (*see* **Note 10**).
2. Use the thermocycling program according to **Table 9.3**.
3. Apply 5 µL of the reaction mix to electrophoresis on a 1% (w/v) agarose gel to check for correct size of the amplicon.
4. For subsequent purification of the amplicon from the reaction mix, use the Qiagen PCR Purification Kit according to the user manual, and elute in 30 µL of H_2O.

3.5. Digest and Ligation

1. For digestion of the pGL3-Basic vector and the PCR amplicon, prepare the following reaction mixtures according to **Table 9.4** and incubate overnight at 37°C.
2. Incubate for 20 min at 80°C to heat-inactivate restriction enzymes.
3. For dephosphorylation of the linearized pGL3-Basic, add 1 µL of CIAP and incubate at 37°C for 30 min. Inactivate CIAP by incubating at 85°C for another 15 min.
4. For subsequent purification of digested vector and amplicon from the reaction mix, use the Qiagen PCR Purification Kit according to the user manual.
5. For ligation, prepare the following reaction mix according to **Table 9.5**, including a re-ligation reaction of the vector as a control. Incubate for 2 h at RT or overnight at 16°C (*see* **Note 11**).

Table 9.2
Cycling conditions for PCR amplification of promoter sequences

	Volume	
Bacmid DNA	1 µL	—
Primer forward	1 µL	1 µL
Primer reverse	1 µL	1 µL
dNTP	1 µL	1 µL
TMSO	1 µL	1 µL
Reaction buffer	2.5 µL	2.5 µL
Pfu polymerase	0.25 µL	0.25 µL
H_2O	17.25 µL	18.25 µL
Total volume	25 µL	25 µL

Table 9.3
Pipetting instructions for endonuclease digestion

Step	Temperature	Time	Cycles
Initial denaturation	95°C	3 min	1
Denaturation	95°C	1 min	
Annealing	58°C	1 min	40
Extension	72°C	2 min/kb	
Final extension	72°C	10 min	1

Table 9.4
Pipetting instructions for endonuclease digestion

	Volume	
pGL3-Basic (100 ng/μL)	5 μL	—
Amplicon	—	25 μL
*Xho*I	1 μL	1 μL
*Hin*dIII	1 μL	1 μL
10× Digestion buffer R+	3 μL	3 μL
H_2O	20 μL	20 μL
Total volume	30 μL	30 μL

Table 9.5
Pipetting instructions for DNA ligation

	Ligation volume	Religation volume
pGL3-Basic	1 μL	1 μL
Amplicon	6.5 μL	—
T4 ligase buffer	1.5 μL	1.5 μL
T4 DNA ligase	1 μL	1 μL
H_2O	5 μL	11.5 μL
Total volume	15 μL	15 μL

3.6. Transformation

1. Thaw bacteria (DH5α subcloning efficiency) on ice.
2. Transfer 50 μL of bacteria in a pre-chilled reaction tube. Then add 5 μL of the ligation mix and resuspend gently.
3. Incubate on ice for 30 min.

4. Heat shock for 20 sec at 37°C.
5. Incubate on ice for 2 min.
6. Add 500 µL SOC medium and incubate on a shaker at 300 rpm for 1 h at 37°C.
7. Centrifuge 30 sec with full speed in a tabletop microfuge. Discard 400 µL of the supernatant.
8. Resuspend the pellet in the remaining supernatant and streak on a selective plate containing ampicillin.

3.7. Identification of Clones

1. For identification of bacterial clones that carry the pGL3-Basic vector with the correctly inserted amplicon, prepare a PCR mastermix for as many bacterial clones as you intend to screen according to **Table 9.6**. Prepare two additional reactions for negative and positive controls. Transfer 25 µL of the mix to a PCR tube, pick a bacterial clone from the selective plate, streak the clone on a masterplate, and resuspend it in the PCR tube. As positive control, pick a bacterial clone from the re-ligation plate.
2. Use the thermocycling program according to **Table 9.7**.
3. Subject 5 µL of each PCR to agarose gel electrophoresis. Clones that did not incorporate the insert will produce an amplicon of 170 bp, e.g. the positive control clone, whereas clones carrying the correct insert from the MMP7 promoter will produce an amplicon of 1,028 bp (*see* **Note 12**).
4. Prepare plasmid DNA from a positive clone from the masterplate by using the miniprep kit from Qiagen and subject it to sequencing reaction. When the region of interest is inserted correctly, perform luciferase assay analysis (*see* **Chapter 8**) (**Volume 1**, Biechele and Moon).

Table 9.6
Pipetting instructions for screening PCR

	Volume
Screening primer forward	1 µL
Screening primer reverse	1 µL
dNTP	1 µL
TMSO	1 µL
$MgCl_2$	0.75 µL
Reaction buffer	2.5 µL
Taq polymerase	0.2 µL
H_2O	17.55 µL
Total volume	25 µL

Table 9.7
Cycling conditions for screening PCR

Step	Temperature	Time	Cycles
Initial denaturation	94	2 min	1
Denaturation	94	30 sec	
Annealing	56.9	30 sec	10
Extension	72	1 min/kb	
Denaturation	94	30 sec	
Annealing	51.9	30 sec	25
Extension	72	1 min/kb	
Final extension	72	10 min	1

4. Notes

1. Mine the PubMed database for the gene of interest and if there are published results on the promoter region and a reporter plasmid is already described, request it from the author.
2. Depending on the localization of the coding sequence of the GOI on the plus or minus strand of DNA, withdraw 5,000 bp in the *from* text box or add 5,000 bp in the *to* text box, respectively, for investigation of the 5′ region.
3. Be mindful to check your GOI for splice variants displaying different transcription start sites and thus different upstream regulatory sequences.
4. Be mindful that in most publications on β–catenin/TCF target genes, the prevalent TCF site identified is the WWCAAAG sequence; there are many fewer publications on the other TCF sites listed in PubMed. Be aware that additional TCF sites might be identified.
5. Depending on the localization and frequency of TCF sites in your specific 5′ region, you may decide to clone a shorter and a longer region of the promoter.
6. For simplicity of the subsequent cloning procedure, always try to choose restriction enzymes that work with high efficiency in a double digest.
7. Genomic DNA may also be used as a template; however, efficiency of the PCR strongly improves when a BAC is used as template.
8. When clicking on the clone designation button, you get additional data, including the international clonename (e.g.

RP11-315O6). Using this name, you should be able to order the clone from different distributors (e.g. BACPAC Resources at CHORI Oakland, CA; http:// bacpac.chori.org).

9. If the pellet is dried for too long, it becomes insoluble.
10. When cloning promoter regions, one often has to deal with GC-rich sequences, so addition of TMSO *(10)* or other reagents such as dimethylsulfoxide (DMSO) or Betaine usually strongly improves PCR efficiency. Use a proofreading polymerase, e.g. Pfu polymerase from Fermentas, to ensure correct amplification of the target sequence.
11. One may also try different ratios of Insert/Vector in the ligation reaction mix. In the case of digestion with blunt-end-generating restriction enzymes, e.g. *Sma*I, use polyethylene glycol (PEG), which is supplied with T4 ligase from Fermentas, according to the user manual, to increase ligation efficiency.
12. If it was difficult to amplify the region of interest in the first place, the screening PCRs may not produce an amplicon, except for the positive control. Nevertheless, the screened clones may carry the insert. It is reasonable to characterize them further by performing minipreps and subsequent restriction controls.

Acknowledgments

This work was supported by grants to TB from the German Research Ministry BMBF (NGFN2 project no. 01GS0436). SS was funded by the German Research Council DFG (grant no. BR1399/4-3) and OS was funded by the Deutsche Krebshilfe (grant no. 106958).

References

1. Giles, R.H., van Es, J.H., Clevers, H. (2003) Caught up in a Wnt storm: Wnt signaling in cancer. *Biochim Biophys Acta* **1653**, 1–24.
2. van de Wetering, M., Sancho, E., Verweij, C., de Lau, W., Oving, I., Hurlstone, A., van der Horn, K., Batlle, E., Coudreuse, D., Haramis, A.P., Tjon-Pon-Fong, M., Moerer, P., van den Born, M., Soete, G., Pals, S., Eilers, M., Medema, R., Clevers, H. (2002) The beta-catenin/TCF-4 complex imposes a crypt progenitor phenotype on colorectal cancer cells. *Cell* **111**, 241–250.
3. Zirn, B., Samans, B., Wittmann, S., Pietsch, T., Leuschner, I., Graf, N., Gessler, M. (2006) Target genes of the WNT/beta-catenin pathway in Wilms tumors. *Genes Chromosomes Cancer* **45**, 565–574.
4. Huang, S., Li, Y., Chen, Y., Podsypanina, K., Chamorro, M., Olshen, A.B., Desai, K.V., Tann, A., Petersen, D., Green, J.E., Varmus, H.E. (2005) Changes in gene expression during the development of mammary tumors in MMTV-Wnt-1 transgenic mice. *Genome Biol* **6**, R84.

5. Sambrook, J., Fritsch, E.F., Maniatis, T. (1989) Molecular cloning: a laboratory manual. 2nd edn. Cold Spring Harbor Laboratory Press.
6. Brabletz, T., Jung, A., Dag, S., Hlubek, F., Kirchner, T. (1999) beta-catenin regulates the expression of the matrix metalloproteinase-7 in human colorectal cancer. *Amer J Pathol* **155**, 1033–1038.
7. Crawford, H.C., Fingleton, B.M., Rudolph-Owen, L.A., Goss, K.J., Rubinfeld, B., Polakis, P., Matrisian, L.M. (1999) The metalloproteinase matrilysin is a target of beta-catenin transactivation in intestinal tumors. *Oncogene* **18**, 2883–2891.
8. Korinek, V., Barker, N., Morin, P.J., van Wichen, D., de Weger, R., Kinzler, K.W., Vogelstein, B., Clevers, H. (1997) Constitutive transcriptional activation by a beta-catenin-Tcf complex in APC-/- colon carcinoma. *Science* **275**, 1784–1787.
9. Ghiselli, G., Coffee, N., Munnery, C.E., Koratkar, R., Siracusa, L.D. (2003) The cohesin SMC3 is a target for beta-catenin/TCF4 transactivation pathway. *J Biol Chem* **278**, 20259–20267.
10. Chakrabarti, R., Schutt, C.E. (2002) Novel sulfoxides facilitate GC-rich template amplification. *Biotechniques* **32**, 866, 868, 870–862, 874.
11. Rockman, S.P., Currie, S.A., Ciavarella, M., Vincan, E., Dow, C., Thomas, R.J., Phillips, W.A. (2001) Id2 is a target of the beta-catenin/T cell factor pathway in colon carcinoma. *J Biol Chem* **276**, 45113–45119.
12. Gradl, D., Kuhl, M., Wedlich, D. (1999) The Wnt/Wg signal transducer beta-catenin controls fibronectin expression. *Mol Cell Biol* **19**, 5576–5587.

Section B

Non-canonical Signaling

Chapter 10

β-Catenin-Independent Wnt Pathways: Signals, Core Proteins, and Effectors

Richard G. James, William H. Conrad, and Randall T. Moon

Abstract

Wnt signaling activates several distinct intracellular pathways, which are important for cell proliferation, differentiation, and polarity. Wnt proteins are secreted molecules that typically signal across the membrane via interaction with the transmembrane receptor Frizzled. Following interaction with Frizzled, the downstream effect of the most widely studied Wnt pathway is stabilization and nuclear translocation of the cytosolic protein, β-catenin. In this chapter, we discuss two β-catenin-independent branches of Wnt signaling: 1) Wnt/planar cell polarity (PCP), a Wnt pathway that signals through the small GTPases, Rho and Rac, to promote changes in the actin cytoskeleton, and 2) Wnt/Ca^{2+}, a Wnt pathway that promotes intracellular calcium transients and negatively regulates the Wnt/β-catenin pathway. Finally, during the course of our discussion, we highlight areas that require future research.

Key words: Wnt, Wnt/Ca, Wnt/PCP.

1. Introduction

The Wnt signaling pathway regulates cell proliferation, differentiation, and polarity in several models of development and disease. Wnt proteins are secreted glycoproteins that activate transduction cascades upon binding to specific cell surface receptors. In the most extensively studied Wnt pathway (Wnt/β-catenin signaling), Wnt proteins bind to members of the Frizzled family of membrane-bound receptors proteins and the co-receptor LRP5/6. The stimulation of these receptors activates a series of intracellular events that result in stabilization of cytoplasmic β-catenin, its transport to the nucleus, and transcription of downstream target genes (see the left cascade in **Fig. 10.1**; reviewed in **Chapter 1** (**Volume 1**) and ref. *(1)*).

Elizabeth Vincan (ed.), *Wnt Signaling, Volume I: Pathway Methods and Mammalian Models, vol. 468*

Book doi: 10.1007/978-1-59745-249-6

Unlike the Wnt-β-catenin pathway that follows one linear signaling cascade, the β-catenin-independent Wnt signaling pathways have multiple branches. Whether these are largely overlapping or distinct is still a mystery. One branch of the pathway modulates intracellular Ca^{2+} levels. This pathway is commonly known as the Wnt/Ca^{2+} pathway (middle cascade, **Fig. 10.1**). A second branch of the pathway directs cytoskeletal rearrangements through several small GTPases as well as the jun kinase (JNK) signaling pathway. This pathway is often called the Wnt/PCP pathway (right cascade, **Fig. 10.1**). The following chapters in this section describe the current experimental approaches utilized to monitor each branch. This chapter puts these assays into a broader context of the β-catenin-independent Wnt pathways. We hope to contextualize the β-catenin-independent Wnt signaling readouts described in future chapters as well as demonstrate the current gaps in the field.

1.1. Glossary of Terms

Wnt—A secreted glycoprotein that binds Frizzled receptors and activates several intracellular pathways, including Wnt/β-catenin, Wnt/PCP, and Wnt/Ca^{2+}.

Planar cell polarity (PCP)—The organization of protein complexes found in individual cells within the plane of a single layered sheet of cells, which occurs orthogonal to the apical–basal axis.

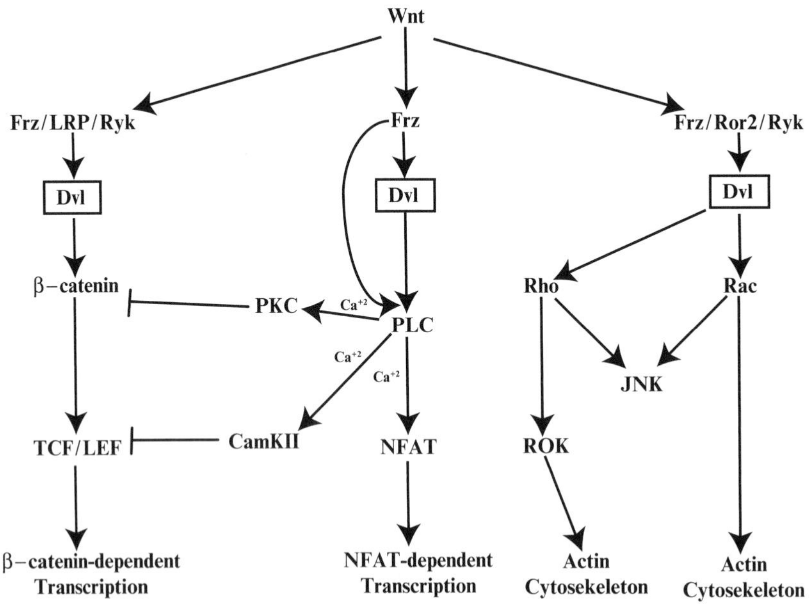

Fig. 10.1. The Wnt signaling pathways. Wnt signaling branches into distinct downstream pathways. Here, we simply diagram Wnt/β-catenin signaling (*left pathway*), Wnt/Ca^{2+} signaling (*middle pathway*), and Wnt/PCP signaling (*right pathway*). The downstream effect of Wnt/β-catenin signaling is TCF-dependent transcription. Downstream of Wnt/Ca^{2+} signaling is negative regulation of Wnt/β-catenin signaling and NFAT-dependent transcription. Finally, the Wnt/PCP pathway promotes changes in the Actin cytoskeleton. *Frz*, Frizzled; *DVL*, Dishevelled; *PLC*, phospholipase C.

Convergent extension (CE)—During zebrafish and *Xenopus* development, tissues narrow (convergence) and lengthen (extension) along the anteroposterior axis. On a cellular level, CE involves cell shape/adhesion changes, and directed migration.

Keller explant—Explants taken from the dorsal mesoderm of *Xenopus* embryos can be cultured in vitro between glass coverslips so that they exhibit cellular movements very similar to those seen during convergent extension during development. Keller explants have been utilized extensively to study the role of β-catenin-independent Wnt signaling in convergent extension.

Wnt/β-catenin signaling—Several Wnt proteins activate signaling via stabilization and nuclear translocation of β-catenin, which results in transcriptional activation of specific target β-catenin target genes.

Upstream activating proteins—These proteins initiate Wnt signaling. For the purposes of this review, these proteins include the Wnt ligands and receptors for Wnt that act upstream of Frizzled.

Core transduction proteins—Frizzled, Dishevelled, Diversin, Prickle, Strabismus, and Flamingo are the core transduction proteins of β-catenin-independent Wnt signaling. These proteins serve the following diverse functions: receiving signal from upstream activating proteins, establishing subcellular localization of other core transduction proteins, and activating the downstream effector pathways of Wnt.

Effector proteins—These proteins signal via the Wnt/β-catenin, Wnt/Ca^{2+}, or Wnt/PCP pathways downstream of Dishevelled.

Wnt/PCP signaling—This β-catenin-independent Wnt pathway directs cytoskeletal rearrangement via JNK and the small GTPases, Rac and Rho.

Wnt/Ca2+ signaling—In some contexts, Wnt signaling promotes changes in cellular Ca^{2+} signaling, which results in activation of protein kinase C (PKC), Ca^{2+}–calmodulin-dependent protein kinase II (CamK2), and nuclear factor of activated T cells (NFAT). Intriguingly, following Wnt/Ca^{2+} signaling, PKC, CamKII, and NFAT separately repress Wnt/β-catenin signaling.

2. β-Catenin-Independent Wnt Signaling Components

When considering the different branches of β-catenin-independent Wnt signaling, it is important to consider what signaling molecules these pathways share as well as what signaling molecules are specific to each branch. Each of the β-catenin-independent Wnt signaling pathways shares a suite of *core transduction proteins*. The *upstream activating proteins* that initiate signaling vary significantly between the branches of β-catenin-independent Wnt signaling. Likewise, each β-catenin-independent Wnt pathway activates different *effector pathways*. Here, we briefly describe several proteins that fall into each of the *core transduction protein* and *upstream activating protein* categories and introduce how these proteins were found to be involved in β-catenin-independent Wnt signaling. Then, we will discuss the various effector pathways of β-catenin-independent Wnt signaling.

2.1. Core Transduction Proteins

Initially, investigators characterized the *core transduction proteins* of the β-catenin-independent Wnt pathway using fly models of planar cell polarity (PCP). PCP is the organization of protein complexes found in individual cells within fields. The wing sensory hairs of *Drosophila* serve as an excellent model of PCP (*see* **Chapters 11** and **12** of **Volume 2**). In the normal wing epithelium, the sensory hairs of an individual cell are always located on the side of the cell that is distal to the animal's body. Mutations

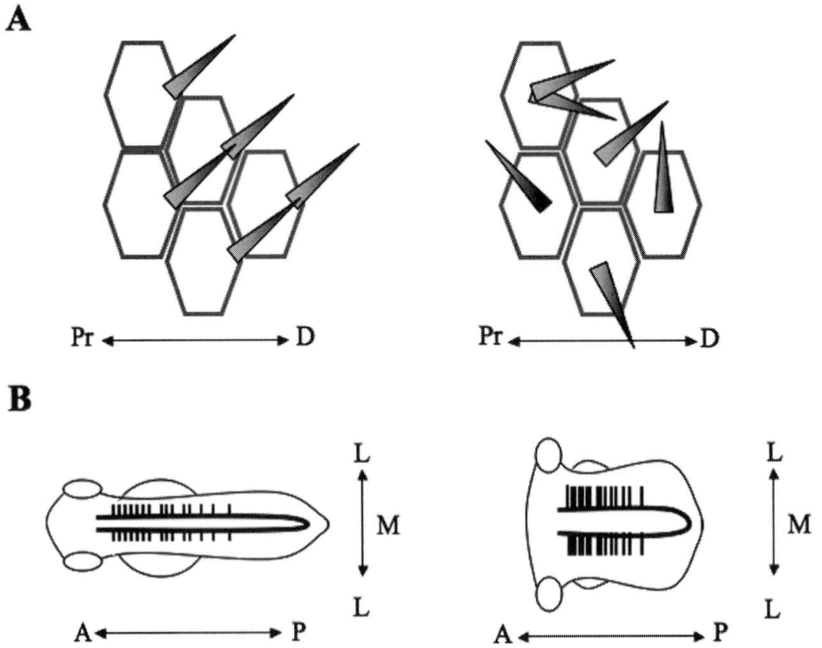

Fig. 10.2. Representation of planar cell polarity (*PCP*) defects in *Drosophila* wing hairs and convergent extension (*CE*) defects in zebrafish. **A** During normal PCP, cells of the *Drosophila* wing produce a single hair on the distal end of the cell (*left*). When PCP is disrupted, the hair pattern becomes disorganized (*right*). **B** Dorsal view of a wild-type zebrafish embryo at the 15-somite stage (*left*). Dorsal view of a zebrafish embryo with a CE defect (*right*). Note the shortened anteroposterior axis and widened mediolateral axis. *Pr*, proximal; *D*, distal; *A*, anterior; *P*, posterior; *L*, lateral; *M*, medial.

of several genes result in a characteristic phenotype in which cells of the wing become disorganized and the uniform distal location of the hairs in individual cells is lost (**Fig. 10.2a**); genes that are required for normal PCP are referred to as PCP genes.

The establishment of correct sensory hair polarity requires the precise organization of protein complexes to both the proximal and distal side of the cell. Among the first proteins that were shown to be required for fly PCP were the Frizzled receptor and the cytoplasmic protein Dishevelled, both of which are localized to the distal part of a wing cell prior to sensory hair formation. The Frizzled/Dishevelled complex activates cytoskeletal rearrangement via indirect modulation of actin fibers and microtubules.

In addition to Dishevelled and Frizzled, there are several other conserved proteins required for *Drosophila* PCP: Diego, Strabismus, Prickle, and Flamingo. Let us look at the fly wing to briefly discuss how these proteins polarize (this has been extensively reviewed in ref. *(2)*). At the beginning of cell polarization, the *core transduction proteins* are localized generally to the apical half of the cell *(3)*. Following stimulation by *upstream activating proteins (4, 5)*, Frizzled and Dishevelled migrate together as a complex to the distal portion of the cell where the future sensory hair will form *(6–8)*. The distal migration of Frizzled and Dishevelled is, in part, catalyzed by the activities of Strabismus and Prickle, which act on the proximal side of the cell to antagonize the interaction between Frizzled and Dishevelled, thus disrupting their complex, and preventing inappropriate proximal localization *(9, 10)*. Diego acts as a positive regulator of Frizzled/Dishevelled distal localization as it complexes with Frizzled and negatively regulates any Prickle present on the distal side of the cell *(11)*. The transmembrane protein Flamingo, while present on both the proximal and distal side of the cell, is required downstream of Frizzled and Dishevelled to direct PCP during sensory hair formation *(12)*. Following proper localization, Frizzled, Flamingo, and Dishevelled direct effector PCP machinery to promote differential formation of a sensory hair *(2, 13–15)*.

In vertebrates, cellular processes that involve directed migration and adhesion require orthologous protein complexes to those that establish fly PCP. The most prevalent vertebrate assay used to investigate the protein complexes utilized by β-catenin-independent Wnt signaling has been the process of convergent extension (CE) during development (*see* **Chapters 17, 21** and **30** of **Volume 2**). CE occurs during embryonic development in zebrafish, and *Xenopus* via a concerted process of directed migration and cellular intercalation. Embryos that exhibit defects in CE are short and fat relative to normal embryos (**Fig. 10.2b**). The short/fat phenotype can be caused by defects in directed cell migration, aberrant division plane polarity, or increased/decreased adhesion during embryonic gastrulation. The β-catenin-independent *core transduction proteins* conserved between fly PCP and vertebrate CE include Diego (or Diversin *(16)*), Strabismus (or Vangl2 *(17)*), Dishevelled *(18, 19)*, Flamingo (or Celsr *(20)*), and Prickle *(21, 22)*. While these genetic linkages have been determined, conservation of molecular function of the *core transduction proteins* between fly PCP and vertebrate CE has not been established.

2.2. Upstream Activating Proteins

For the purposes of this review, we will define *upstream activating proteins* as proteins that act upstream to the previously described *core transduction proteins*. In this section, we discuss both the receptors and the ligands that activate the β-catenin-independent Wnt pathways.

The link between the PCP proteins and the Wnt ligand was originally suggested following zebrafish loss-of-function experiments that demonstrated that Wnt5a *(23)* and Wnt11 *(24)* are required for normal CE. This link was cemented by a series of experiments in zebrafish and *Xenopus* that showed genetic linkage in CE of Wnt5a/11 and several of the β-catenin-independent Wnt signaling *core transduction proteins* including Prickle *(25)*, Flamingo *(20)*, and Diversin *(26)*.

Although there is strong genetic data linking Wnt with Frizzled, Prickle, Flamingo, and Diversin, the biochemical mechanism by which Wnt activates Dishevelled is still unclear. A few aspects of Dishevelled regulation by Wnt are clear: 1) Wnt signaling causes Frizzled-mediated recruitment of Dishevelled to the membrane *(4, 27–29)*, and 2) Wnt signaling promotes serine/threonine phosphorylation of Dishevelled *(29, 30)*. Wnt-directed phosphorylation of Dishevelled can take place in the absence of LRP6 *(30)*, and with or without stabilization of β-catenin *(30)*, indicating that Dishevelled phosphorylation may be a property of β-catenin-independent Wnt signaling. Another clue about the mechanism by which Wnt regulates Dishevelled at the membrane comes from pharmacological inhibition of G protein-coupled signaling, which prevents Wnt-mediated stabilization of β-catenin *(31)*. Additionally, studies in vertebrates *(32–34)* and flies *(35, 36)* have implicated the kinase, CK1ε as a Wnt-activated kinase that phosphorylates Dishevelled during the activation of both the β-catenin-dependent and β-catenin-independent pathways. The precise biochemical mechanism by which Frizzled receptors activate CK1ε via G protein signaling remains to be worked out.

Emerging research indicates that, in addition to Frizzled, several heretofore-uncharacterized receptors and/or co-receptors may transduce a Wnt signal (β-catenin-dependent and β-catenin-independent). Several of these have been linked to the Wnt/β-catenin pathway, including Frl1 *(37)* and receptor-like tyrosine kinase (Ryk) *(38)*, and many with the β-catenin-independent pathways including receptor tyrosine kinase-like orphan receptor 2 (Ror2) *(39–41)*, Ryk *(42, 43)*, and PTK7 *(44)*. Adding to the complexity that is already described here, a number of reports demonstrate that receptor context likely plays a large role in determining whether a particular Wnt activates β-catenin-dependent or β-catenin-independent pathways. For example, Wnt5a can promote β-catenin nuclear translocation and *Xenopus* axis duplication when co-expressed with Frizzled5 *(45)*. Additionally, treatment of several specific mammalian cell lines with the Wnt/β-catenin ligand, Wnt3, can promote activation of Rho kinase *(46)*, a response typical of the β-catenin-independent pathways.

In the fly, the situation is more confusing as it is not clear what upstream activating molecule establishes PCP. Wnt has been ruled out in the fly wing, eye, and abdomen as no Wnt mutation to date shows defective PCP. However, in other contexts in the fly, Wnt is required for cell polarity and has been genetically linked to members of the β-catenin-independent Wnt pathway. Consistent with the idea that Wnt can regulate *Drosophila* cell polarity, *wingless* is required for cell shape reorganization and polarity during denticle formation *(47)*, and dWnt3 (the fly ortholog of Wnt5a) is essential for axon pathfinding *(43)* and directed migration of salivary gland cells *(48)*.

3. Effector Proteins

While the Wnt/β-catenin signaling pathway exerts its effects specifically through the stabilization of β-catenin and activation of target gene transcription, the β-catenin-independent Wnts signal through several pathways. To add further complication, some of these signaling pathways may be context dependent. Below, we describe how Wnt exerts effects through intracellular Ca^{2+} modulation as well as cytoskeletal rearrangement. Future chapters will discuss interpretation of β-catenin-independent Wnt effectors in greater detail.

3.1. Cytoskeletal Rearrangement

We know that Wnt regulates the cytoskeleton polarization thanks, in large part, to genetic studies of PCP in *Drosophila*. Work by Wong and Adler *(49)* as well as Eaton et al. *(14)* characterized the dramatic changes that occur in the actin cytoskeleton during the development of a fly hair cell. Initially, immature hair cells display circumferential arrays of actin, as is the case in non-PCP epithelial cells. As hair formation initiates, actin accumulates at the distal vertex of each cell, where filamentous actin outgrowths ultimately form the mature hair *(49)*.

These cytoskeletal rearrangements are under the direct control of the core PCP machinery in *Drosophila*. As mentioned above, mutation of the cytosolic protein Dishevelled results in a defect in wing hair orientation of *Drosophila (50)*. Of course, Dishevelled does not polarize the cytoskeleton itself. Instead, Dishevelled directs actin reorganization via the activation of the small GTPases, Rho and Rac, which have long been known to be regulators of cytoskeletal rearrangement in a variety of contexts *(51)*. In mammalian cell culture, Rho activation leads to the assembly of contractile actin–myosin filaments known as stress fibers, which ultimately leads to the formation of focal adhesion complexes *(52)*, whereas Rac activation produces a meshwork of actin filaments at the cell periphery that produce lamellipodia and membrane ruffles *(53)*.

The mechanism by which Dishevelled mediates actin cytoskeletal remodeling during fly PCP involves both Rho and Rac. Conditional overexpression of both RhoA and Rac1 recovers PCP eye defects exhibited in Fzd or Dvl loss of function in *Drosophila (15)*. Conversely, in the wing, conditional loss of function of Rac1 and Rac2 recovers the PCP phenotype caused by Dvl overexpression in the *Drosophila* wing *(54)*. RhoA and Rac1 are also important effectors for Wnt/PCP signaling in vertebrates. Like mutations for Wnt5a and Wnt11, loss of function for Rho or Rac will result in CE defects in both *Xenopus* and zebrafish *(55)*. Furthermore, Rho can rescue CE defects present in the zebrafish mutants for Wnt5a and Wnt11 *(55)*. This evidence indicates that the GTPases utilized by Dishevelled to modulate the actin cytoskeleton are conserved between fly PCP and vertebrate CE.

Much work has been done to explain the distinct ways that Rho and Rac mediate cell polarity. While activated RhoA will rescue CE defects caused by loss of function of RhoA, Rac will not *(56)*. Likewise, RhoA will not rescue CE defects caused by loss of function of Rac *(56)*. Consistent with the idea that Rho and Rac act separately, experiments in *Xenopus* Keller explants show that during CE, activated RhoA promotes the formation of cellular protrusions, whereas activated Rac results in increased numbers of filopodia *(56)*. One specific downstream target of RhoA is RhoA-associated kinase (ROK) (*see* **Chapter 15, Volume 1**), which can also rescue CE defects caused by genetic deficiency of zebrafish Wnt5a and Wnt11 *(57)*. Ultimately, ROK modulates cytoskeletal rearrangement by direct phosphorylation of myosin II light chain actin *(58)*. The mechanism by which Rac directs cytoskeletal rearrangement downstream of Wnt remains unknown.

In addition to cytoskeletal rearrangement, Rho and Rac activate Dishevelled-mediated JNK signaling *(15, 54, 59, 60)*. For instance, it has been shown that dominant-negative XRhoA represses JNK signaling induced by overexpressed Dishevelled or Wnt11 *(59)*. Separately Rac also activates JNK: in *Drosophila*, loss of JNK signaling recovers the phenotype induced by the overexpression of Rac *(15)* and overexpression of Rac leads to upregulation of JNK transcriptional targets *(15)*. Finally, JNK activation is necessary for normal progression of fly PCP as well as vertebrate CE *(15, 54, 60, 61)* (*see* **Chapter 14, Volume 1**).

3.2. Wnt/Calcium Pathway

Modulation of intracellular Ca^{2+} levels is a second effector pathway of β-catenin-independent Wnt signaling (reviewed in refs. *(62–64)*) (*see* **Chapters 11–13**). Wnt/Ca^{2+} signaling has roles in negatively regulating the Wnt/β-catenin pathway, and CE movements in zebrafish and *Xenopus.* Importantly, elevation of calcium transients has been observed by the gold standard: treating cells with purified Wnt5a *(65, 66)*. In this section, we briefly discuss embryonic Ca^{2+} signaling, how Wnt molecules regulate Ca^{2+} signaling, and the mechanism by which Wnt/Ca^{2+} regulates downstream targets.

Unlike the Wnt/cytoskeletal pathway, characterization of Wnt/Ca^{2+} signaling has been accomplished primarily during embryonic gastrulation in the zebrafish and xenopus vertebrate embryonic models. Recent breakthroughs in cellular Ca^{2+} visualization has enabled live cell Ca^{2+} imaging of intact and healthy vertebrate embryos. The cells of zebrafish embryos normally exhibit significant fluctuation of cellular Ca^{2+} levels starting at the 32-cell stage and continuing throughout development *(67–69)*. In zebrafish, embryonic Ca^{2+} fluctuations occur in two flavors: isolated Ca^{2+} flux in individual or small groups of cells *(67)*, and waves of Ca^{2+} flux that involve large groups of cells along the entire embryonic axis *(70)*. During the process of CE, zebrafish embryos exhibit large, fast (5 μM/s) waves of Ca^{2+} signaling that travel from the dorsal to the ventral side of the embryo *(70)*. Ca^{2+} signaling is also evident in a more isolated model of CE, the *Xenopus* Keller explant, where rapid Ca^{2+} flux moves through actively intercalating cells *(71)*. Remarkably, CE requires Ca^{2+} signaling, as general pharmacological depletion of intracellular Ca^{2+} stores prevents normal CE movements yet fails to affect patterning in *Xenopus (71)*. In total, these data demonstrate that waves of Ca^{2+} signaling happen during development, and that they are required for proper CE movements.

The link between β-catenin-independent Wnt signaling and Ca^{2+} was initially proposed following gain-of-function experiments that demonstrated that Wnt5a, and Frizzled2 can synergize, in a G protein- and phospholipase C-dependent manner, to increase the rate of spontaneous Ca^{2+} waves during zebrafish development *(68, 69)*. Consistent with the idea that Wnts can regulate embryonic Ca^{2+} waves, the zebrafish mutant fish, *pipetail*, exhibits a decreased frequency of Ca^{2+} transients during gastrulation *(72)*. In addition to Wnt5a, several of the *core transduction proteins* that regulate Wnt/PCP and *Drosophila* PCP also regulate the Wnt/Ca^{2+} pathway. For instance, the Frizzled receptors Fz2, Fz3, Fz4, and Fz6 specifically activate downstream effectors of intracellular Ca^{2+} signaling *(73)*. Additional evidence that the *core transduction proteins* have a conserved role in Wnt/Ca^{2+} signaling come from gain-of-function experiments for zebrafish Dishevelled, which show that Dishevelled is sufficient in regulating the frequency of embryonic Ca^{2+} waves *(74)*. More recently, morpholino experiments in zebrafish implicated Prickle *(22)* and Strabismus *(17)* as required members of the Wnt/Ca^{2+} signal transduction cascade. Beyond describing the requirement of the *core transduction proteins* for normal frequency of embryonic Ca^{2+} waves, the mechanism of action of transduction of the Wnt/Ca^{2+} pathway at the membrane has been poorly characterized. Most importantly, in individual breast cells *(65)* and mouse embryonic stem cells *(66)*, Wnt5a promotes Ca^{2+} transients.

In relation to what is happening at the membrane, the downstream pathways of Wnt/Ca^{2+} signaling are studied more extensively *(62, 63)*. In the earlier studies performed to characterize the effectors of Wnt/Ca^{2+} signaling, overexpression of Wnt5a and Frizzled activated the Ca^{2+}-responsive kinases, PKC *(75)* and CamKII *(73)*, which presumably catalyze signaling via phosphorylation of substrates. NFAT activation comprises a distinct type of Wnt/Ca^{2+}-dependent target *(76)*, because upon Wnt5a treatment, NFAT translocates to the nucleus and activates calcineurin-dependent transcription. The role of Wnt5a in NFAT signaling is complicated by the fact that in addition to activating NFAT, the Wnt/PCP pathway negatively regulates NFAT activity *(65)*.

Intriguingly, the Wnt/Ca^{2+} pathway (reviewed in ref. *(62)*) and Ca^{2+} signaling in general *(77)* antagonizes the activity of the Wnt/β-catenin pathway. Negative regulation of the Wnt/β-catenin activity by Wnt/Ca^{2+} can be achieved via the disparate action of several downstream intermediates including PKC *(18)*, CamKII *(73)*, and NFAT *(76)*. PKC inhibits Wnt/β-catenin signaling by directly phosphorylating cytoplasmic β-catenin and targeting it for β-transducin repeat-containing protein (TRCP)-independent degradation by the proteasome *(78)*. CamKII utilizes a second mechanism whereby it phosphorylates Tak1, which activates Nlk. In turn, Nlk directly phosphorylates Lef1 and catalyzes the dissociation of the Lef1–β-catenin protein complex from DNA, thus inhibiting target gene transcription *(79–81)*. Presumably, which effector pathway that the Wnt/Ca^{2+} pathway utilizes to negatively regulate Wnt/β-catenin signaling depends on cell type and context.

4. β-Catenin-Independent Wnt Signaling in the Future

While the genetic branches of the β-catenin-independent Wnt signaling pathway have been carefully characterized, large questions remain. First, what are the mechanisms by which the various branches of Wnt signaling are activated at the membrane? Second, as each pathway regulates Dishevelled, how does Dishevelled decide what downstream effector pathway to activate? Also, there are several suggestions in the literature that the vertebrate Wnt/Ca and Wnt/PCP pathways are integrated. To test whether that is true, the field requires assays to monitor both Wnt/PCP and Wnt/Ca in parallel during real time. Finally, vertebrate CE is a complex process that involves proper cell adhesion, directed migration, and cell proliferation rates. As the β-catenin-independent Wnt pathways have been implicated in each of the processes separately, in vitro cellular models of adhesion, migration, and proliferation will enhance our knowledge of β-catenin-independent signaling greatly. Hopefully, each of the above questions will be clarified by complementing the extensive genetic data in the literature with improved biochemistry and proteomic approaches.

Acknowledgments

This work was supported by the Howard Hughes Medical Institute and the National Institutes of Health (NIH) RO1 GM073887-03 to RTM.

References

1. Moon, R.T., A.D. Kohn, G.V. De Ferrari, and A. Kaykas. (2004) WNT and beta-catenin signalling: diseases and therapies. *Nat Rev Genet.* **5**, 691–701.
2. Seifert, J.R. and M. Mlodzik. (2007) Frizzled/PCP signalling: a conserved mechanism regulating cell polarity and directed motility. *Nat Rev Genet.* **8**, 126–138.
3. Wu, J., T.J. Klein, and M. Mlodzik. (2004) Subcellular localization of frizzled receptors, mediated by their cytoplasmic tails, regulates signaling pathway specificity. *PLoS Biol.* **2**, E158.
4. Amonlirdviman, K., N.A. Khare, D.R. Tree, W.S. Chen, J.D. Axelrod, and C.J. Tomlin. (2005) Mathematical modeling of planar cell polarity to understand domineering nonautonomy. *Science.* **307**, 423–426.
5. Yang, C.H., J.D. Axelrod, and M.A. Simon. (2002) Regulation of Frizzled by fat-like cadherins during planar polarity signaling in the Drosophila compound eye. *Cell.* **108**, 675–688.
6. Axelrod, J.D. (2001) Unipolar membrane association of Dishevelled mediates Frizzled planar cell polarity signaling. *Genes Dev.* **15**, 1182–1187.
7. Das, G., A. Jenny, T.J. Klein, S. Eaton, and M. Mlodzik. (2004) Diego interacts with Prickle and Strabismus/Van Gogh to localize planar cell polarity complexes. *Development.* **131**, 4467–4476.
8. Strutt, D.I. (2001) Asymmetric localization of frizzled and the establishment of cell polarity in the Drosophila wing. *Mol Cell.* **7**, 367–375.
9. Bastock, R., H. Strutt, and D. Strutt. (2003) Strabismus is asymmetrically localised and binds to Prickle and Dishevelled during Drosophila planar polarity patterning. *Development.* **130**, 3007–3014.
10. Tree, D.R., J.M. Shulman, R. Rousset, M.P. Scott, D. Gubb, and J.D. Axelrod. (2002) Prickle mediates feedback amplification to generate asymmetric planar cell polarity signaling. *Cell.* **109**, 371–381.

11. Jenny, A., J. Reynolds-Kenneally, G. Das, M. Burnett, and M. Mlodzik. (2005) Diego and Prickle regulate Frizzled planar cell polarity signalling by competing for Dishevelled binding. *Nat Cell Biol.* 7, 691–697.
12. Usui, T., Y. Shima, Y. Shimada, S. Hirano, R.W. Burgess, T.L. Schwarz, et al. (1999) Flamingo, a seven-pass transmembrane cadherin, regulates planar cell polarity under the control of Frizzled. *Cell.* **98**, 585–595.
13. Adler, P.N., J. Charlton, K.H. Jones, and J. Liu. (1994) The cold-sensitive period for frizzled in the development of wing hair polarity ends prior to the start of hair morphogenesis. *Mech Dev.* **46**, 101–107.
14. Eaton, S., R. Wepf, and K. Simons. (1996) Roles for Rac1 and Cdc42 in planar polarization and hair outgrowth in the wing of Drosophila. *J Cell Biol.* **135**, 1277–1289.
15. Fanto, M., U. Weber, D.I. Strutt, and M. Mlodzik. (2000) Nuclear signaling by Rac and Rho GTPases is required in the establishment of epithelial planar polarity in the Drosophila eye. *Curr Biol.* **10**, 979–988.
16. Schwarz-Romond, T., C. Asbrand, J. Bakkers, M. Kuhl, H.J. Schaeffer, J. Huelsken, et al. (2002) The ankyrin repeat protein Diversin recruits Casein kinase Iepsilon to the beta-catenin degradation complex and acts in both canonical Wnt and Wnt/JNK signaling. *Genes Dev.* **16**, 2073–2084.
17. Park, M. and R.T. Moon. (2002) The planar cell-polarity gene stbm regulates cell behaviour and cell fate in vertebrate embryos. *Nat Cell Biol.* **4**, 20–25.
18. Wallingford, J.B., B.A. Rowning, K.M. Vogeli, U. Rothbacher, S.E. Fraser, and R.M. Harland. (2000) Dishevelled controls cell polarity during Xenopus gastrulation. *Nature.* **405**, 81–85.
19. Wang, J., N.S. Hamblet, S. Mark, M.E. Dickinson, B.C. Brinkman, N. Segil, et al. (2006) Dishevelled genes mediate a conserved mammalian PCP pathway to regulate convergent extension during neurulation. *Development.* **133**, 1767–1778.
20. Formstone, C.J. and I. Mason. (2005) Combinatorial activity of Flamingo proteins directs convergence and extension within the early zebrafish embryo via the planar cell polarity pathway. *Dev Biol.* **282**, 320–335.
21. Takeuchi, M., J. Nakabayashi, T. Sakaguchi, T.S. Yamamoto, H. Takahashi, H. Takeda, et al. (2003) The prickle-related gene in vertebrates is essential for gastrulation cell movements. *Curr Biol.* **13**, 674–679.
22. Veeman, M.T., D.C. Slusarski, A. Kaykas, S.H. Louie, and R.T. Moon. (2003) Zebrafish prickle, a modulator of noncanonical Wnt/Fz signaling, regulates gastrulation movements. *Curr Biol.* **13**, 680–685.
23. Kilian, B., H. Mansukoski, F.C. Barbosa, F. Ulrich, M. Tada, and C.P. Heisenberg. (2003) The role of Ppt/Wnt5 in regulating cell shape and movement during zebrafish gastrulation. *Mech Dev.* **120**, 467–476.
24. Heisenberg, C.P., M. Tada, G.J. Rauch, L. Saude, M.L. Concha, R. Geisler, et al. (2000) Silberblick/Wnt11 mediates convergent extension movements during zebrafish gastrulation. *Nature.* **405**, 76–81.
25. Carreira-Barbosa, F., M.L. Concha, M. Takeuchi, N. Ueno, S.W. Wilson, and M. Tada. (2003) Prickle 1 regulates cell movements during gastrulation and neuronal migration in zebrafish. *Development.* **130**, 4037–4046.
26. Moeller, H., A. Jenny, H.J. Schaeffer, T. Schwarz-Romond, M. Mlodzik, M. Hammerschmidt, et al. (2006) Diversin regulates heart formation and gastrulation movements in development. *Proc Natl Acad Sci U S A.* **103**, 15900–15905.
27. Axelrod, J.D., J.R. Miller, J.M. Shulman, R.T. Moon, and N. Perrimon. (1998) Differential recruitment of Dishevelled provides signaling specificity in the planar cell polarity and Wingless signaling pathways. *Genes Dev.* **12**, 2610–2622.
28. Boutros, M., J. Mihaly, T. Bouwmeester, and M. Mlodzik. (2000) Signaling specificity by Frizzled receptors in Drosophila. *Science.* **288**, 1825–1828.
29. Yanagawa, S., F. van Leeuwen, A. Wodarz, J. Klingensmith, and R. Nusse. (1995) The dishevelled protein is modified by wingless signaling in Drosophila. *Genes Dev.* **9**, 1087–1097.
30. Gonzalez-Sancho, J.M., K.R. Brennan, L.A. Castelo-Soccio, and A.M. Brown. (2004) Wnt proteins induce dishevelled phosphorylation via an LRP5/6- independent mechanism, irrespective of their ability to stabilize beta-catenin. *Mol Cell Biol.* **24**, 4757–4768.
31. Liu, T., A.J. DeCostanzo, X. Liu, H. Wang, S. Hallagan, R.T. Moon, et al. (2001) G protein signaling from activated rat frizzled-1 to the beta-catenin-Lef-Tcf pathway. *Science.* **292**, 1718–1722.
32. Cong, F., L. Schweizer, and H. Varmus. (2004) Casein kinase Iepsilon modulates the signaling specificities of dishevelled. *Mol Cell Biol.* **24**, 2000–2011.
33. Peters, J.M., R.M. McKay, J.P. McKay, and J.M. Graff. (1999) Casein kinase I transduces Wnt signals. *Nature.* **401**, 345–350.
34. Swiatek, W., I.C. Tsai, L. Klimowski, A. Pepler, J. Barnette, H.J. Yost, et al. (2004) Regulation

of casein kinase I epsilon activity by Wnt signaling. *J Biol Chem.* **279**, 13011–13017.

35. Klein, T.J., A. Jenny, A. Djiane, and M. Mlodzik. (2006) CKIepsilon/discs overgrown promotes both Wnt-Fz/beta-catenin and Fz/PCP signaling in Drosophila. *Curr Biol.* **16**, 1337–1343.
36. Strutt, H., M.A. Price, and D. Strutt. (2006) Planar polarity is positively regulated by casein kinase Iepsilon in Drosophila. *Curr Biol.* **16**, 1329–1336.
37. Tao, Q., C. Yokota, H. Puck, M. Kofron, B. Birsoy, D. Yan, et al. (2005) Maternal wnt11 activates the canonical wnt signaling pathway required for axis formation in Xenopus embryos. *Cell.* **120**, 857–871.
38. Lu, W., V. Yamamoto, B. Ortega, and D. Baltimore. (2004) Mammalian Ryk is a Wnt coreceptor required for stimulation of neurite outgrowth. *Cell.* **119**, 97–108.
39. Hikasa, H., M. Shibata, I. Hiratani, and M. Taira. (2002) The Xenopus receptor tyrosine kinase Xror2 modulates morphogenetic movements of the axial mesoderm and neuroectoderm via Wnt signaling. *Development.* **129**, 5227–5239.
40. Mikels, A.J. and R. Nusse. (2006) Purified Wnt5a protein activates or inhibits beta-catenin-TCF signaling depending on receptor context. *PLoS Biol.* **4**, e115.
41. Schambony, A. and D. Wedlich. (2007) Wnt-5A/Ror2 regulate expression of XPAPC through an alternative noncanonical signaling pathway. *Dev Cell.* **12**, 779–792.
42. Keeble, T.R., M.M. Halford, C. Seaman, N. Kee, M. Macheda, R.B. Anderson, et al. (2006) The Wnt receptor Ryk is required for Wnt5a-mediated axon guidance on the contralateral side of the corpus callosum. *J Neurosci.* **26**, 5840–5848.
43. Yoshikawa, S., R.D. McKinnon, M. Kokel, and J.B. Thomas. (2003) Wnt-mediated axon guidance via the Drosophila Derailed receptor. *Nature.* **422**, 583–588.
44. Lu, X., A.G. Borchers, C. Jolicoeur, H. Rayburn, J.C. Baker, and M. Tessier-Lavigne. (2004) PTK7/CCK-4 is a novel regulator of planar cell polarity in vertebrates. *Nature.* **430**, 93–98.
45. He, X., J.P. Saint-Jeannet, Y. Wang, J. Nathans, I. Dawid, and H. Varmus. (1997) A member of the Frizzled protein family mediating axis induction by Wnt-5A. *Science.* **275**, 1652–1654.
46. Habas, R., Y. Kato, and X. He. (2001) Wnt/Frizzled activation of Rho regulates vertebrate gastrulation and requires a novel Formin homology protein Daam1. *Cell.* **107**, 843–854.
47. Price, M.H., D.M. Roberts, B.M. McCartney, E. Jezuit, and M. Peifer. (2006) Cytoskeletal dynamics and cell signaling during planar polarity establishment in the Drosophila embryonic denticle. *J Cell Sci.* **119**, 403–415.
48. Harris, K.E. and S.K. Beckendorf. (2007) Different Wnt signals act through the Frizzled and RYK receptors during Drosophila salivary gland migration. *Development.* **134**, 2017–2025.
49. Wong, L.L. and P.N. Adler. (1993) Tissue polarity genes of Drosophila regulate the subcellular location for prehair initiation in pupal wing cells. *J Cell Biol.* **123**, 209–221.
50. Gubb, D. and A. Garcia-Bellido. (1982) A genetic analysis of the determination of cuticular polarity during development in Drosophila melanogaster. *J Embryol Exp Morphol.* **68**, 37–57.
51. Hall, A. (1998) Rho GTPases and the actin cytoskeleton. *Science.* **279**, 509–514.
52. Ridley, A.J. and A. Hall. (1992) The small GTP-binding protein rho regulates the assembly of focal adhesions and actin stress fibers in response to growth factors. *Cell.* **70**, 389–399.
53. Ridley, A.J., H.F. Paterson, C.L. Johnston, D. Diekmann, and A. Hall. (1992) The small GTP-binding protein rac regulates growth factor-induced membrane ruffling. *Cell.* **70**, 401–410.
54. Boutros, M., N. Paricio, D.I. Strutt, and M. Mlodzik. (1998) Dishevelled activates JNK and discriminates between JNK pathways in planar polarity and wingless signaling. *Cell.* **94**, 109–118.
55. Zhu, S., L. Liu, V. Korzh, Z. Gong, and B.C. Low. (2006) RhoA acts downstream of Wnt5 and Wnt11 to regulate convergence and extension movements by involving effectors Rho kinase and Diaphanous: use of zebrafish as an in vivo model for GTPase signaling. *Cell Signal.* **18**, 359–372.
56. Tahinci, E. and K. Symes. (2003) Distinct functions of Rho and Rac are required for convergent extension during Xenopus gastrulation. *Dev Biol.* **259**, 318–335.
57. Marlow, F., J. Topczewski, D. Sepich, and L. Solnica-Krezel. (2002) Zebrafish Rho kinase 2 acts downstream of Wnt11 to mediate cell polarity and effective convergence and extension movements. *Curr Biol.* **12**, 876–884.
58. Winter, C.G., B. Wang, A. Ballew, A. Royou, R. Karess, J.D. Axelrod, et al. (2001) Drosophila Rho-associated kinase (Drok) links Frizzled-mediated planar cell polarity signaling to the actin cytoskeleton. *Cell.* **105**, 81–91.

59. Kim, G.H. and J.K. Han. (2005) JNK and ROKalpha function in the noncanonical Wnt/RhoA signaling pathway to regulate Xenopus convergent extension movements. *Dev Dyn.* **232**, 958–968.
60. Strutt, D.I., U. Weber, and M. Mlodzik. (1997) The role of RhoA in tissue polarity and Frizzled signalling. *Nature.* **387**, 292–295.
61. Yamanaka, H., T. Moriguchi, N. Masuyama, M. Kusakabe, H. Hanafusa, R. Takada, et al. (2002) JNK functions in the non-canonical Wnt pathway to regulate convergent extension movements in vertebrates. *EMBO Rep.* **3**, 69–75.
62. Kohn, A.D. and R.T. Moon. (2005) Wnt and calcium signaling: beta-catenin-independent pathways. *Cell Calcium.* **38**, 439–446.
63. Slusarski, D.C. and F. Pelegri. (2007) Calcium signaling in vertebrate embryonic patterning and morphogenesis. *Dev Biol.* **307**, 1–13.
64. Webb, S.E. and A.L. Miller. (2003) Calcium signalling during embryonic development. *Nat Rev Mol Cell Biol.* **4**, 539–551.
65. Dejmek, J., A. Safholm, C. Kamp Nielsen, T. Andersson, and K. Leandersson. (2006) Wnt-5a/Ca2+-induced NFAT activity is counteracted by Wnt-5a/Yes-Cdc42-casein kinase 1alpha signaling in human mammary epithelial cells. *Mol Cell Biol.* **26**, 6024–6036.
66. Ma, L. and H.Y. Wang. (2006) Suppression of cyclic GMP-dependent protein kinase is essential to the Wnt/cGMP/Ca2+ pathway. *J Biol Chem.* **281**, 30990–31001.
67. Reinhard, E., H. Yokoe, K.R. Niebling, N.L. Allbritton, M.A. Kuhn, and T. Meyer. (1995) Localized calcium signals in early zebrafish development. *Dev Biol.* **170**, 50–61.
68. Slusarski, D.C., V.G. Corces, and R.T. Moon. (1997) Interaction of Wnt and a Frizzled homologue triggers G-protein-linked phosphatidylinositol signalling. *Nature.* **390**, 410–413.
69. Slusarski, D.C., J. Yang-Snyder, W.B. Busa, and R.T. Moon. (1997) Modulation of embryonic intracellular Ca2+ signaling by Wnt-5A. *Dev Biol.* **182**, 114–120.
70. Gilland, E., A.L. Miller, E. Karplus, R. Baker, and S.E. Webb. (1999) Imaging of multicellular large-scale rhythmic calcium waves during zebrafish gastrulation. *Proc Natl Acad Sci U S A.* **96**, 157–161.
71. Wallingford, J.B., A.J. Ewald, R.M. Harland, and S.E. Fraser. (2001) Calcium signaling during convergent extension in Xenopus. *Curr Biol.* **11**, 652–661.
72. Westfall, T.A., R. Brimeyer, J. Twedt, J. Gladon, A. Olberding, M. Furutani-Seiki, et al. (2003) Wnt-5/pipetail functions in vertebrate axis formation as a negative regulator of Wnt/beta-catenin activity. *J Cell Biol.* **162**, 889–898.
73. Kuhl, M., L.C. Sheldahl, C.C. Malbon, and R.T. Moon. (2000) Ca(2+)/calmodulin-dependent protein kinase II is stimulated by Wnt and Frizzled homologs and promotes ventral cell fates in Xenopus. *J Biol Chem.* **275**, 12701–12711.
74. Sheldahl, L.C., D.C. Slusarski, P. Pandur, J.R. Miller, M. Kuhl, and R.T. Moon. (2003) Dishevelled activates Ca2+ flux, PKC, and CamKII in vertebrate embryos. *J Cell Biol.* **161**, 769–777.
75. Sheldahl, L.C., M. Park, C.C. Malbon, and R.T. Moon. (1999) Protein kinase C is differentially stimulated by Wnt and Frizzled homologs in a G-protein-dependent manner. *Curr Biol.* **9**, 695–698.
76. Saneyoshi, T., S. Kume, Y. Amasaki, and K. Mikoshiba. (2002) The Wnt/calcium pathway activates NF-AT and promotes ventral cell fate in Xenopus embryos. *Nature.* **417**, 295–299.
77. Westfall, T.A., B. Hjertos, and D.C. Slusarski. (2003) Requirement for intracellular calcium modulation in zebrafish dorsal-ventral patterning. *Dev Biol.* **259**, 380–391.
78. Gwak, J., M. Cho, S.J. Gong, J. Won, D.E. Kim, E.Y. Kim, et al. (2006) Protein-kinase-C-mediated beta-catenin phosphorylation negatively regulates the Wnt/beta-catenin pathway. *J Cell Sci.* **119**, 4702–4709.
79. Ishitani, T., S. Kishida, J. Hyodo-Miura, N. Ueno, J. Yasuda, M. Waterman, et al. (2003) The TAK1-NLK mitogen-activated protein kinase cascade functions in the Wnt-5a/Ca(2+) pathway to antagonize Wnt/beta-catenin signaling. *Mol Cell Biol.* **23**, 131–139.
80. Ishitani, T., J. Ninomiya-Tsuji, and K. Matsumoto. (2003) Regulation of lymphoid enhancer factor 1/T-cell factor by mitogen-activated protein kinase-related Nemo-like kinase-dependent phosphorylation in Wnt/beta-catenin signaling. *Mol Cell Biol.* **23**, 1379–1389.
81. Ishitani, T., J. Ninomiya-Tsuji, S. Nagai, M. Nishita, M. Meneghini, N. Barker, et al. (1999) The TAK1-NLK-MAPK-related pathway antagonizes signalling between beta-catenin and transcription factor TCF. *Nature.* **399**, 798–802.

Chapter 11

Image Analysis of Calcium Release Dynamics

Christina M. Freisinger, Douglas W. Houston, and Diane C. Slusarski

Abstract

Many aspects of animal development are dependent on the dynamic release of calcium (Ca^{2+}) ions. Although Ca^{2+} release within a cell is tightly controlled, how the release dynamics result in a specific biological output in embryonic development is less clear. The integration of pharmacological and molecular-genetic studies with in vivo imaging in zebrafish and *Xenopus* has linked endogenous Ca^{2+} release to the Wnt signaling network. Our data suggests that distinct Ca^{2+} release dynamics lead to antagonism of the developmentally important Wnt/β-catenin pathway while sustained Ca^{2+} release modulates polarized cell movements.

Key words: Calcium, Wnt-5, Zebrafish, *Xenopus*, Fura-2, Wnt/calcium, Wnt/beta-catenin, Ratiometric calcium imaging.

1. Introduction

Calcium (Ca^{2+}) release is a key signal for many cellular processes including neuronal synapse, muscle contraction, cell division, and fertilization. As an essential second messenger molecule, the dynamics of Ca^{2+} release inside a cell are tightly regulated. Ca^{2+} levels are predominantly controlled by a gradient of Ca^{2+} concentration either across the plasma membrane or across the membrane of intracellular Ca^{2+} stores. The opening of specialized ion channels and the release from intracellular organelles generates bursts of Ca^{2+} into the cytosol. The location, extent, and duration of the ion channel opening can result in a local or global signaling event. During embryogenesis, growth factors such as the Wnts, a large family of secreted cysteine-rich glycoproteins, and their

Elizabeth Vincan (ed.), *Wnt Signaling, Volume I: Pathway Methods and Mammalian Models, vol. 468*

Book doi: 10.1007/978-1-59745-249-6

associated receptors, can modulate Ca^{2+} release, leading to activation of Ca^{2+}-binding proteins and influence patterning.

Ca^{2+} indicators have proven to be useful tools in the investigation of the role of Ca^{2+} in the Wnt signal transduction network. This method utilizes microscope-based applications to highlight Ca^{2+} activity in intact embryos and embryo explants. The basic technical approach is to detect endogenous Ca^{2+} release activity with a light source to excite a Ca^{2+} indicator in a sample, coupled with a detector that monitors indicator emission. Variants of Fura, aequorin, Fluo, Oregon Green, and Calcium Green are among the most common forms of Ca^{2+} indicators to date, and, in general, neither the dye nor the fluorescence imaging has any detrimental effect on the developing embryo *(1, 2)*. Changes in Ca^{2+} can be detected by single wavelength and "ratiometric" approaches. Single-wavelength excitation allows imaging in the visible range and the use of confocal microscopy. However, one disadvantage of single wavelength measurements is the potential to interpret signal artifacts as Ca^{2+}-dependent changes. Luminescence studies use a Ca^{2+}-stimulated photoprotein, such as aequorin, and have the advantage of rapid measurement and detection of Ca^{2+} gradients and they can be genetically encoded. However, aequorin is photon limited, prone to photobleaching (the irreversible destruction of fluorophores), and requires the incorporation of the cofactor coelenterazine *(3)*. Ratiometric dyes such as Fura-2, a fluorescent derivative of the Ca^{2+}-chelator, EGTA, developed by Tsien and collegues *(4)*, have reduced sensitivity to signal artifacts and enabled quantitative measurement of Ca^{2+} concentrations; this is the approach outlined in this chapter.

The focus of this chapter is on Ca^{2+} imaging and cell–cell interaction in development, in particular, early vertebrate embryonic development in the zebrafish and *Xenopus*. The transparent zebrafish (*Danio rerio*) embryos are ideally suited for fluorescent studies while the ability to add recombinant proteins to *Xenopus* animal caps allows for complimentary studies in that model system. In zebrafish, mis-expression of a subset of Wnts and Frizzleds was found to stimulate an increase in intracellular Ca^{2+} signaling; conversely, mis-expression of Wnts that stimulate Wnt/β-catenin activity did not stimulate Ca^{2+} release in the zebrafish *(5–8)*. Four methods are used in this application: microinjection to deliver reagents to the cytoplasm, expression of exogenous proteins from microinjected messenger RNAs (mRNAs), isolation of animal caps, and fluorescence ratio imaging. The procedure below focuses on the use of fluorescence ratio imaging of Ca^{2+} release as an assay for early signal transduction events in intact zebrafish embryos and dissociated *Xenopus* animal caps.

2. Materials

2.1. Embryo Culture and Pharmacological Reagents

1. Buffers (fish recipes from *The Zebrafish Book*) *(9)* Stock salts: 280 g Instant Ocean Sea Salts (Aquarium Systems, Inc., Mentor, OH) in 2 L distilled water, (dH_2O). Egg water: 1.5 mL stock salts/L of dH_2O (final concentration, 60 µg/mL).
2. Isoproterenol and propranolol (Sigma-Aldrich, St. Louis, MO). Final concentration of 100 µM in egg water (prepare immediately before use).
3. Marc's Modified Ringers (MMR): 100 mM NaCl, 2 mM KCl, 1 mM $MgSO_4$, 2 mM $CaCl_2$, and 5 mM HEPES, pH 7.8.
4. Ca^{2+}–Mg^{2+}-free MMR (CMFM): 15 mM HEPES, pH 7.8, 88 mM NaCl, and 1.8 mM KCl. Store at 4°C in a plastic bottle.
5. *Xenopus* explant materials: Sharp forceps, 2% agarose-coated medium Petri dishes (Falcon 1007), and 67 mM phosphate buffer, pH 7.4.
6. Chamber slides: 0.7-mm aluminum cut to the size of a microscope slide, with two holes 1.5 cm in diameter for the samples. Coverslips are attached to the bottom with nail polish.
7. Glass Pasteur pipettes, wide bore and thinly pulled with fire-polished rims.

2.2. Microinjection Reagents

1. Fura-2 dextran: 10,000 MW (Molecular Probes; Invitrogen, Carlsbad, CA). Working solution of 5 mM in sterile ddH_2O (*see* **Note 1**).
2. Texas Red-conjugated dextran (TxR): 10,000 MW (Molecular Probes). Stock solution of 5 mg/mL in sterile ddH_2O.
3. Synthetic mRNA: Xwnt-5A or rat Frizzled-2 receptor/β_2-adrenergic receptor (Rfz-2/β_2AR) chimera. mRNA concentration of 100–200 ng, mix 1:1 with TxR for a final concentration of 50–100 ng of mRNA in the injection cocktail (*see* **Note 2**).
4. Recombinant mouse Wnt-5a and Wnt-3a protein (R&D Systems, Minneapolis, MN). Stock solution of 10 µg/mL in phosphate-buffered saline with 0.1% (w/v) bovine serum albumin (BSA). Final concentration of 0.2 µg/mL in MMR.

2.3. Zebrafish and Xenopus Embryo Collection

1. Adult zebrafish are maintained at 27.5°C in aquariums (60 mg Instant Ocean Sea Salts/L of water) with biological filters with 14 h light/10 h dark light cycle.
2. Zebrafish embryos are obtained from natural pair matings and staged as detailed in *The Zebrafish Book (9)* (*see* **Note 3**).

3. *Xenopus* eggs are expressed from females ovulated with human chorionic gonadotropin and mixed with sperm in 1× MMR. After 4 min, eggs are washed with 0.1× MMR and kept at room temperature (approximately 24°C). Twenty minutes postinsemination, zygotes were dejellied in 0.1× MMR + 2% cysteine, washed extensively, and cultured at room temperature in 0.1× MMR. Embryos are staged according to Nieuwkoop and Faber *(10)*.

2.4. Instrumentation (Hardware/Software)

1. Inverted epifluorescence microscope (Zeiss Axiovert S100) equipped with coverslip bottomed heating chamber, filters for epifluorescence (Chroma Technology Corp., Rockingham, VT), dual filter wheel and 10× Plan-Neofluar objective (Zeiss, Thornwood, NY; numerical aperture [NA]: 0.5).
2. Xenon arc lamp (75 W) illumination system with Ludl Electronic Products, Ltd (LEP) shutter drive and filter wheel controller.
3. Slow-scan CCD camera at high spatial resolution and high bit depth (12-bit gray scale) (Photometrics Quantix, Tucson, AZ).
4. Computer (LINUX workstation) and image analysis software (ISee and Ratio Tool programs from Inovision Corp., Raleigh, NC) to digitize and manipulate images.

3. Methods

3.1. Microinjection

1. Preparation of synthetic mRNA for microinjection: Linearize the plasmid of interest. In this example, the Rfz-2/β2AR or Xwnt5A cDNA is linearized with NotI and used as template. Synthetic RNA was made by in vitro transcription using T7 RNA polymerase with mMessage mMachine high-yield Capped RNA transcription kit (Ambion/Applied Biosystems, Austin, TX). After DNAse treatment, run the reaction over a mini Quick Spin Column (Roche, Nutley, NJ). Estimate mRNA concentrations by optical density and gel electrophoresis. Just prior to microinjection, mix 0.5 μL of RNA (from 100 to 200 ng working stock) with 0.5 μL of TxR and 2 μL ddH_2O. Store on ice, protect from light.
2. Preparation and calibration of injection micropipette: Pull a tapered needle with a sharp tip from a micropipette (Drummond, 25-λ microcap) on a standard needle puller. Backfill the needle with 0.5–1 μL of Fura-2 or RNA

co-mixed with lineage marker. Mount the filled needle onto the micromanipulator or some other support. Break off the needle tip with fine forceps under a light microscope and lower into a petri dish with oil. Inject a drop of Fura-2 into the oil and measure the diameter with a calibrated eye-piece reticule. Adjust the injection pressure/time to yield an injection volume of ~100–200 pL (*see* **Note 4**).

3. Zebrafish embryos are microinjected with a pressure injector (Harvard apparatus) with approximately 3 nL volumes at the 1-cell stage and ~200 pL at the 8- to 16-cell stage. Place embryos into an injection dish (petri dish with 1.5% agarose that has small depressions formed by 1-mm capillaries and cover with egg water). Gently orient the embryos using fine forceps. Position the needle to penetrate the chorion, gently insert the needle through the yolk into the blastodisc (which is clearer than the yolk region) and inject the Fura-2–dextran. Slowly withdraw the needle using forceps to support the embryo if needed. Allow the injected embryos to develop to the 8- to 16-cell stage.
4. Microinjection into the 8- to 16-cell zebrafish embryo: Load the needle with RNA/TxR cocktail and calibrate for 100–200 pL injection volume. When the Fura-2 injected embryos enter the 8- to 16-cell stage, inject one to two blastomeres with the RNA cocktail as described in the previous step. Maintain the embryos in egg water with β-adrenergic antagonist (propranolol) to prevent endogenous activity before image analysis. XWnt-5A induced Ca^{2+} release does not require additional reagents (*see* **Note 5**).
5. An alternative approach to identify components necessary for Wnt/Fz-induced Ca^{2+} is to co-inject the Ca^{2+} activating reagent (i.e., Wnt-5 RNA) with Fura-2 at the 1-cell stage for uniform distribution. Followed by localized injection at the 8- to 16-cell stage with a pharmacological reagent, antisense oligo, or small molecule inhibitor co-mixed with TxR lineage marker (*see* **Note 6**).
6. In *Xenopus*, dejellied embryos at the 1- to 2-cell stage are transferred into 2% Ficoll/0.5× MMR in a medium Petri dish. Fura-2–dextran is briefly centrifuged and then front-filled into a calibrated needle (ideally 5–10 nL/1 sec injection) using a Harvard Apparatus PLI-100 (Harvard Apparatus, Holliston, MA). Each blastomere is injected with 5 nL Fura-2–dextran near the animal pole, delivering a total of 10 nL per embryo. The final Fura concentration in the embryo should be ~50 μM. Embryos are left in the Ficoll solution until stage 9 and then transferred to 1× CMFM prior to dissection of animal caps.

3.2. Generation of Dissociated Animal Caps from Xenopus

1. Late blastula *Xenopus* embryos (stage 9) are transferred to agarose-coated dishes containing CMFM.
2. The vitelline membranes are removed and animal caps are dissected using sharp forceps.
3. The dissected caps are rinsed for 2 min in 67 mM phosphate buffer and transferred to new agarose-coated dishes containing CMFM.
4. The pigmented outer epithelium layer is peeled away and the dish is gently agitated to separate the cells (blastomeres) (*see* **Note** 7).
5. Groups of cells are transferred to pre-coated imaging chambers using thin bore pipettes coated in BSA (cut yellow pipette tips can also be used).
6. Recombinant Wnt proteins (R&D Systems) can be added directly to cells in the chamber slide.

3.3. Microscopy and Image Analysis

In order to image intracellular Ca^{2+} in living cells, the ratiometric dye Fura-2 will be used. The excitation spectra are different between the Ca^{2+}-bound (340 nm) and Ca^{2+}-free (380 nm) forms. By taking the ratio of the fluorescence intensity at these two wavelengths, an estimate of intracellular free Ca^{2+} can be derived, to some degree, independent of cell thickness and distribution of the fluorescent indicator (which can vary in living cells). There is a wide range of hardware and software that can be used for fluorescence ratio imaging. Our system is briefly outlined below, consult manufacturer instructions for details on other setups.

1. Set up the microscopy workstation: Turn on the hardware by first firing the arc lamp (Xenon). After the arc lamp is powered, the other components can be turned on; the computer to run the software (LINUX), shutter drive and filter wheel controller (LEP), Photometrics CCD camera (Quantix), and the motorized stage controller. Install Fura-2 and TxR filter cubes in a slider.
2. Zebrafish embryos: Orient embryo as desired on the coverslip floor in the thermostated chamber, set at 28.5°C, of the inverted microscope. Analysis is possible at room temperature but ensure that the temperature does not drop below 25°C. Two orientations are useful, one places the blastodisc face down such that the top of the embryo is facing the objective (a small gold wire loop is needed to position the embryo in an inverted orientation). The other orientation places the embryo such that a side view of the blastodisc (one side of the long axis) is facing the objective (**Fig. 11.1a**). Select an embryo with good fluorescence intensity at the 340 nm and 380 nm excitation wavelength (510 nm emission) and determine the optimal exposure time. Change the excitation wavelength to

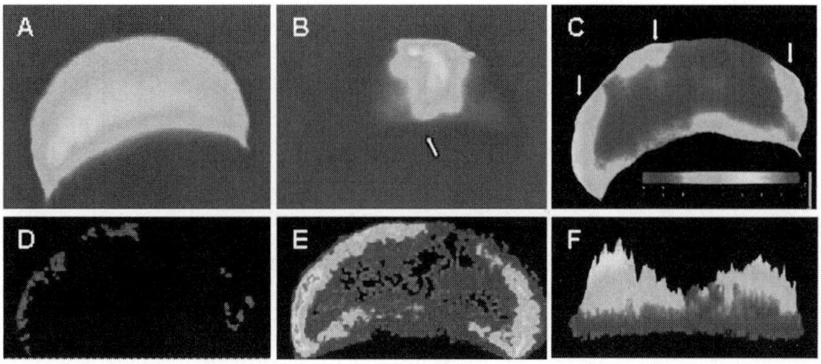

Fig. 11.1. Ca^{2+} imaging and data analysis in zebrafish. **A** Lateral view of a cellular blastoderm stage embryo injected with Fura-2, animal pole to the top. One excitation channel of Fura-2 fluorescence (380 nm) appears as *yellow* in the embryo proper with little or no signal in the underlying yolk. **B** The same embryo as in (**A**) showing the Texas Red (TxR) distribution. The *white arrow* demonstrates the restricted domain of lineage marker (co-mixed with G protein signaling inhibitor). **C** Corresponding ratiometric image of (**A**). *White arrows* denote regions of increased Ca^{2+}, inset is the color bar showing higher Ca^{2+} as *yellow* and lower Ca^{2+} levels as *blue*/*purple*. **D** An example of the subtractive image, highlighting regions of increased Ca^{2+} present in the frame shown in (**C**) that were not in the previous time point. **E** Compiled subtractive images for a time course of 200 frames color coded with a high number of transients as *red* and a low number as *purple*. **I** Conversion of the compiled transients to a topographical representation. Colors and peaks represent numbers of Ca^{2+} transients identified in the embryo during the time course (*purple* is 1 transient and *red* indicates 25 transients). The "dip" in the center of the embryo corresponds to reduced Ca^{2+} release activity in the TxR-positive (inhibitor-injected) region.

540 nm and move the TxR cube into place. Select embryos with TxR (and reagent) distribution appropriate for the experiment (uniform or restricted, **Fig. 11.1b**).

3. *Xenopus* dissociated cells: dissected animal caps are transferred to a chamber with a coverslip floor on an inverted microscope and excited with 340- and 380-nm wavelength to determine the optimal range of exposure.
4. Once the exposure time and orientation is established, move the microscope stage such that the zebrafish embryo (or cluster of *Xenopus* cells) is out of the field of view and collect a background exposure pair at 340 and 380 nm. Center the embryo in the field of view. Collect a sample image pair at 340 and 380 nm and review a ratio calculated by the computer software (in this case, Ratiotool from Inovision). The ratio image is a pixel-by-pixel match of the two excitation wavelengths after the background has been subtracted. Use the sample ratio to determine the proper threshold. Set at a level to exclude from the ratio image areas outside of the embryo such as yolk or chorion (**Fig. 11.1c**). Determine the TxR distribution by collecting a reference exposure at 540-nm excitation.
5. Initiating the time course. Determine the stage of the embryo (i.e., 16 or 32 cell) and time. Initiate the time course by

collecting image pairs at 15-sec intervals (*see* **Note 8**). In zebrafish, the first phase of Ca^{2+} release will be very obvious. There will be long-lived Ca^{2+} elevations associated with the forming cleavage furrows. These will persist until the 64-cell stage. The second phase of Ca^{2+} release will appear as localized Ca^{2+} elevations roughly the size of single cells (**Fig. 11.1c**).

6. For Wnt-5 injected zebrafish embryos, increased Ca^{2+} release will become apparent in the region of the TxR tracer. Conversely, localized injection of a signaling inhibitor would demonstrate reduced activity in the TxR region (**Fig. 11.1b–f**). For the inducible Fz2 receptor, collect baseline activity (4–10 min), then add the β-adrenergic agonist, isoproterenol, to activate the Rfz-2/β2AR chimera. After the solution change, reorient the embryo and collect another TxR reference. Collect 340/380-nm image pairs at 15-sec intervals. After the data collection is completed, return the embryo to egg water (*see* **Note 9**).
7. Data analysis. The ratio image from the archived raw data is processed using computer software (ISee; Inovision). In short, the region of interest (ROI) is defined (ROI, the embryo) and centered in the image frame such that the ratio images from the time course are super-imposable on each other (compare **Fig. 11.1c** with the shifted image in **Fig. 11.1d** and **e**). The images are low-pass filtered and sequential ratio images are subtracted from each other. In the subtractive image, a transient is defined as a feature approximately the size of a cell with an increase in fluorescence intensity (the "features" in **Fig. 11.1c** identified by arrows are detected by the algorithm and shown in **Fig. 11.1d**). The numerical output of the total number of transients is subjected to statistical analyses using Student's *t* test or analysis of variance to determine if there is a significant change of activity from endogenous and/or control levels, and can be plotted as a function of time. The compiled features for the entire time course can also be superimposed into one image (**Fig. 11.1e**) and converted to a topographical representation showing the spatial distribution of the total number of transients along the embryo (**Fig. 11.1f**). The color and peak height indicate the total number of Ca^{2+} transients present in that particular region of the embryo.
8. Xenopus animal cap imaging: After collection of baseline activity, add recombinant Wnt protein to the chamber while continuously imaging. Reorient if cells shift (*see* **Note 10**). Recombinant Wnt-3a-treated cells show Ca^{2+} release activity similar to that of untreated cells (**Fig. 11.2a, b**); whereas addition of recombinant Wnt-5a protein to dissociated caps

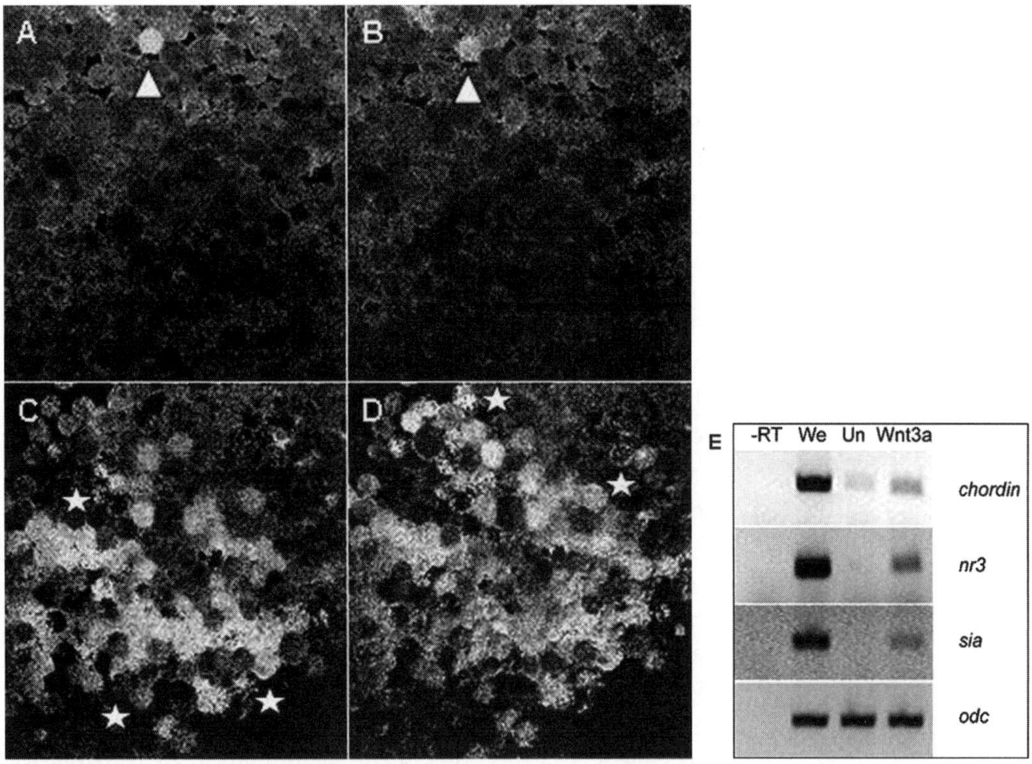

Fig. 11.2. Ca^{2+} imaging in *Xenopus*. Dissociated cells from Fura-2-injected *Xenopus* animal caps were treated with Wnt-3a (**A**, **B**) and Wnt-5a (**C**, **D**) recombinant protein. **A** Ratiometric image of Wnt-3a-treated cells. A *white arrowhead* indicates a cell that has increased Ca^{2+} levels. **B** The same tissue 15 sec later, showing a decrease in Ca^{2+} levels in the one cell. **C** Ratiometric image after application of recombinant Wnt-5a. *Stars* indicate groups of cells with increased Ca^{2+} levels. **D** The same tissue 15 sec later shows a decrease in active cells from the previous frame but increased activity in adjacent cells. **E** RT-PCR analysis in whole embryo (*We*), untreated animal caps (*Un*), animal caps cultured in Wnt-3a recombinant protein, and a negative control (*-RT*). *Ornithine decarboxylase* (*odc*) is used as an internal control for the relative amount of RNA used for each sample. Wnt target genes are activated in Wnt-3a-treated caps. *nr3*, *nodal-related 3*; *sia*, *siamois*.

stimulates Ca^{2+} release in clusters of cells in a wave across the tissue (**Fig. 11.2c, d**) (*see* **Note 11**).

9. The activity of the Wnt-3a protein can be confirmed with analysis of Wnt target genes. Treatment of intact caps at the same concentration used in image analysis is followed by reverse transcriptase (RT) polymerase chain reaction (PCR). Changes in Wnt target genes (such as *nodal-related 3*, *siamois*, and *chordin*) can determine the potency of the recombinant protein (**Fig. 11.2e**). Co-culture of animal caps with Wnt-5a and Wnt-3a should suppress Wnt-3a-mediated transcriptional changes.

4. Notes

1. Unless stated otherwise, all injection solutions are prepared under RNAse-free conditions. Either purchase nuclease-free solutions (i.e., from Ambion) or treat with diethylpyrocarbonate (DEPC) and autoclave. Care is taken in handling tips and tubes and gloves are worn.
2. The concentration of injected RNA may vary with the investigator as different setups may deliver different volumes. Mis-expression of Wnt-5 leads to cell movement defects. Using the embryonic phenotype as an output, determine the optimal RNA injection concentration prior to imaging.
3. Adult zebrafish need to be well cared for in order to obtain eggs reliably. Consult *The Zebrafish Book (9)* for details on stock maintenance.
4. Test several different parameters on the needle puller and the breaking back of the tip to generate a needle thin enough to avoid damage to the embryo, yet strong enough to penetrate the chorion.
5. Introducing the mRNA (or other reagent) into a subset of cells is a useful control for manipulation of Ca^{2+} release because changes should only be observed in or near the lineage tracer region (TxR positive signal). Globin mRNA, as well as Wnt-8, which does not generate Ca^{2+} release, can serve as negative controls for Ca^{2+} activation *(6)*.
6. Wnt-5- and Fz2-induced Ca^{2+} fluxes can be suppressed with inhibitors to G protein and phosphoinositide (PI) cycle signaling *(5, 6, 11, 12)*. Introduction of the inhibitor into a portion of the embryo allows for identification of molecules necessary for Ca^{2+} stimulation. The adjacent region of the embryo (non-inhibitor injected) will show activated/increased activity. This type of analysis sets the stage for in vivo physiological testing of small molecule inhibitors of Wnt signaling.
7. Prior to imaging of *Xenopus* animal caps, chamber slides are coated with CMFM plus 1 mg/mL BSA and rinsed in CMFM without BSA.
8. When imaging a developmental stage or tissue for the first time, use shorter intervals (1–5 sec) to determine the optimal collection frequency. Keep in mind that increased rate of collection and long duration of the time course will require more storage space on the computer. Typically, 40 min to 1h time courses are sufficient to identify trends in Ca^{2+} modulation in whole embryos, while 10–20 min time courses suffice for cultured animal caps.

9. The manipulations described result in alterations in embryonic development. There are several molecular markers that can be used to assess alterations in embryonic patterning. Culture the injected embryos to the appropriate stage and evaluate by morphology, marker gene expression via whole mount in situ hybridization, or protein distribution via immunohistochemistry.
10. Endogenous activity in dissociated cells includes increases at the edges of cells that are actively moving on the coverslip. Other release activity, more relevant to Wnt signaling, involves increases in cytoplasmic Ca^{2+} levels, typically filling the entire cell (**Fig. 11.2a**, arrowhead) but decreases within 10–20 sec (**Fig. 11.2b**, arrowhead). This type of Ca^{2+} release activity is rare and usually only observed in one cell in untreated or Wnt-3a-treated disassociated animal caps.
11. The animal cap imaging discussed in this chapter was performed shortly after caps were isolated. If interested in neural induction and calcium dynamics, culture the tissue for 4 h after to dissociating the cells.

Acknowledgments

This work was supported by RO1 CA112369 to DCS, the Roy J. Carver Charitable Trust to DWH, and An American Heart Association predoctoral fellowship to CMF.

References

1. Tsien, R. Y., Pozzan, T., and Rink, T. J. (1982) Calcium homeostasis in intact lymphocytes: cytoplasmic free calcium monitored with a new, intracellularly trapped fluorescent indicator. *J Cell Biol* **94**, 325–334.
2. Thomas, D., Tovey, S. C., Collins, T. J., Bootman, M. D., Berridge, M. J., and Lipp, P. (2000) A comparison of fluorescent Ca2+ indicator properties and their use in measuring elementary and global Ca2+ signals. *Cell Calcium* **28**, 213–223.
3. Shimomura, O. (2005) The discovery of aequorin and green fluorescent protein. *J Microsc* **217**, 1–15.
4. Grynkiewicz, G., Poenie, M., and Tsien, R. Y. (1985) A new generation of Ca2+ indicators with greatly improved fluorescence properties. *J Biol Chem* **260**, 3440–3450.
5. Slusarski, D. C., Corces, V. G., and Moon, R. T. (1997) Interaction of Wnt and a Frizzled homologue triggers G-protein-linked phosphatidylinositol signalling. *Nature* **390**, 410–413.
6. Slusarski, D. C., Yang-Snyder, J., Busa, W. B., and Moon, R. T. (1997) Modulation of embryonic intracellular Ca2+ signaling by Wnt-5A. *Dev Biol* **185**, 114–120.
7. Westfall, T. A., Brimeyer, R., Twedt, J., Gladon, J., Olberding, A., Furutani-Seiki, M., and Slusarski, D. C. (2003) Wnt-5/pipetail functions in vertebrate axis formation as a negative regulator of Wnt/{beta}-catenin activity. *J Cell Biol* **162**, 889–898.
8. Liu, X. X., Liu, T., Slusarski, D. C., Yang-Snyder, J., Malbon, C. C., Moon, R. T., and Wang, H. Y. (1999) Activation of a Frizzled-2/beta-adrenergic receptor chimera

promotes Wnt signaling and differentiation of mouse F9 teratocarcinoma cells via G alpha o and G alpha t. *Proc Natl Acad Sci USA* **96**, 14383–14388.

9. Westerfield, M. (1995) *The zebrafish book: A guide for the laboratory use of zebrafish.* University of Oregon Press, Eugene.
10. Nieukoop, P. D., and Faber, J. (1994) *Normal table of Xenopus laevis (Daudin): a systematical and chronological survey of the development from the fertilized egg till the end of metamorphosis.* Garland Publishing, New York, London.
11. Westfall, T. A., Hjertos, B., and Slusarski, D. C. (2003) Requirement for intracellular calcium modulation in zebrafish dorsal-ventral patterning. *Dev Biol* **259**, 380–391.
12. Ahumada, A., Slusarski, D. C., Liu, X. X., Moon, R. T., Malbon, C. C., and Wang, H. Y. (2002) Signaling of rat Frizzled-2 through phosphodiesterase and cyclic GMP. *Science* **298**, 2006–2010.

Chapter 12

Detecting PKC Phosphorylation as Part of the Wnt/Calcium Pathway in Cutaneous Melanoma

Samudra K. Dissanayake and Ashani T. Weeraratna

Abstract

Signaling networks play crucial roles in the changes leading to malignancy. In melanoma, increased Wnt5A expression increases melanoma cell motility via activation of protein kinase C (PKC). PKC isoforms comprise a family of serine/threonine kinases that are involved in the transduction of signals for cell proliferation, differentiation, and metastasis. The important role of PKC in processes leading to carcinogenesis and tumor cell invasion would render PKC a suitable target for cancer therapy, if not for its ubiquitous nature. Thus, targeting pathways leading to PKC activation that are more tumor specific, such as the non-canonical Wnt pathway, may prove to be the key to targeting PKC in cancer. Here we summarize the current understanding of the Wnt/calcium pathway and discuss methods of detecting activated/phosphorylated PKC as a result of Wnt signaling in malignant melanoma. We have shown that overexpression of Wnt5A results in the activation of PKC, while inhibition of Wnt5A via small interfering RNA (siRNA) treatment results in its inactivation. In addition, the use of PKC activators and inhibitors has allowed us to study Wnt5A effects on downstream genes that may prove to be key targets for molecular therapy.

Key words: Melanoma, Wnt5A, Protein kinase C, PKC.

1. Introduction

The Wnt/calcium pathway is one of the three major pathways by which Wnt proteins exert their intracellular signaling events *(1)*. Dysregulation of Wnt signaling can cause developmental defects and is implicated in the genesis of several human cancers. In melanoma, complementary DNA (cDNA) microarray analysis identified Wnt5A

Elizabeth Vincan (ed.), *Wnt Signaling, Volume I: Pathway Methods and Mammalian Models, vol. 468*

Book doi: 10.1007/978-1-59745-249-6

as the gene that best discriminated highly aggressive melanomas from their less invasive counterparts *(2)*. In a follow-up study, we demonstrated that introducing Wnt5A into less aggressive melanomas resulted in an increase in their metastatic potential most likely via the activation of PKC and rises in intracellular calcium *(3)*. The most recent study from our laboratory has combined the use of siRNA technology with microarray analysis to identify the pathways and mechanisms by which Wnt5A might be mediating motility in melanoma cells. In this study, we used recombinant Wnt5A protein as well as siRNA to validate our array observations, and furthermore more fully assessed the role of PKC in this process using activation and inhibition studies *(4)*. In addition to our laboratory, many others have highlighted the importance of G protein-mediated signaling and the resultant activation of PKC and increases in intracellular calcium in melanoma progression *(5–7)*.

The PKC family consists of a number of serine/threonine kinases, which are divided into three major groups based on their activating factors (**Table 12.1**). PKC isoforms have been linked to carcinogenesis because PKC activators can act as tumor promoters,

Table 12.1
PKC isoforms

Isoform	Type	Calcium dependence	Phorbol stimulation	Amino acids	Predicted molecular weight (kDa)
Alpha (α)	Conventional	Yes	Yes	672	76.8
Beta I (β_I)	Conventional	Yes	Yes	673	76.9
Beta II (β_{II})	Conventional	Yes	Yes	671	76.8
Gamma (γ)	Conventional	Yes	Yes	697	78.4
Delta (δ)	Novel	No	Yes	673	77.5 (r)
Epsilon (ε)	Novel	No	Yes	737 (r)	83.5 (r)
Eta (η)	Novel	No	Yes	680	77.6
Mu (μ)	Novel	No	Yes	912 (m)	115 (m)
Theta (θ)	Novel	No	Yes	706	82 (r)
Zeta (ζ)	Atypical	No	No	592 (r)	67.7 (r)
Lambda (λ)	Atypical	No	No	586	67.2

All forms are monomeric. The β_I and β_{II} isoforms differ in their C-terminal amino acid residues, and the forms are the result of alternative splicing of the 3′ exon and different PKC isoforms, especially PKC α and β, have been often linked to a malignant phenotype in melanoma *(8, 9)*. We have confirmed this observation in cutaneous melanoma, and have shown that the PKC isoforms affected by Wnt5A are predominantly the conventional PKCs, PKC α, β, and γ *(3)*. In its nonphosphorylated state, PKC resides in the cytosol. Binding of a hormone or other effector molecule to the membrane receptor results in the activation of phospholipase C (PLC) via a G protein-dependent phenomenon. The activated PLC hydrolyzes phosphatidylinositol-4,5-bisphosphate (PIP2) to produce diacylglycerol (DAG) and inositol-1,4,5-triphosphate (IP_3) from plasma membrane phospholipids. DAG activates conventional PKC isoforms, while IP3 causes the releases of Ca^{2+} from intracellular stores, which in turn potentiates the activation of these PKC isoforms. The binding of Ca^{2+} causes PKC translocation to the plasma membrane, where it interacts with DAG and is transformed into a fully active enzyme that is capable of phosphorylating specific substrates on serine/threonine residues, as reviewed in ref. *(8)*. Commercial antibodies that recognize the individual isoforms of PKC, both in phosphorylated and nonphosphorylated forms, are readily available. These antibodies give us the ability to determine activity of PKC via Western blot analysis. These phospho-antibodies are very important tools, because of some caveats that should be noted when working with antibodies against nonphosphorylated proteins, specifically, as we discuss here, antibodies against total PKC. These antibodies are marketed as antibodies to "total" protein, as they are often designed to peptides from the protein sequence. Thus, these antibodies should technically recognize both phosphorylated and nonphosphorylated proteins. However, our current data indicate that, in fact, these antibodies may recognize phosphorylation-sensitive epitopes, resulting in a decreased ability to recognize phosphorylated protein. For example, when treating Wnt5A low cells with recombinant Wnt5A, we see that as the phosphorylated form of a protein increases, the intensity of the band recognized by the antibody to the corresponding nonphosphorylated isoform decreases *(4)*. The converse is true for knocking down Wnt5A in Wnt5A high cells—decreases in Wnt5A correspond to decreases in PO_4-PKC, but to increases in PKC α, β, and γ. This is only true for molecules that activate the existing pool of PKC, such as Wnt5A, and not for chemical agents such as phorbol 12-myristate 13-acetate (PMA) that appear to act via an increase in the protein expression as well as in its activation. This is also true for the chemical inhibitors we have tested that decrease the overall expression of PKC. In order to determine if these antibodies were indeed phosphorylation sensitive, duplicate sets of samples were subjected to sodium dodecyl sulfate (SDS) polyacrylamide gel electrophoresis (PAGE), then transferred to polyvinyl difluoride (PVDF). According to the protocol by Maya et al. *(10)*, the membranes were cut in half, separating the duplicate blots, and one set of samples was blocked and probed as usual, but the other was first incubated in alkaline phosphatases for a half hour. Phosphatase treatment significantly increased the ability of antibodies against PKC α, β, and γ to recognize these isoforms *(4)*. These data, some of which are included as examples below, should provide a strong caution to researchers, as the use of the antibodies to the nonphosphorylated form of PKC could result in a misinterpretation of data, even when doing what are considered "gold standard" assays for PKC activation, such as Western analysis of membrane extracts as compared with cytosolic extracts. In such studies, for example, these antibodies could indicate a depletion in membrane-bound PKC, where there may in fact be an increase in membrane-bound phosphorylated PKC. Thus, antibodies to both the phosphorylated and nonphosphorylated isoforms should be used when assessing the effects of signaling molecules on PKC activation. Additionally, we suggest the use of other techniques such as substrate-based assays, or the use of commercially available green fluorescent protein (GFP)-tagged PKC isoforms followed by confocal microscopy to determine if activation or deactivation of a pathway of interest can affect the translocation of PKC to the membrane. The methods we use, including transfection of GFP-PKC, confocal microscopy, and Western analysis of PKC and PO_4-PKC are provided in detail below.

2. Materials

2.1. Cell Culture and Lysis Reagents

1. RPMI (UACC 903, UACC 647, M93-047, 1273, and 1205LU melanoma cell lines), EMEM (C-32 cells), or McCoy's (G361 cells) medium (all from Invitrogen, Carlsbad, CA) supplemented with 10% (v/v) fetal bovine serum (FBS; Invitrogen), 100 U/mL penicillin and streptomycin, and 4 mM L-glutamine. C-32 cells only were additionally supplemented with 1% non-essential amino acids (Invitrogen). All cell lines were cultured at 37°C in 5% CO_2/95% air, and the medium was replaced every 2 to 3 days.
2. TripleE Solution (Invitrogen) containing 0.25% (w/v) ethylenediamine tetraacetic acid (EDTA) (1 mM).
3. Phorbol 12-myristate 13-acetate (PMA, Sigma-Aldrich, St. Louis, MO) is dissolved at 200 μM in dimethyl sulfoxide (DMSO) and stored in single use aliquots at –20°C.
4. Gö 6983 and GF109203X (Calbiochem, San Diego, CA) are dissolved at 1 mM in DMSO and stored in single use aliquots at –20°C.

5. Recombinant Wnt3A and Wnt5A (R&D Biosystems, Minneapolis, MN) are reconstituted in sterile phosphate-buffered saline (PBS) containing 0.1% (w/v) BSA to a stock concentration of 10 μg/mL and stored in single use aliquots at -20°C.
6. RIPA buffer for cell lysis: 20 mM Tris-HCl, pH 7.5, 150 mM NaCl, 5 mM EDTA, 1% (v/v) Triton X-100, 1 tablet protease inhibitor cocktail (Complete, Roche, Indianapolis, IN) per 50 mL total buffer solution, plus 1 mM Na_3VO_4, and 1 mL of phosphatase inhibitor cocktail 1 (Sigma-Aldrich) per 100 mL total buffer. This buffer is stored in single use aliquots at -80°C (*see* **Note 1**).
7. BCA protein assay kit (Pierce, Rockford, IL) for quantifying protein lysates.
8. Spectrophotometer set to 570 nM.

2.2. Transfection Reagents

1. PDest-Wnt5A was designed using the Gateway cloning system (The Harvard Gene Therapy Initiative), and is available from the authors upon request.
2. Wnt5A siRNA was designed using Invitrogen's online design tools, which designs 21-nt siRNA according to the Tushcl rules of siRNA design. SiRNAs are ordered from Qiagen (Valencia, CA), and are ordered in both rhodamine tagged (3′-UTR) and untagged versions (*see* **Note 2**). The sequence that works most efficiently as demonstrated in Dissanayake et al. *(4)* is AAGACCTGGTCTACATCGACC.
3. GFP-PKCβII was obtained from Invitrogen.
4. Lipofectamine Plus system, from Invitrogen.

2.3. Reagents for Confocal Microscopy

1. Glass slide chambers, single well.
2. Coverslips.
3. Prolong Gold (Invitrogen).

2.4. Sds-page

1. 4–20% Tris-glycine pre-cast Novex mini gels (Invitrogen). Store at 4°C.
2. Running buffer (10×): Tris/glycine buffer (Bio-Rad, Hercules, CA). Store at room temperature.
3. Prestained molecular weight markers: Kaleidoscope markers (Bio-Rad). Store at –20°C.
4. 4× Loading buffer: 12% (w/v) SDS, 40% (v/v) glycerol, 0.2 M Tris-HCl, pH 7.6, 0.004% (w/v) bromo phenol blue (BPB), and 0.05% (v/v) BME. Store at 4°C.
5. XCell SureLock Novex Mini-Cell (Invitrogen) for running mini-gels.

2.5. Western Blotting for Active PKC

1. Transfer buffer: 25× Novex Tris/Gly transfer buffer (Invitrogen) and 20% (v/v) methanol. Store at room temperature, but cool to 4°C for use.
2. Filter paper sandwich: 0.2-μm PVDF membranes (Invitrogen).
3. Tris-buffered saline (TBS) with Tween (TBS-T): Prepare from 10× TBS (Quality Biological Inc, Gaithersburg, MD), 1% (v/v) Tween-20 (Sigma-Aldrich).
4. Blocking buffer: 5% (w/v) nonfat dry milk (BioRad nonfat blocking solution) in TBS-T.
5. Primary antibody dilution buffer: 5% (w/v) nonfat dry milk in TBS-T.
6. Secondary antibody: anti-rabbit IgG conjugated to horseradish peroxidase (HRP) (Amersham Biosciences, Buckinghamshire, UK).
7. ECL-Plus reagent and high-performance chemiluminescence film (Hyperfilm™ ECL; Amersham Biosciences).
8. Mini-PROTEAN 3 cell transfer system (Bio-Rad) for transfer of gel.

2.6. Antibodies

PO_4-Pan-PKC (1:1,000) and β-tubulin (1:1,000) antibodies were obtained from Cell Signaling Technology, Inc. (Danvers, MA); and the Wnt5A-biotinylated antibody (1:100) was obtained from R&D Biosystems (*see* **Note 3**). Antibodies to PKC isoforms (nonphosphorylated) were obtained from BD Biosciences (San Jose, CA).

2.7. Stripping and Reprobing Blots

1. Stripping buffer: 50 mM Tris-HCl, pH 6.8, 2% (w/v) SDS, and 2% (v/v) β-mercaptoethanol (BME).
2. Wash Buffer: Tris-buffered saline (TBS) with 0.1% (v/v) Tween CTBS-T)
3. Sealed plastic box.
4. Water bath with shaker at 55°C in fume hood.

3. Methods

PKC is considered active when at the membrane. Because Wnt5A has been shown to increase the phosphorylation of PKC *(3, 11)*, we recently used siRNA against Wnt5A to determine if inhibition of Wnt5A corresponded to an inhibition of PKC phosphorylation, and recombinant Wnt5A to determine if increasing Wnt5A could increase PO_4-PKC levels *(4)*. Subsequently, we investigated whether PKC translocated from its active site at the membrane of the cell after inhibition of Wnt5A, for further confirmation of PKC de-activation. To perform this experiment, we examined the conventional PKC, PKCβII. Cells were transiently transfected with a GFP-tagged PKCβII vector and a rhodamine-tagged control siRNA, and then examined using confocal microscopy. Cells with endogenously high Wnt5A showed PKCβII expression predominantly at the membrane (**Fig. 12.1a**, GFP-PKC, double arrows; rhodamine-tagged siRNA, single arrow). When co-transfected with rhodamine-tagged A2-RNAi (Wnt5A-A2Rh) against Wnt5A, GFP-PKCβII moved from the membrane into the cytoplasm (**Fig. 12.1b**, GFP-PKC, double arrows; rhodamine tagged siRNA, single arrow).

3.1. Assessment of PKC Translocation Using GFP-Tagged PKC and Confocal Microscopy

1. Passage all cells when approaching confluence with Trypsin/EDTA (TripleE) to provide new maintenance cultures in T-75 tissue culture dishes and experimental cultures on 1-well glass slide chambers (*see* **Note 4**). One slide chamber is required for each experimental data point. Typically, we seed between 0.3×10^5 and 7.5×10^5 cells per slide chamber for experimental cultures that would require 60–70% confluency after 16 h. This should be calculated for each cell line depending on its doubling time. Rinse the cultures twice with cell culture media (without serum) and incubate for a further 1–3 h in serum-free media before transfection.

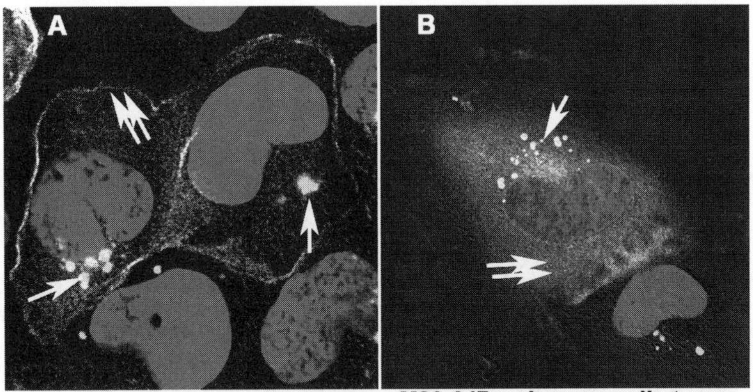

Fig. 12.1. Confocal microscopy for PKC activation. Using a GFP-tagged PKCβII expression vector, cells were treated with either control siRNA or siRNA against Wnt5A. In cells co-transfected with a control siRNA (*single arrow*), PKC (*double arrows*) remains at its active site in the membrane (**A**). When these cells are cotransfected with siRNA against Wnt5A (*single arrow*), PKC (*double arrows*) moves into the cytoplasm (**B**).

2. Transfect Wnt5A siRNA or plasmid DNA into cells (60–70% confluency) using Lipofectamine Plus (Invitrogen). Allow the Plus reagent to incubate with DNA/siRNA for 15 min, and then incubate this complex with Lipofectamine for a further 15 min, prior to adding to the serum-starved cells (*see* **Note 5**). After 5 to 6 h of transfection, replace the medium with fresh serum-containing medium. Transfection efficiencies are usually upward of 90% for siRNA oligos as gauged by transfection with rhodamine-tagged siRNAs, and siRNAs are routinely transfected at a concentration of 150 nM. For the PCDNA-Wnt5A vector and GFP-tagged PKC II vector, transfect approximately 1 μg of DNA per 35-mm dish to achieve transfection efficiencies around 75%. Interestingly, transfection efficiency is highly dependent on melanoma cell confluency, and cell densities higher than 80% result in inefficient transfection (**Fig. 12.2**).
3. To test effects of Wnt5A on GFP-tagged PKC, either cotransfect Wnt5A-A2 siRNA (Wnt5A high cells), or add recombinant Wnt5A to transfected cells (Wnt5A low cells). For recombinant Wnt5A treatments, after testing a range of concentrations and time points (**Fig. 12.3**), a concentration of 0.2 μg/mL for 16 h in serum-free medium was decided upon as ideal for our melanoma cells (*see* **Note 6**). Researchers attempting this in other cell types should assess the ideal times and concentrations for their cells. Additionally, assess whether 24 h posttransfection or 48 h posttransfection gives the best results in each individual case. We find (for our cells) that when co-transfecting GFP-tagged PKC and Wnt5A siRNA, 48 h is an effective time point. For Wnt5A treatments, we prefer to treat cells 24 h posttransfection for maximal results.

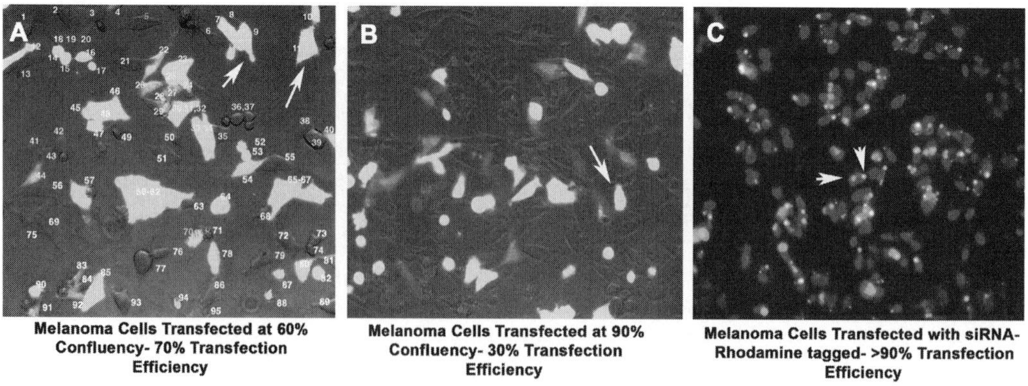

Fig. 12.2. Transfection efficiency based on cell confluency. When melanoma cells are transfected at efficiencies of 50–60% (**A**), the efficiency of transfection of a GFP-plasmid (*arrows*) is around 70–75%. Cells transfected at a higher confluence (**B**), lose efficiency of transfection dramatically, dropping down to about 25–30%. siRNA transfection efficiency, as gauged by the rhodamine fluorescence (*arrows*) in this image (**C**) is consistently high, around 90–95%.

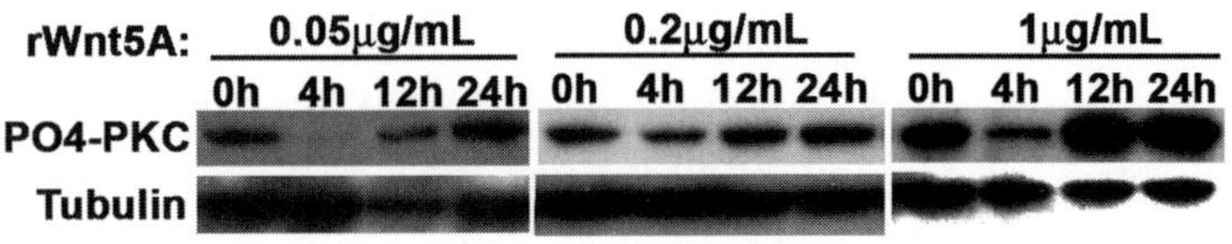

Fig. 12.3. Concentration and time course determination for rWnt5A treatment. Cells were assayed for their ability to increase PO_4-PKC in response to rWnt5A treatment at differing concentrations and over differing times. The ideal time course for Wnt5A treatment was ultimately determined to be 16–24 h. Note that at 4 h, Wnt5A effects are consistently dampened, perhaps due to receptor recycling.

PKC inhibitors and activators are also added at 24 h posttransfection (*see* **Notes 7** and **8**).

4. After treatment with agents of interest, fix the cells using either 95% methanol or 4% formaldehyde. We prefer to use 95% methanol, as this effectively permeabilizes membranes while fixing proteins that are in the membrane to the membrane, without allowing leaching into the cytoplasm. Rinse chambers once with PBS, use ice-cold methanol, and fix the cells for 20 min at room temperature. Break off the chambers and store the slides in PBS, pH 7.4, at 4°C until ready to stain.
5. Use Prolong Gold, with or without DAPI (we prefer with DAPI for easier visualization). Drop 100 μL onto the slide, then coverslip. Allow the slides to cure in the dark at room temperature for at least 24 h (*see* **Note 9**). Our data indicate that Wnt5A high cells will have GFP-PKC at the membranes, with some cytoplasmic expression, while Wnt5A low cells will have predominantly cytoplasmic PKCβII. Treatment with siRNA against Wnt5A reduces membrane-associated PKC in Wnt5A high cells (**Fig. 12.1**), whereas treatment with recombinant Wnt5A increases membrane-associated PKC, as does treatment with phorbol ester. Image cells using confocal microscopy to accurately determine membrane association. Expect that about 20% of your cells will not have membrane or cytoplasmic staining as expected, i.e., in Wnt5A high cells, about 20% of cells will have some significant cytoplasmic staining, and in Wnt5A low cells, a proportion will have some membrane staining.

3.2. Assessment of PKC Activation by Western Blotting

The activation of PKC is under the control of three distinct phosphorylation events, specifically threonine 500 at the activation loop, the threonine 641 autophosphorylation site, and the serine 660 hydrophobic site at the carboxy terminus of PKCβII *(12)*. The pan PO_4-PKC antibody used in our studies detects endogenous levels of PKC, α β, and γ isoforms only when phosphorylated at a carboxy-terminal residue homologous to serine 660 at PKCβII. The integrity of the total protein loaded per sample was assessed by reprobing blots for β-tubulin, a conventional house-keeping gene widely used in many protein expression analysis studies. Using the PO_4-PKC antibody as well as antibody to non-phosphorylated PKC, we see that Wnt5A appears to be activating only the existing pool of PKC, because inhibiting Wnt5A decreases PO4-PKC, but increases nonphosphorylated PKC (**Fig. 12.4a**), and increasing Wnt5A produces the reverse effect. PKC antibodies are phosphorylation sensitive; this is determined by probing Western blots with antibodies against PKC isoforms before and after phosphatase treatment (**Fig. 12.4b, c**). Perhaps due to the technique used (phosphatase treatment of PVDF, where phosphatase treatment may be incomplete), phosphatase treatment does not result in complete ablation of the ability of the phospho-antibody to recognize PO4-PKC (**Fig. 12.4c**).

1. Prepare cells as described in Section 3.1, step 1, with the exception that they should be grown in 35-mm dishes or 6-well plates. For our cells to achieve 60–70% confluency

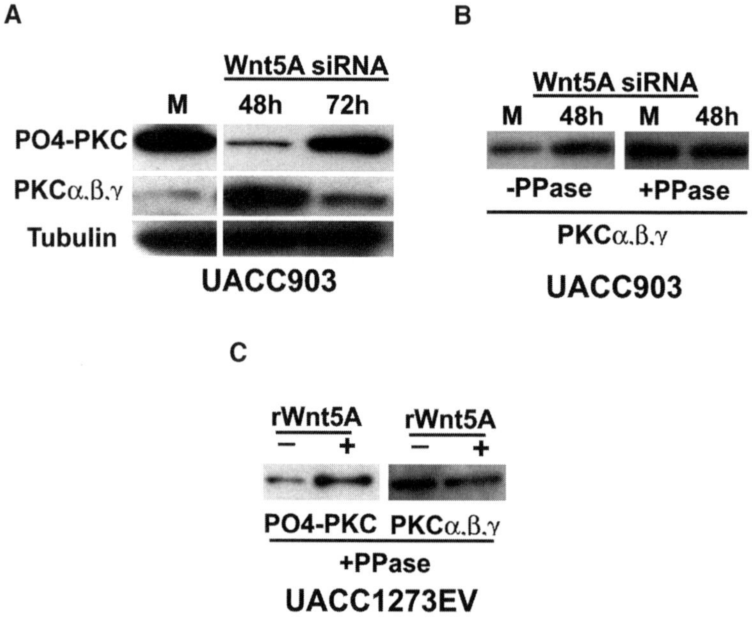

Fig. 12.4. Wnt5A affects the existing pool of PKC. **A** Wnt5A inhibition results in a decrease in PO_4-PKC, which is maximal at 24–48 h. At 72 h posttransfection, Wnt5A siRNA effects begin to decrease, and PO_4-PKC levels start to increase again. Conversely, nonphosphorylated PKC levels increase with Wnt5A knockdown as PO_4-PKC decreases, and when PO4-PKC starts to increase again, nonphosphorylated PKC levels decrease. **B** This is attributable to the fact that the PKC antibodies recognize phosphorylation-sensitive epitopes. This is shown by probing two identical membranes with antibodies against PKC, and subjecting one membrane to phosphatase treatment (*PPase*; *right panel*), but not the other (*left panel*). **C** Treating cells with rWnt5A increases PO_4-PKC (shown, *left panel*) but decreases nonphosphorylated PKC (shown in ref. *(4)*). Phosphatase treatment of the membrane results in an increase of the ability of the nonphospho-antibody to recognize PKC (*right panel*).

the following day, the required seeding densities in these dishes are 0.75×10^6 to 1×10^6. Researchers should determine this for each cell line. Transfect or treat cells as described in **Section 3.1**, steps 1 and 2.

2. Label microfuge tubes for each sample and have ready a 22-gauge 1½-inch needle attached to a 1-mL syringe, cell lysis buffer (RIPA), an ice bucket, and a centrifuge set at 4°C. For adherent cells, rinse flasks once with PBS, then add RIPA directly onto the dish. We use 150 µL of RIPA per well of a 6-well plate. Incubate the plate on ice for 5 min. Using a cell scraper, scrape the cells, and place the cells in a microfuge tube on ice. Shear the cells using a syringe by pushing them through the syringe seven times. Incubate the cells on ice for 30 min and centrifuge at 15,000×*g* for 10 min. Quantitate using the Pierce BCA protein quantitation assay (*see* **Note 10**).
3. Mix 50 µg of each protein lysate with 4× loading dye, heat denature at 95°C for 10 min in a heat block, and run out on a 4–20% Tris-glycine precast Novex mini gel at 100 V for approximately 2 h.
4. Transfer as follows (these directions assume the use of a Bio-Rad Mini-PROTEAN 3 cell transfer system): Prepare a tray of transfer buffer that is large enough to lay out a transfer cassette with 1 piece of foam and 3-mm-thick filter paper submerged on one side. Cut the PVDF membrane on one corner to allow the orientation to be tracked, and immerse the membrane in methanol followed by two washes in distilled water. Submerge the wet membrane in the transfer buffer on top of the filter paper.
5. Disconnect the gel unit from the power supply and dissemble. Remove and discard stacking gel, and lay the separating gel on top of the PVDF membrane. Wet another sheet of filter paper in the transfer buffer and carefully lay it on top of the gel, ensuring that no bubbles are trapped in the resulting sandwich, which can be done by firmly rolling a 15-mL centrifuge tube or pipette across the sandwich. Lay the second wet foam sheet on top, and close the transfer cassette.
6. Place the cassette into the transfer tank such that the PVDF membrane is between the gel and the anode. It is critical that this orientation is maintained or the proteins will be lost from the gel into the buffer rather than transferred to the PVDF membrane.
7. Place the lid on the tank and activate the power supply. Transfers are done at 4°C, preferably with a magnetic stir-bar in the tank activated to maintain a temperature between 10°C and 15°C in the tank. Transfers can be accomplished at either 30 mA overnight or 100 mA for 2 h, but the overnight transfer is preferable.

8. Once the transfer system is complete, take the cassette out of the tank and carefully dissemble, removing the top sponge and sheet of filter paper. Discard the gel and immerse the PVDF membrane containing the transferred protein in TBS-T. The colored molecular weight markers should be clearly visible on the membrane.
9. Incubate the PVDF membrane in 10 mL of blocking buffer for 1 h at room temperature on a rocking platform (*see* **Note 11**).
10. Discard the blocking buffer and add a primary antibody made up in fresh blocking buffer (*see* **Note 12**). The membrane is incubated overnight at 4°C on a rocking platform. Alternately, the membrane can be incubated with antibody in a 50-mL tube on a rotating wheel at 4°C.
11. Remove the primary antibody after the overnight incubation step and wash the membrane three times for 5 min each with 10 mL of TBS-T.
12. Freshly prepare the secondary antibody for each experiment at a 1:2,000 dilution in blocking buffer and add this to the membrane for 1 h at room temperature, with gentle agitation on a rocking platform. Alternately, as in the case of the primary antibody, the membrane can be incubated with secondary antibody in a 50-mL tube on a rotating wheel at 4°C.
13. Discard the secondary antibody and wash the membrane three times for 5 min each with TBS-T.
14. During the final wash, mix 1 mL of solution A of ECL-Plus reagent with 25 µL of solution B and apply immediately to the membrane removed from the final wash. Rotate the ECL-Plus reagents by hand for 1 min to ensure even coverage.
15. Remove the membrane from the electrochemiluminescence (ECL) reagents, blot with Kim-Wipes, and then place the membrane between the leaves of a polythene sheet protector, and place in an X-ray film cassette.
16. Expose the membrane is to film for a suitable exposure time, typically 1–2 min.

3.3. Stripping and Reprobing Blots

1. The PKC and β-tubulin antibodies work on stripped and blocked membranes. The Wnt5A antibody works only on fresh blots.
2. For stripping, warm the stripping buffer (10 mL per membrane—*see* **Note 13**) to 55°C and add the warm buffer to the membrane. Incubate the membrane for 20 min with gentle agitation in a water bath in the fume hood. Once the blot is stripped and washed extensively (3× with 10 mL TBS-T, each wash for 10 min), it can be reprobed with the desired primary

antibody immediately. Further blocking with milk for 1 h is not necessary. Incubation with primary antibody is followed by incubation with the secondary antibody and ECL-Plus detection, as above.

4. Notes

1. The phosphorylation and activation of PKC occurs on its threonine and serine residues, and PKC in turn is a serine/threonine kinase. Therefore it is important that the RIPA buffer used in the extraction of protein contains inhibitors of serine/threonine protein phosphatase such as PP1 and PP2A (Phosphatase Inhibitor Cocktail 1; Sigma-Aldrich) and sodium orthovanadate (Na_3VO_4), which protect these PKC phosphorylation sites. Protease and phosphatase inhibitors are very unstable in aqueous solution and are usually degraded with repeated thaw–freeze cycles. Therefore, RIPA buffer should not be subjected to such repeated thaw–freeze cycles and should be stored in single-use aliquots.
2. We have tried other siRNAs (such as Dharmacon smartpool siRNA) against Wnt5A, and found them ineffective as determined by Western analysis. Also, when using rhodamine-tagged siRNA, fluorescence can be overwhelming when using confocal microscopy, so we dilute our tagged siRNA with nontagged siRNAs 1:3.
3. When performing Western analysis of Wnt5A in melanoma, researchers should note that there are often two bands present, one of which is nonspecific in some lines, and in some lines represents the glycosylated or unglycosylated version of Wnt5A. To ascertain which band is Wnt5A, it is advisable to perform a glycosylase assay. To do this, use PNGase F. Set up the following three conditions—in the example shown here we also subjected the recombinant protein to the assay as a positive control—untreated protein, protein lysate subjected to PNGase F treatment, and protein lysate subjected to identical treatment with no enzyme. We use PNGase F from New England Biolabs (Ipswitch, MA), which allows the use of protein lysates extracted in RIPA. The enzyme comes with denaturing buffer at 10×, NP-40, and a 10× G7 reaction buffer. Incubate 50–100 μg of protein (RIPA lysates are fine, but NP-40 *must* be added to counteract the inhibitory effects of SDS) with 1× Denaturing Buffer at 100°C for 10 min. After the addition of NP-40 and G7 Buffer, add 2 μL of PNGase F per 10 μL of reaction volume, and incubate the reaction for

1 h at 37°C. Visualize the results by Western analysis, probing for Wnt5A as described in the Methods section (**Section 3**). A sample gel is shown in **Fig. 12.5**.

4. Many researchers like to use multiwell chamber slides and treat different wells on the same slide. We appreciate that this should be the ideal experimental situation, however it has been our experience that when treating with reagents that are in solution in the medium, the evaporation and condensation that occurs on the lids of these chambers can sometimes cause a cross reaction, affecting untreated cells.
5. It is vital to test the effects of the Plus reagent on each cell line that will be used. It can be toxic to some cells. This applies to other agents also. For example, we find that Lipofectamine 2000 is highly toxic to many of our cells, and Lipofectamine alone results in an inefficient transfection when using DNA vectors. siRNA, however, can be transfected very efficiently using only Lipofectamine with no additional reagents.
6. Some notes on the use of recombinant Wnt5A: Early time points of Wnt5A treatment, such as 4 h, should be avoided due to the consistent deactivation of PKC across a wide range of concentrations of Wnt5A that may be indicative of receptor internalization prior to recycling. As with Wnt5A siRNA treatments, it should be noted that increases in PO4-PKC upon treatment of melanoma cells with recombinant Wnt5A result in a decrease in the nonphosphorylated forms of PKC. Finally, R&D Systems estimates that Wnt5A activity, as gauged by its ability to inhibit Wnt3A, is 5–25 times lower than that of Wnt3A. Thus, when using assays where Wnt3A is used as a control, cells should be treated with Wnt3A at a concentration of 0.04 μg/mL for the same time points.

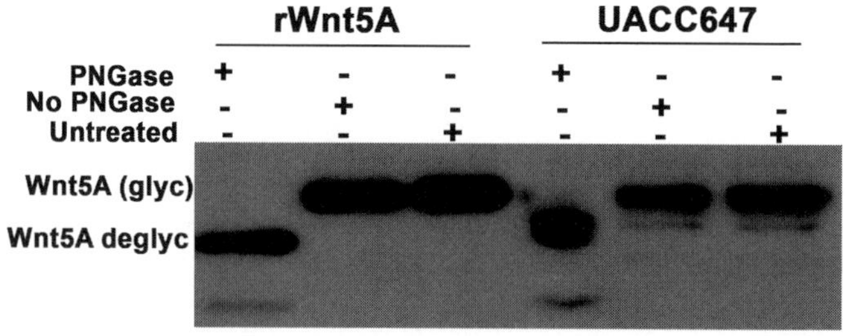

Fig. 12.5. Wnt5A is glycosylated in melanoma cells. UACC647 melanoma cells have high levels of Wnt5A and are highly metastatic. To determine if the Wnt5A in these cells was glycosylated, protein lysates were subjected to treatment with PNGase F, a deglycosylase. After treatment with PNGase F, Wnt5A migrated at around 38–40 kDa as compared with its migration at 42–44 kDa before treatment. Recombinant Wnt5A was also run as control, and demonstrated the same deglycosylation.

7. To study the role of PKC phosphorylation in the Wnt/Ca^{2+} pathway in melanoma, we also employed PKC-specific activators and inhibitors to determine if we could mimic Wnt5A effects via PKC activation, and also if Wnt5A could still mediate its effects in the presence of PKC inhibitors. Our recent data implicating Wnt5A in the epithelial-to-mesenchymal transition in melanoma cells indicate that, for many of the Wnt5A-mediated effects that we see, Wnt5A requires PKC. To assess PKC activation effects, melanoma cell lines low in Wnt5A (UACC1273EV and C-32) were incubated with 200 nM PMA over a 24-h time period. Increases in PKC could be detected as early as 30 min, peaking at 4–12 h in both cell lines, but with sustained effects to 24 h. Many studies have suggested that prolonged PMA activation can result in PKC depletion, but this is not the case in our cells *(4, 13)*.
8. For PKC inhibition studies, a range of times, concentrations, and inhibitors were tested, and ultimately, Gö 6983 (a specific inhibitor of PKC α, β, γ, δ, and ζ, but not μ) and GF 109203X (an inhibitor of PKC α, β, δ, and ε) were used in an attempt to use two different inhibitors of the conventional PKC pathway, at concentration of 1 μM each. Cells were pretreated with the inhibitor over a 24-h time course and assayed by Western blot for phospho-PKC. A decrease in phospho-PKC can be observed as early as 30 min after addition of inhibitor, with maximum inhibition occurring at 12–24 h. For studies in which the effect of PKC inhibition on rWnt5A treatment was examined, cells were pretreated with inhibitor for 12 h, then media was changed, and fresh inhibitor and rWnt5A were added for an additional 16 h.
9. Placement at 4°C "uncures" the slides, making the coverslips slide around, so always leave them out for at least one night (manufacturer recommends 80 h) prior to examining the slides.
10. If cells are frozen at –80°C in RIPA and need to be homogenized at a later date, do the following: Quick thaw the cell lysates in RIPA in a 37°C water bath, and homogenize them with a 22-gauge, 1½-inch needle immediately. Let the lysates sit on ice for 30 min and spin the samples down at 12,000×*g* for 10 min at 4°C. Transfer the supernatant to a new tube and measure the concentration using a Pierce BCA Protein Assay kit.
11. The source of milk is very important. We prefer Biorad nonfat blocking solution, and switching to another company caused a loss of signal on our Western blots.
12. The PKC antibodies from BD Biosciences are provided as a sampler kit of all the different isoforms. Use the antibodies

at the dilutions recommended by the manufacturer, but be aware that although each antibody is provided in a 40-μL aliquot, the required concentrations are very different, causing some antibodies in the kit to be depleted faster than others.

13. Western stripping buffer can be reused until the effect of BME is gone (this is evident when the solution no longer smells potent), if stored at 4°C.

Acknowledgments

We thank Dr. Arya Biragyn and Dr. Teresa D'Souza for helpful comments on this manuscript. Any data represented in this chapter was generated with the support of funds from the Intramural Research Program of the National Institute on Aging, Baltimore, MD.

References

1. Kuhl, M., Sheldahl, L. C., Park, M., Miller, J. R., and Moon, R. T. (2000) The Wnt/Ca2+ pathway: a new vertebrate Wnt signaling pathway takes shape. *Trends Genet* **16**, 279–283.
2. Bittner, M., Meltzer, P., Chen, Y., Jiang, Y., Seftor, E., Hendrix, M., et al. (2000) Molecular classification of cutaneous malignant melanoma by gene expression profiling. *Nature* **406**, 536–540.
3. Weeraratna, A. T., Jiang, Y., Hostetter, G., Rosenblatt, K., Duray, P., Bittner, M., et al. (2002) Wnt5a signaling directly affects cell motility and invasion of metastatic melanoma. *Cancer Cell* **1**, 279–288.
4. Dissanayake, S. K., Wade, M. S., Johnson, C. E., O'Connell, M. P., Leotlela, P. D., French A.D., et al. (2007) The Wnt5a/Pkc pathway mediates motility in melanoma cells via the inhibition of metastasis suppressors, and initiation of an epithelial to mesenchymal transition. *J Biol Chem* **282**, 17259–17271.
5. Parker, C., and Sherbet, G. V. (1992) Modulators of intracellular Ca2+ and the calmodulin inhibitor W-7 alter the expression of metastasis-associated genes MTS1 and NM23 in metastatic variants of the B16 murine melanoma. *Melanoma Res* **2**, 337–343.
6. Li, S., Huang, S., and Peng, S. B. (2005) Overexpression of G protein-coupled receptors in cancer cells: involvement in tumor progression. *Int J Oncol* **27**, 1329–1339.
7. Fink-Puches, R., Helige, C., Kerl, H., Smolle, J., and Tritthart, H. A. (1993) Inhibition of melanoma cell directional migration in vitro via different cellular targets. *Exp Dermatol* **2**, 17–24.
8. Oka, M. and Kikkawa, U. (2005) Protein kinase C in melanoma. *Cancer Metastasis Rev* **24**, 287–300.
9. Lahn, M. M. and Sundell, K. L. (2004) The role of protein kinase C-alpha (PKC-alpha) in melanoma. *Melanoma Res* **14**, 85–89.
10. Maya, R. and Oren, M. (2000) Unmasking of phosphorylation-sensitive epitopes on p53 and Mdm2 by a simple Western-phosphatase procedure. *Oncogene* **19**, 3213–3215.
11. Sheldahl, L. C., Park, M., Malbon, C. C., and Moon, R. T. (1999) Protein kinase C is differentially stimulated by Wnt and Frizzled homologs in a G-protein-dependent manner. *Curr Biol* **9**, 695–698.
12. Keranen, L. M., Dutil, E. M., and Newton, A. C. (1995) Protein kinase C is regulated in vivo by three functionally distinct phosphorylations. *Curr Biol* **5**, 1394–1403.
13. Leotlela, P. D., Wade, M. S., Duray, P. H., Rhode, M. J., Brown, H. F., Rosenthal, D. T., et al. (2007) Claudin-1 overexpression in melanoma is regulated by PKC and contributes to melanoma cell motility. *Oncogene* **26**, 3846–3856.

Chapter 13

Measuring CamKII Activity in *Xenopus* Embryos as a Read-out for Non-canonical Wnt Signaling

Michael Kühl and Petra Pandur

Abstract

It has been known for quite some time that not all members of the Wnt family induce the formation of a secondary body axis when ectopically expressed in *Xenopus* embryos. An ingenious hypothesis led to the discovery that some Wnt ligands have the capacity to elicit intracellular Ca^{2+} signaling. This finding has been studied in more detail in the past years, which has revealed an intriguing complexity of Wnt signaling. The significance of a Wnt-induced Ca^{2+}-mediated pathway during development has been demonstrated in various model systems so far and includes processes such as dorsal–ventral patterning, regulation of the canonical Wnt/β-catenin signaling pathway, tumor formation, bone formation, and regulation of epithelial–mesenchymal transitions. Here we describe two assays to measure the activation of the Ca^{2+}/calmodulin-dependent kinase (CamK)-II, a Ca^{2+}-sensitive molecule described as a mediator of a non-canonical Wnt signaling pathway.

Key words: *Xenopus*, Wnts, Non-canonical signaling, Calcium, Ca^{2+}/calmodulin-dependent kinase II, CamKII, Enzyme activity assay.

1. Introduction

In the past years, Wnt signaling has become increasingly complex due to the findings that Wnt ligands can also activate intracellular pathways that do not signal through β-catenin to activate target genes. These pathways are called non-canonical Wnt signaling pathways. Some members of the Wnt gene family can regulate developmental processes by activating Jun N-terminal kinase (JNK) signaling and some Wnt ligands can mobilize intracellular Ca^{2+} *(1–5)*. With the availability of purified Wnt-5A or peptides derived from the Wnt-5A protein, it was recently shown that the effect of Wnts on Ca^{2+} release is a rather rapid response

Elizabeth Vincan (ed.), *Wnt Signaling, Volume I: Pathway Methods and Mammalian Models, vol. 468*

Book doi: 10.1007/978-1-59745-249-6

of the cell, indicating that the effect is direct *(6–9)*. The Wnt-induced intracellular increase in Ca^{2+} leads to the activation of various Ca^{2+}-dependent proteins, such as protein kinase C (PKC) *(10–13)*, calcineurin *(14)*, and Ca^{2+}/calmodulin-dependent kinase (CamK)-II *(6, 15–19)*. The assays described in this chapter address the analysis of CamKII activation. CamKII is a complex molecule composed of 8 to 12 subunits. Each subunit has an autoinhibitory domain, a catalytic domain, and an association domain, which is required for multimerization. In the absence of Ca^{2+}/calmodulin, the autoinhibitory domain blocks the catalytic site and the enzyme is inactive. Binding of Ca^{2+}/calmodulin induces a conformational rearrangement of the subunit and renders the catalytic site amenable to the substrate. After binding of Ca^{2+}/calmodulin, the enzyme becomes autophosphorylated. This event maintains an active state of CamKII well beyond the initial Ca^{2+} signal, hence the activity of CamKII is now independent of Ca^{2+}. We describe two experimental approaches to investigate whether a particular factor is able to activate CamKII: the enzyme activity assay allows measuring the Ca^{2+}-independent CamKII activity, whereas the Western blotting technique reveals the increase of the phosphorylated form of CamKII.

2. Materials

2.1. Enzyme Activity Assay

1. Cell lysis buffer (pH 7.5): 20 mM Tris-HCl, pH 7.5, 0.1% (v/v) β-mercaptoethanol, 2 mM phenylmethylsulfonyl fluoride (PMSF) (*see* **Note 1**), 100 nM okadaic acid (*see* **Note 2**), and 40 μM EGTA. Prepare 1 or 1.5 mL of lysis buffer fresh before use and keep it on ice.
2. Reaction mixture without inhibitor: 50 mM HEPES, pH 7.5, 10 mM Mg acetate, 0.6 μM ATP, 20 μM Syntide-2 (*see* **Note 3**), 100 pM okadaic acid, 2 mM sodium orthovanadate (*see* **Note 4**), 200 μM EGTA (or 1 mM $CaCl_2$ as a positive control), and 5–10 μCi ^{32}P-γ-ATP (*see* **Note 5**). Additionally, prepare a reaction mixture containing 1–2 μM of the CamKII inhibitor Autocamtide-2 (AiP) (*see* **Note 3**). Prepare the reaction mixture fresh before use and keep it on ice. The final volume of the reaction mixture is 50 μL (45 μL reaction mixture + 5 μL lysate). For an example, *see* **Table 13.1**.
3. P81 cellulose phosphate paper. We have been using P81 phosphocellulose disk sheets from Invitrogen (Karlsruhe, Germany), however the company does not distribute this product anymore. A similar product can be obtained from Whatman (Dassel, Germany). These are circular P81 phosphocellulose sheets with a diameter of 2.1 cm.

Table 13.1
An example for the preparation of the reaction mixture to measure CamKII activity

Solutions and reagents (stock concentration)	Amount for one reaction	
	– Inhibitor	+ Inhibitor
HEPES, pH 7.5 (500 mM) + Mg acetate (100 mM)[a]	5 μL	5 μL
ATP (10 μM)	3 μL	3 μL
Syntide-2 (1 mg/mL)	1.5 μL	1.5 μL
Okadaic acid (100 μM)[b]	1 μL	1 μL
Sodium orthovanadate (0.1 M)	1 μL	1 μL
EGTA (2 mM) *or* $CaCl_2$ (10 mM)	5 μL	5 μL
AiP (1 mg/mL)[c]	—	1 μL
H_2O	27.5 μL	26.5 μL
10 μCi ^{32}P-γ-ATP	1 μL	1 μL
Total volume	45 μL	45 μL

The reaction is started by adding 5 μL of lysate
[a]Note that the HEPES and the Mg acetate solution have been combined (they can also be stored as such) and 5 μL of the mixture is used
[b]Dilute the okadaic acid stock solution 1:20 before use and take 1 μL from this dilution.
[c]Use 1 μL from the diluted stock solution (*see* **Note 3**)

4. Phosphoric acid solution for washing the phosphocellulose squares. Add 5 mL 85% phosphoric acid (H_3PO_4) to 995 mL H_2O.
5. Scintillation vials and a scintillation cocktail, for example, Rotiszint Eco-Plus from Carl Roth, Karlsruhe, Germany.
6. For determining the protein concentration by the Bradford method, prepare a 1 mg/mL bovine serum albumin (BSA; cat. #A4503 Sigma, Hamburg, Germany) stock solution and use the Protein Assay Dye Reagent Concentrate from Bio-Rad (München, Germany) or any other Coomassie brilliant blue solution.

2.2. Immunoprecipitation and Western Blotting for Active CamKII

1. Lysis and immunoprecipitation (IP) buffer: 20 mM Tris-HCl, pH 7.5, 150 mM NaCl, 1 mM EDTA, 1 mM EGTA, 0.1% (v/v) Triton X-100, 2 mM PMSF, 100 nM okadaic acid, and 2 mM sodium orthovanadate (*see* **Note 4**).
2. 1,1,2-Trichloro-1,2,2-trifluoroethane solution (Freon 113) from Sigma.

3. Electrophoresis buffer: 25 mM Tris, 192 mM glycine, and 1% (w/v) sodium dodecyl sulfate (SDS), pH 8.3.
4. Buffer for Western blotting: 25 mM Tris, 192 mM glycine, and 10% (v/v) methanol, pH 8.3.
5. Stock solutions for polyacrylamide gels: Solution A (keep at 4°C): 38% (w/v) acrylamide, 2% (w/v) bis-acrylamide (*see* **Note 6**). Solution B: 1.5 M Tris-HCl, pH 8.8. Solution C: 10% (w/v) SDS. Solution D: 0.5 M Tris-HCl, pH 6.8. Keep solutions B, C, and D at room temperature.
6. Ammonium persulfate (APS): prepare 1 mL of a 10% (w/v) solution in water. Although single-use aliquots can be frozen immediately at –20°C, we prefer to prepare the solution fresh before use.
7. *N,N,N,N'*-Tetramethylethylenediamine (TEMED; Sigma; cat. #T9281).
8. SDS sample buffer: 126 mM M Tris-HCl, pH 6.8, 0.005% (w/v) bromophenol blue or Pyronin Y, 4% (w/v) SDS, 4% (v/v) glycerol, and 5% (v/v) β-mercaptoethanol.
9. TBS (10×): dissolve 80 g NaCl, 2 g KCl, and 30 g Tris-base in 800 mL H_2O, use concentrated HCl (at least 15 mL) to adjust the pH to 7.4. Add H_2O to obtain a final volume of 1 L.
10. TBST is a 1× solution of 10× TBS with 10% (v/v) Tween-20.
11. Blocking buffer: 2% (w/v) BSA in 1× TBST. Optional: add 1 μL of okadaic acid (*see* **Note 2**) to 20 mL of the BSA/TBST solution.
12. Antibodies for total and active CamKII: mouse anti-CamKIIα monoclonal antibody, clone 6G9 (Sigma; cat. #C265) for immunoprecipitation; rabbit anti-CamKII polyclonal antibody (R&D Systems, Minneapolis, MN; cat. #PPS003) (*see* **Note 8**) for Western blot; and rabbit anti-ACTIVE CamKII pAb (pT^{286}) (Promega, Mannheim, Germany) for Western blot.
13. Secondary antibody: anti-rabbit IgG peroxidase conjugate (Sigma; cat. #A0545).
14. Whatman 3MM Chr paper and nitrocellulose membrane (Bio-Rad; cat. #162-0115) for blotting.
15. Protein A–Sepharose 4B is available as aqueous ethanol suspension (Sigma; cat. #P9424).
16. Blot staining solution: 0.1% (w/v) Poinceau red and 1% (v/v) acetic acid in H_2O.
17. Protein standard for blotting: Kaleidoscope prestained standard from Bio-Rad (cat. #161-0324).
18. For signal detection: Western blotting luminal reagent from Santa Cruz Biotechnology (Santa Cruz, Heidelberg, Germany). Hyperfilm ECL from Amersham (Buckinghamshire, UK; cat. #RPN 2103K).
19. Stripping solution: 4 M $MgCl_2$ solution, pH 4.

3. Methods

Measuring CamKII activity using a radioactive isotope to label its substrate is highly susceptible to erroneous pipetting. To obtain reliable and reproducible results, it is mandatory that the steps to determine the enzymatic activity of CamKII are performed as precisely as possible and in duplicates. This applies to all steps once the reaction is started by adding the cell lysate to the reaction mixture. Therefore it is very important that the workplace is set up in a way that each step can be executed without any time delay. The principle of the assay is to measure CamKII activity every 10 seconds in a total time period of 40 seconds. The reaction is stopped by pipetting an aliquot of the reaction mixture onto the phosphocellulose disk sheets. If there is even a slight time variation in taking the aliquots or if the volume of the aliquot varies, for example, when a drop of the reaction mixture sticks on the outside of the pipette tip and ends up on the phosphocellulose paper, it will perturb the measurement. To determine full activation of CamKII, prepare additional reaction mixtures that contain $CaCl_2$ instead of EGTA (*see* **Table 13.1**). This sample then represents a positive control and helps to evaluate the performance of the assay (**Fig. 13.1**).

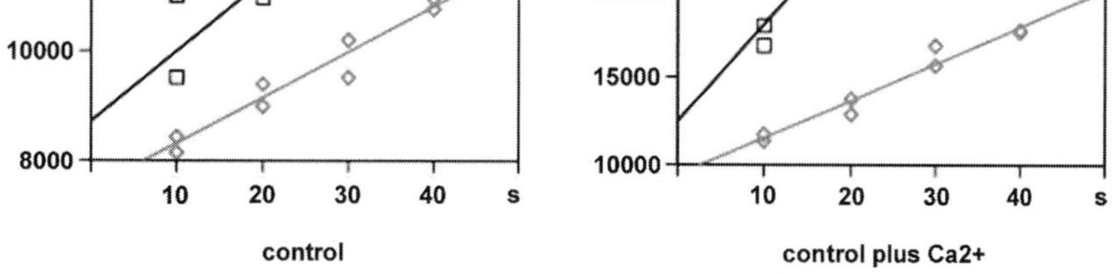

Fig. 13.1. An example of results obtained from the CamKII activity assay. The graph on the left depicts CamKII activity in untreated tissue, the graph on the right depicts full activation of CamKII after adding $CaCl_2$ to the reaction mixture. CamKII activity in the absence of inhibitor is shown in *black*; CamKII activity in the presence of the inhibitor AiP is shown in *grey*. To determine the fold activation of CamKII by Ca^{2+}, subtract the value of the slope of the *grey line* (with inhibitor) from the value of the slope of the *black line* (without inhibitor). In the example shown, the subtraction yields 45 for the control reaction and 336 for the reaction with $CaCl_2$. Hence, the result is a 7.5-fold activation of CamKII. When testing the CamKII activating potential of a factor of interest, the activity typically increases between twofold and threefold. The data was plotted using the Cricket Graph software.

3.1. Enzyme Activity Assay

3.1.1. Preparation of Samples

The following protocol has been designed to identify factors, in this case Wnt ligands and components of the Wnt signaling pathway, that can activate CamKII in *Xenopus* tissue. The ability of a factor to activate CamKII can be tested by either injecting in vitro-synthesized RNA or by treating explanted tissue with the protein if available (*see* **Note 7**). The in vitro synthesis of RNA, the microinjection technique, and the composition of the solution for culturing *Xenopus* embryos are described in **Chapter 29, Volume 2**.

1. In vitro-synthesized RNA is injected bilaterally into the animal pole of a two-cell stage *Xenopus* embryo. The optimal amount of RNA injected needs to be determined for each investigated factor. In previous assays, injection of 1 ng RNA of various Wnts and Frizzled (Fz) has worked well *(15)*. The embryos are then cultured in 0.1× Modified Barth Solution High-Salt (MBSH) until the early blastula stage (stage 7), which is before the onset of zygotic transcription *(20)*. The cell extract for the assay can be prepared from either whole embryos or dissected animal halves (*see* **Fig. 3** in **Chapter 29, Volume 2**).
2. Prepare the lysis buffer and the reaction mixture in 1.5-mL microcentrifuge tubes (do not add ^{32}P-γ-ATP yet) just before the embryos or explants are ready to proceed with. When preparing the reaction mixture, remember that the measurements of CamKII activity are performed as duplicates. For example, a total volume of 200 μL of reaction mixture (4× 50 μL with CamKII inhibitor and 4× 50 μL without CamKII inhibitor) is needed for determining CamKII activity in a batch of control and injected embryos.

3.1.2. Setting Up the Work Place in the Radiation Work Area

1. Set a heating block at 37°C. Pour the phosphoric acid solution in a 500-mL glass beaker, add a stir bar, and place the beaker on a magnetic stirrer.
2. Number the P81 phosphocellulose disk sheets using a black pencil (labeling with any other pen will be washed off during the procedure!), arrange the phosphocellulose disks on a sheet of paper towel in a tray (do not forget a timer and forceps for picking up the phosphocellulose disks), and place the tray behind the Plexiglas shield.
3. Make sure a waste container for radioactive waste is next to the tray so that the pipette tips can be rapidly exchanged during the assay.
4. Prior to lysis, transfer the embryos or animal halves into 1.5-mL microcentrifuge tubes and wash five times with Ca^{2+}/Mg^{2+}-free 0.1× MBSH to remove Ca^{2+} ions from the previous medium.

After the last wash step, remove as much of the 0.1× MBSH as possible and add the lysis buffer. Use 20 µL lysis buffer per embryo (400 µL for 20 embryos) or a minimum of 100 µL for 15 animal halves.

5. Centrifuge the samples at 15,870 ×*g* for 5 minutes at 4°C. Transfer approximately half of the initial lysis buffer volume to a new 1.5-mL microcentrifuge tube. Avoid contamination of the lysate with yolk and cell debris as much as possible.
6. Centrifuge the samples again at 15,870 ×*g* for 5 minutes at 4°C. Transfer the clear lysate into a new 1.5-mL microcentrifuge tube. Make sure to recover at least 30 µL of clear lysate after the second centrifugation.
7. Move to the radiation work area and add ^{32}P-γ-ATP to each reaction mixture. For each sample and its duplicate, pipette 45 µL of the reaction mixture into a new 1.5-mL microcentrifuge tube and preincubate the tubes at 37°C in the heating block (once the lysate is added, the final reaction temperature will be approximately 30°C).
8. Start the reaction by adding 5 µL of the lysate to the reaction mixture. Discard the pipette tip and *immediately* start the timer. At 8 seconds, start with pipetting a 5-µL aliquot onto the phosphocellulose disk sheet (it will be 10 seconds by the time you actually place the aliquot onto the phosphocellulose disk and the reaction is stopped), discard the pipette tip, pick up the phosphocellulose disk and put it into the stirring phosphoric acid solution. At 18 seconds (20 seconds time point), repeat this procedure and again at 28 seconds (30 seconds time point), and a last time at 38 seconds (40 seconds time point). Then, start the next reaction by adding 5 µL of lysate to the next reaction mixture and repeat the complete procedure.
9. After the last phosphocellulose disk has been put into the phosphoric acid solution, wash for an additional 3 minutes.
10. Carefully discard the radioactive phosphoric acid solution into a waste container for liquid radioactive waste and pour fresh phosphoric acid solution onto the phosphocellulose disks. Swirl the beaker and discard most of the solution. Leave approximately 100 mL of phosphoric acid solution in the beaker because the residual liquid makes it easier to pick up the disks with the forceps.
11. Place the phosphocellulose disks onto a paper towel in the tray and let them dry briefly. The time should be used to label the scintillation vials according to the number of the phosphocellulose disks.

12. Transfer the phosphocellulose disks into the scintillation vials, add 4–5 mL of the scintillation cocktail, and place the vials into a scintillation counter. The β-radiation emitted by the incorporated ^{32}P is measured as counts per minute (cpm). An increase in CamKII activity is given by an increase in counts per minute over the 40-second time course (**Fig. 13.1**).
13. For normalization, determine the protein concentration of your samples using the Bradford method.

3.2. Immuno-precipitation and Western Blotting for Phosphorylated CamKII

The following protocol describes the determination of phosphorylated CamKII in *Xenopus* embryo extracts but can be easily adapted to determine phosphorylated CamKII in cultured cell lines. When using cells, omit the Freon extraction, which is required to remove lipids from *Xenopus* extracts.

3.2.1. Sample Preparation

1. Lyse 10 embryos in 300 µL of IP buffer using a pipette tip that fits on a 20–200-µL pipette. If lysing cells derived from cell culture, use a 26-gauge syringe needle. Centrifuge the lysate in a refrigerated microtube centrifuge at 15,870×*g* for 5 minutes.
2. Remove the supernatant and extract with an equal volume of Freon. Use 100 µL of the upper phase for the subsequent immunoprecipitation.

3.2.2. Immunoprecipitation

1. When comparing the change in CamKII phosphorylation between different samples, make sure to use equal amounts of proteins. This requires determination of protein concentration by the Bradford method.
2. Preincubate lysates with 10 µL of the protein A–Sepharose suspension for 60 minutes at 4°C with gentle agitation.
3. Centrifuge at low speed (845×*g*) for 2 minutes to separate the protein A–Sepharose from the sample. Transfer the supernatant to a new microcentrifuge tube. Be careful not to loose any of the sample, and to treat all samples alike.
4. Add the mouse anti-CamKIIα monoclonal antibody for immunoprecipitating CamKII. The amount of antibody required needs to be determined individually since it depends on the batch and on the company from which it was purchased. We have been using 2 µg for immunoprecipitation. Incubate the sample for 2 hours at 4°C with gentle agitation.
5. Add 20 µL of the protein A–Sepharose suspension to the samples and incubate for an additional hour at 4°C under gentle agitation.
6. Pellet the protein A–Sepharose at 845×*g* for 2 minutes at 4°C. Discard the supernatant unless you are interested in

checking the efficiency of the immunoprecipitation with subsequent Western blotting.

7. Wash the protein A–Sepharose three times with 100–200 µL of 1× TBST by adding the buffer to the pelleted protein A–Sepharose and snipping the tube.

8. Finally, remove as much of the washing buffer as possible and add an appropriate amount of SDS sample buffer (e.g., 10 µL). Heat the samples for 5–10 minutes at 95°C, and place and load the gel. If removal of the protein A–Sepharose is desired before loading the gel, puncture the tube with a 26-gauge syringe needle at the bottom and the top (*see* **Note 9**), put the tube into another microcentrifuge tube, and spin the nested microcentrifuge tubes in a table top centrifuge at 845 × *g* for 2 minutes. The protein A–Sepharose remains in the inner microcentrifuge tube and the sample is drained into the bottom microcentrifuge tube.

3.2.3. Polyacrylamide Gel Electrophoresis

The following protocol is according to Laemmli *(21)* and has been adapted to the Mini-Protean III unit of Bio-Rad. The protocol needs to be modified if an apparatus of a different supplier is used.

1. We routinely prepare the required number of gels during the incubation times of the immunoprecipitation procedure. It is critical that the glass plates for the gels are very clean. Clean the glass plates with ethanol and acetone and rinse well with distilled H_2O.

2. First pour a gel for separating the proteins in the samples according to their molecular weight. Prepare a solution containing: 1.5 mL of solution A, 1.875 mL of solution B, 75 µL of solution C, 4.1 mL distilled H_2O, 25 µL of 10% (w/v) APS, and 12.5 µL TEMED. This solution should be sufficient to pour two gels. Mix the solution well and pour it between the assembled glass plates. Leave space for the stacking gel (approximately 1 cm below the pouches from the comb that will be inserted into the stacking gel).

3. Cover the top rim of the gel with H_2O-saturated isobutanol. The gel should polymerize in about 15 minutes.

4. After polymerization, remove the isobutanol and wash with distilled water or running buffer using a syringe with a 26-gauge needle. Carefully dry the top rim of the gel with a small piece of 3MM Whatman paper without actually touching the gel surface. It is crucial that the top rim of the gel is even.

5. Prepare the stacking gel by mixing: 0.55 mL of solution A, 1.25 mL of solution D, 50 µL of solution C, 3.2 mL distilled H_2O, 25 µL of 10% (w/v) APS, and 15 µL TEMED. This

is sufficient for two stacking gels. Mix the components well, pour the stacking gel onto the separating gel, and insert the comb. Avoid the generation of air bubbles.

6. After the stacking gel has polymerized, remove the comb and clean the wells with distilled H_2O using a syringe.
7. Assemble the gel apparatus and pour the electrophoresis buffer into the upper and lower chamber.
8. Load each sample (from the immunoprecipitation) in a well. Also load a prestained molecular weight marker in one of the wells.
9. Finish assembling the gel unit, put the lid on the unit, and start the gel. Run the gel at 100 V until the dye front reaches the bottom of the gel.

3.2.4. Western Blotting

The samples that have been separated are subsequently electrophoretically transferred to nitrocellulose. The following protocol is adapted to the Mini-Protean II Trans-Blot system from Bio-Rad, but can be adapted to other models.

1. While the separation of the samples is in progress, cut six sheets of 3MM Whatman paper and a sheet of nitrocellulose to the size of the separating gel. Wet the Whatman papers and the nitrocellulose with blotting buffer.
2. Disconnect the gel unit and disassemble the gel. Carefully lift off one of the glass plates and cut off one corner of the gel. This will help to orient the gel later on.
3. Place the gel onto the nitrocellulose, avoiding the inclusion of any air bubbles. Place three 3MM Whatman papers on each side of the gel/nitrocellulose combination, again avoiding the inclusion of any air bubbles. Place this sandwich into the blotting cassette and introduce the cassette into the blotting apparatus. Note that the proteins migrate toward the anode and therefore the nitrocellulose membrane has to be placed between the gel and the anode; otherwise the proteins will be lost in the buffer. Pour blotting buffer and a small stir bar into the chamber and place the apparatus into a Styrofoam box filled with ice. The Bio-Rad Mini-Protean II system provides an additional cooling device that is placed into the chamber. Put the box onto a magnetic stirrer. This will ensure a uniform cooling of the buffer.
4. Put the lid on the blotting apparatus and connect it with the power supply. We routinely perform the transfer for 1 hour at 111 V (a number that can be remembered easily).
5. Disassemble the blotting apparatus and open the blotting cassette. *Do not* touch the nitrocellulose without gloves! Use forceps to remove the nitrocellulose and put the nitrocellulose

in a plastic dish of appropriate size. Add the Poinceau Red solution for several seconds and subsequently wash several times with 1× TBST. The staining should uncover the prestained marker as well as the transferred proteins. Ideally, you should only see the light and heavy chains of the antibodies used. Label the size of the prestained marker bands with a waterproof pen and indicate the orientation and numbering of lanes.

6. Incubate the nitrocellulose in blocking buffer for 30 minutes at room temperature to block binding sites for proteins. We do not use milk powder for blocking because the milk powder might contain phosphatases that might remove the phosphate from CamKII!
7. Wash the nitrocellulose three times with 1× TBST and incubate the nitrocellulose with the rabbit anti-ACTIVE CamKII pAb (pT^{286}) overnight at 4°C with gentle agitation.
8. Remove the primary antibody and wash three times with 1 × TBST for 30 minutes at room temperature. The diluted primary antibody can be saved and reused several times.
9. Dilute the secondary antibody, which in this example is an anti-rabbit IgG peroxidase conjugate, freshly before each experiment. We use a dilution of 1:20,000 and incubate for 2 hours at room temperature with gentle agitation.
10. Replace the solution containing the secondary antibody with 1 × TBST and repeat the washes three times with 1 × TBST for 30 minutes at room temperature.
11. Incubate the nitrocellulose in the solutions supplied with the electrochemiluminescence (ECL) detection kit according to the manufacturer's guidelines. Wrap the nitrocellulose in Saran wrap, place it into an X-ray film cassette, and put the hyperfilm ECL on top of it. The time of exposure has to be determined individually but in general takes several minutes.
12. Once a satisfactory exposure for the result of the activated CamKII has been obtained, the nitrocellulose is stripped of this signal and then reprobed with the rabbit anti-CamKII polyclonal antibody that detects total CamKII independent of its phosphorylation status. This serves as a loading control to confirm that equal amounts of total CamKII were recovered in each sample.
13. Incubate the nitrocellulose in 4 M $MgCl_2$, pH 4, for 3 minutes at room temperature followed by at least three washes for 15 minutes each in 1 × TBST.
14. The nitrocellulose can then be reprobed with the rabbit anti-CamKII polyclonal antibody. Prepare a 1:250 dilution of the

antibody fresh before use. After the washes, add the HRP-conjugated secondary antibody and perform the ECL detection as above.

4. Notes

1. PMSF is an irreversible inhibitor of serine proteases. Dissolve PMSF (Sigma) in ethanol to yield a stock solution of 200 mM and keep it at 4°C. PMSF is toxic and has to be handled with great care.
2. Okadaic acid, *Prorocentrum* sp. (Darmstadt, Germany), is a reversible inhibitor of serine/threonine-specific protein phosphatases 1. Following reconstitution in ethanol at a concentration of 100 µM, it has to be kept at –20°C. Okadaic acid is toxic and should be handled with great care.
3. The 1 mg Syntide-2 (CamKII substrate) and 1 mg AiP (both from Calbiochem) are each dissolved in 1 mL H_2O. Single-use aliquots have to be kept at -20°C. The AiP stock solution has to be diluted prior to adding to the reaction mixture (3.8 µL AiP in a final volume of 50 µL H_2O).
4. For a 0.1 M sodium orthovanadate (Sigma) stock solution, dissolve 183.9 mg of sodium orthovanadate in 8 mL H_2O and add 360 µL of 1 M HCl, which will turn the solution yellow. Heat the solution until it is colorless, let it cool to room temperature, and adjust the pH to 10. Keep the stock solution at 4°C.
5. The radionuclide ^{32}P-γ-ATP is a high-energy beta-emitter and it is important to follow good laboratory practice in addition to specific precautions regarding this radionuclide. The range of beta particles in air can be several meters, so make sure to perform the assay behind protective shielding.
6. In the unpolymerized state, the acrylamide and bis-acrylamide solutions are neurotoxins and any unprotected contact with these substances has to be avoided.
7. The described protocol can be adapted to the use of protein instead of injecting RNAs. In this case, it is recommended that animal halves rather than whole embryos be incubated in the protein-containing solution. To obtain a sufficient amount of protein for the assay, a minimum of 15 animal halves should be dissected. The incubation should be performed in a 1% BSA-coated well of, for example, a 24-well plate. The duration of incubation needs to be determined individually for the protein of interest.

8. To detect total CamKII we have been using a rabbit anti-CamKII polyclonal antibody from Transduction Laboratories (Lexington, KY). However, this product is no longer available. A similar product can be obtained from R&D Systems as suggested in the Materials section (**Section 2**), but this antibody needs to be tested first.
9. Be careful while puncturing the tube not to poke your fingers with the needle. Also make sure that the hole is not too large, otherwise the protein A–Sepharose will also be collected in the bottom microcentrifuge tube and you have to redo the separation. Do not recap used needles but dispose of them in an appropriate waste container.

Acknowledgments

MK and PP are funded by the Deutsche Forschungsgemeinschaft (DFG).

References

1. Slusarski, D. C., Corces, V. G., and Moon, R. T. (1997a) Interaction of Wnt and a Frizzled homologue triggers G-protein-linked phosphatidylinositol signaling. *Nature* **390**, 410–413.
2. Slusarski, D. C., Yang-Snyder, J., Busa, W. B., and Moon, R. T. (1997b) Modulation of embryonic intracellular Ca^{2+} signaling by Wnt-5A. *Dev. Biol.* **182**, 114–120.
3. Eisenberg, L. M. and Eisenberg, C. A. (2002) *Onset of a cardiac phenotype in the early embryo. Cardiovascular molecular morphogenesis: myofibrillogenesis.* New York: Springer Verlag.
4. Kühl, M. (2002) Non-canonical Wnt signaling in Xenopus: regulation of axis formation and gastrulation. *Semin. Cell. Dev. Biol.* **13**, 243–249.
5. Westfall, T. A., Brimeyer, R., Twedt, J., Gladon, J., Olberding, A., Furutani-Seiki, M., and Slusarski, D. C. (2003) Wnt-5/pipetail functions in vertebrate axis formation as a negative regulator of Wnt/beta-catenin activity. *J. Cell. Biol.* **162**, 889–898.
6. Kremenevskaja, N., von Wasielewski, R., Rao, A. S., Schofl, C., Andersson, T., and Brabant, G. (2005) Wnt-5a has tumor suppressor activity in thyroid carcinoma. *Oncogene* **24**, 2144–2154.
7. Säfholm, A., Leandersson, K., Dejmek, J., Nielsen, C. K., Villoutreix, B. O., and Andersson, T. (2006) A formylated hexapeptide ligand mimics the ability of Wnt-5a to impair migration of human breast epithelial cells. *J. Biol. Chem.* **281**, 2740–2749.
8. Dejmek, J., Säfholm, A., Kamp Nielsen, C., Andersson, T., and Leandersson, K. (2006) Wnt5a/Ca^{2+}-induced NFAT activity is counteracted by Wnt-5a/Yes-Cdc42-Casein kinase 1α signaling in human mammary epithelial cells. *Mol. Cell. Biol.* **26**, 6024–6036.
9. Ma, L. and Wang H. Y. (2006) Suppression of cyclic GMP-dependent protein kinase is essential to the Wnt/cGMP/Ca^{2+} pathway. *J Biol Chem.* **281**, 30990–31001.
10. Sheldahl, L. C., Park, M., Malbon, C. C., and Moon, R. T. (1999) Protein kinase C is differentially stimulated by Wnt and Frizzled homologs in a G-protein-dependent manner. *Curr. Biol.* **9**, 695–698.
11. Winklbauer, R., Medina, A., Swain, R. K., and Steinbeisser, H. (2001) Frizzled-7 signalling

controls tissue separation during Xenopus gastrulation. *Nature* **413**, 856–860.

12. Kinoshita, N., Iioka, H., Miyakoshi, A., and Ueno, N. (2003) PKC delta is essential for Dishevelled function in a non-canonical Wnt pathway that regulates Xenopus convergent extension movements. *Genes Dev.* **17**, 1663–1676.
13. Tu, X., Joeng, K. S., Nakayama, K. I., Nakayama, K., Rajagopal, J., Carroll, T. J., Mc Mahon, A. P., and Long, F. (2007) Non-canonical Wnt signaling through G-protein-linked PKC delta activation promotes bone formation. *Dev. Cell* **12**, 113–127.
14. Saneyoshi, T., Kume, S., Amasaki, Y., and Mikoshiba, K. (2002) The Wnt/calcium pathway activates NF-AT and promotes ventral cell fate in Xenopus embryos. *Nature* **417**, 295–299.
15. Kühl, M., Sheldahl, L. C., Malbon, C. C., and Moon, R. T. (2000) Ca^{2+}/Calmodulin-dependent protein kinase II is stimulated by Wnt and Frizzled homologs and promotes ventral cell fates in Xenopus. *J. Biol. Chem.* **275**, 12701–12711.
16. Ishitani, T., Kishida, S., Hyodo-Miura, J., Ueno, N., Yasuda, J., Waterman, M., Shibuya, H., Moon, R. T., Ninomiya-Tsuji, J., and Matsumoto, K. (2003) The TAK1-NLK mitogen-activated protein kinase cascade functions in the Wnt-5a/Ca(2+) pathway to antagonize Wnt-beta catenin signaling. *Mol. Cell. Biol.* **23**, 131–139.
17. Topol, L., Jiang, X., Choi, H., Garrett-Beal, L., Carolan, P. J., and Yang, Y. (2003) Wnt-5a inhibits the canonical Wnt pathway by promoting GSK-3-independent beta-catenin degradation. *J. Cell. Biol.* **162**, 899–908.
18. Ouko, L., Ziegler, T. R., Gu, L. H., Eisenberg, L. M., and Yang, V. W. (2004) Wnt11 signaling promotes proliferation, transformation, and migration of IEC6 intestinal epithelial cells. *J. Biol. Chem.* **279**, 26707–26715.
19. Garriock, R. J. and Krieg, P. A. (2007) Wnt11-R signaling regulates a calcium sensitive EMT event essential for dorsal fin development of Xenopus. *Dev. Biol.* **304**, 127–140.
20. Nieuwkoop, P. D. and Faber, J. (1967) Normaltabelle von *Xenopus laevis.* Elsevier North-Holland Biomedical Press, Amsterdam.
21. Laemmli, U. K. (1970) Cleavage of structural proteins during the assembly of bacteriophage T4. *Nature* **227**, 680–685.

Chapter 14

Analysis of Wnt7a-Stimulated JNK Activity and cJun Phosphorylation in Non-Small Cell Lung Cancer Cells

Lynn E. Heasley and Robert A. Winn

Abstract

The cJun N-terminal kinases (JNKs) are activated in response to diverse growth factors and morphogens, including specific Wnt proteins. Genetic approaches have defined key roles for the JNKs as mediators of the Wnt-regulated epithelial cell programs including planar cell polarity and convergent extension. Moreover, our recent studies demonstrate that the JNK pathway is activated by Wnt-7a and Fzd9 to promote an epithelial differentiation program in lung cancer cells. In comparison to cell stresses such as hypertonicity or ultraviolet irradiation, which strongly activate JNKs, morphogens and growth factors induce activation of the pathway that is more modest and that may be difficult to assess by immunoblotting approaches with anti-phospho-JNK antibodies. We find that the tight binding of JNKs by their substrate, cJun, provides the basis for a simple and reliable assay for measuring JNK activity in cells stimulated with Wnt proteins and growth factors.

Key words: JNK, Protein kinase assay, Protein phosphorylation, cJun, Phospho-cJun (Ser 73), GST-cJun (1–79).

1. Introduction

The cJun N-terminal kinases (JNKs), encoded by three distinct genes (*jnk*1, *jnk*2, and *jnk*3), are activated by diverse cell stresses, morphogens, growth factors, and oncogenes *(1–4)*, implying broad roles of this mitogen-activated protein (MAP) kinase family in differentiation, development, and transformation. As indicated by their name, the JNKs were initially identified, in part, by their ability to phosphorylate cJun on serine 63 and 73, leading to increased transcriptional activation of the cJun protein *(5, 6)*. The substrate specificity of JNKs for cJun derives from their binding to a high-affinity docking motif that resides N-terminal to the phosphorylated residues *(7)*. Thus,

Elizabeth Vincan (ed.), *Wnt Signaling, Volume I: Pathway Methods and Mammalian Models, vol. 468*

Book doi: 10.1007/978-1-59745-249-6

cJun amino acids 1–79 fused to glutathione *S*-transferase (GST), GST-cJun(1–79), functions both as an affinity matrix for JNK purification from cytosolic extracts and a specific substrate following addition of ATP and forms the basis for a sensitive and specific JNK assay *(8)*.

The literature supports the JNK pathway as a required element in development, morphogenesis, and cell differentiation *(2)*. Genetic analyses in *Drosophila* reveals Wnt-dependent regulation of two epithelial programs, planar polarity and dorsal closure, that proceed through JNK activation, but independently of β-catenin *(9)*. Recent studies have established that the mammalian counterpart of planar polarity, termed convergent extension, is also a Wnt-dependent, β-catenin-independent program *(10, 11)*. Moreover, studies in *Xenopus* reveal that Wnt-regulated convergent extension requires the JNK pathway *(10)*. We have recently demonstrated a role for the JNK pathway in epithelial differentiation in response to Wnt-7a and Fzd-9 coexpression in non-small cell lung cancer cells *(12)*.

Cell stresses such as increased tonicity or ultraviolet radiation serve as powerful stimuli for JNKs, such that activation can be easily monitored by immunoblotting for JNKs that are dually phosphorylated at the threonine and tyrosine residues within their activation loops. By contrast, JNK activation by morphogens such as specific Wnt proteins, is invariably more modest (twofold to threefold) and reliable detection may be better assessed by direct assay of JNK activity or phosphorylation of endogenous cJun. Protocols for these assays are outlined herein.

2. Materials

2.1. Isolation of GST-cJun (1–79)

1. LB Medium: In a total volume of 1 L, dissolve 10 g Bacto-tryptone, 5 g Bacto-yeast extract, and 10 g NaCl. Sterilize by autoclaving for 20 min at 15 lb/square inch on the liquid cycle.
2. Ampicillin dissolved at 100 mg/mL in distilled water and sterilized through a 0.2-μm filter. Store in 1-mL aliquots at -20°C.
3. Isopropyl β-D-thiogalacto-pyranoside (IPTG) dissolved in distilled water to 100 mM, sterilized through a 0.2-μm filter. Store in 1-mL aliquots at -20°C.
4. NETN lysis buffer: 20 mM Tris-HCl, pH 7.6, 100 mM NaCl, 1 mM EDTA, 2 mM DTT, 0.5% NP-40, and 1 mM phenylmethanesulfonylfluoride.

5. Lysozyme (egg white) freshly dissolved in distilled water at 1 mg/mL.
6. Glutathione (GSH)–agarose (Sigma-Aldrich, St. Louis, MO; #G4510) rehydrated in NETN buffer and suspended to 50% (v/v). Store at 4°C.

2.2. Assay of JNK Activity

1. JNK assay buffer: Prepare 40 μL per assay of 50 mM β-glycerophosphate, pH 7.2, 0.1 mM sodium orthovanadate, 10 mM $MgCl_2$, 1 mM EGTA, and 20 μM (γ-^{32}P]ATP (~25,000 cpm/pmol).
2. Sample Buffer (5×): 3.78 g Tris base, 25 mL glycerol, 10 g sodium dodecyl sulfate (SDS), 5 mg bromophenol blue, and 2.5 mL β-mercaptoethanol. Bring to 100 mL with water, adjust pH to 6.8.
3. 10% polyacrylamide SDS gel with 4% stack.
4. Coomassie staining solution: 0.1% (w/v) Coomassie Blue R-250 in 25% methanol, 10% acetic acid, and deionized water (v/v/v).
5. Destain solution: 10% ethanol, 10% acetic acid, and deionized water (v/v/v).

2.3. Preparation of Cell Extracts

1. Phosphate-buffered saline (PBS): In 1 L distilled water, 8 g NaCl, 0.2 g KCl, 1.44 g Na_2HPO_4, and 0.24 g KH_2PO_4. Adjust pH to 7.4 and store at 4°C.
2. MAP kinase lysis buffer (MKLB): 0.5% (v/v) Triton X-100, 300 mM NaCl, 50 mM β-glycerophosphate, 0.1 mM sodium orthovanadate, 2 mM $MgCl_2$, 1 mM EGTA, 1 mM DTT, 2 μg/mL leupeptin, 4 μg/mL aprotinin, and 1 mM phenylmethanesulfonylfluoride. Adjust pH to 7.2 and store at 4°C.

2.4. SDS-Polyacrylamide Gel Electrophoresis (SDS-PAGE)

1. Separating buffer (4×): 1.5 M Tris-HCl, pH 8.8, and 0.4% SDS. Store at 4°C.
2. Stacking buffer (4×): 0.5 M Tris-HCl, pH 6.8, and 0.4% SDS. Store at 4°C.
3. Acrylamide/bis solution (30:0.8). Store in a foil-wrapped container at 4°C.
4. *N,N,N,N*-Tetramethyl-ethylenediamine (TEMED).
5. Ammonium persulfate (APS): 10% (w/v) in distilled water, store at 4°C.
6. Electrophoresis Buffer (10×): Dissolve in 4 L deionized water, 121.1 g Tris base, 576.7 g glycine, and 40 g SDS. The pH should be 8.63 at room temperature without adjustment.

7. Prestained, broad-range molecular weight markers for SDS-PAGE.
8. Sample Buffer (5×).

2.5. Immunoblot Analysis for Phospho-cJun (Ser 73)

1. Transfer buffer: 700 mL deionized water, 200 mL methanol, and 100 mL of 10× Electrophoresis Buffer; store at room temperature.
2. Tris-buffered saline (TBS) (20×): 48.4 g Tris base and 327 g NaCl, bring to 2 L with deionized water, adjust pH to 7.4.
3. TBS with Tween-20 (TTBS): 200 mL 20× TBS and 4 mL Tween-20 added to 3.8 L deionized water.
4. NitroPure (GE Water and Process Tech., Watertown, MA) supported nitrocellulose membrane (0.45 μm) and 3MM chromatography paper.
5. Blocking/antibody dilution buffer: 3% (w/v) bovine serum albumin (True Cohn Fraction V; ICN Biomedicals, Inc., Aurora, OH) dissolved in TTBS.
6. Anti-phospho-cJun (Ser73) rabbit polyclonal antibody (Cell Signaling Technology, Beverly, MA).
7. Goat anti-rabbit alkaline phosphatase-conjugated antibody (Santa Cruz Biotechnology, Santa Cruz, CA).
8. Lumi-Phos WB (Pierce Biotechnology, Inc., Rockford, IL).

2.6. Stripping and Reprobing Blots for Total cJun Protein

1. Stripping Buffer: 6.25 mL of 1 M Tris HCl, pH 7.6, 780 μL β-mercaptoethanol, and 2 g SDS, brought to a total volume of 100 mL with deionized water.
2. TTBS.
3. Blocking/antibody dilution buffer: 3% (w/v) nonfat dry milk dissolved in TTBS.
4. Anti-cJun rabbit polyclonal antibody (H-79; Santa Cruz Biotechnology).

3. Methods

3.1. Isolation of GST-cJun (1–79)

1. Grow a 5-mL overnight culture of *Escherichia coli* bacteria transformed with pGEX-cJun (1–79) *(13)* in LB medium containing 100 μg/mL ampicillin.
2. Inoculate 500 mL of LB containing ampicillin (100 μg/mL) with the 5-mL overnight culture and grow at 37°C with vigorous shaking until the OD_{600} is ~0.6 (4–5 h).

3. Add IPTG to 0.4 mM final (2 mL of a 100 mM stock) and continue incubation at 37°C with shaking for 3 h.
4. Collect bacteria by centrifugation at 5,000×*g*, decant supernatant, and freeze pellets at -20°C.
5. Thaw pelleted *E. coli* on ice and resuspend in 10 mL of ice-cold NETN lysis buffer containing 200 μL of freshly prepared 1 mg/mL lysozyme.
6. Incubate in a Falcon 2059 tube (Becton Dickinson Labware, Franklin Lakes, NJ) on ice for 1 h.
7. Sonicate with a probe sonicator for 2 min to shear bacterial DNA and centrifuge at 8,000×*g* for 10 min in a refrigerated centrifuge.
8. Transfer supernatant to 15-mL screwtop tube and add 1.5 mL of 50% GSH–agarose in NETN buffer.
9. Rock at 4°C for 2 h, then wash three times by repetitive centrifugation (5 min, 1,000×*g*) with 10 mL NETN buffer per wash.
10. Suspend beads to a 10% suspension (v/v) in MKLB and store at 4°C. A representative preparation of GST-cJun (1–79) beads resolved on a 10% acrylamide SDS gel and stained with Coomassie is shown in **Fig. 14.1a** (*see* **Note 1**).

3.2. Preparation of Cell Extracts for Analysis of JNK Activity and cJun Phosphorylation

1. Rinse confluent cultures of H2122 NSCLC cells stably transfected with empty LPCX or expressing Fzd9 *(12)* in 10-cm petri dishes on ice with 3 mL PBS. Decant residual PBS, then lyse the attached cells with 0.5 mL MKLB and rapidly collect the cells with a plastic scraper into 1.5-mL microcentrifuge tubes.
2. Centrifuge the extracts in a refrigerated microcentrifuge at 12,000×*g* for 5 min.
3. Transfer the supernatant to new 1.5-mL microcentrifuge tube. Discard the pelleted material.
4. Assay aliquots (10 μL) of the extracts for protein concentration with the Bradford protein assay or similar assay.

3.3. Preparation of 10% Polyacrylamide SDS Gels

1. Prepare a 10% separating gel by mixing 10 mL of 4× separating buffer, 13.3 mL acrylamide/bis solution, 16.5 mL distilled water, 160 μL of 10% APS, and 20 μL TEMED. Pipette a sufficient volume of this solution into a gel-casting device to generate an 11-cm separating gel. Overlay with 100% ethanol.
2. When the separating gel has polymerized (~30 min), decant the ethanol and prepare a 4% stacking gel by mixing 2.5 mL of 4× stacking buffer, 1.3 mL acrylamide/bis solution, 6.2 mL distilled water, 80 μL of 10% APS, and 10 μL TEMED. Pipette

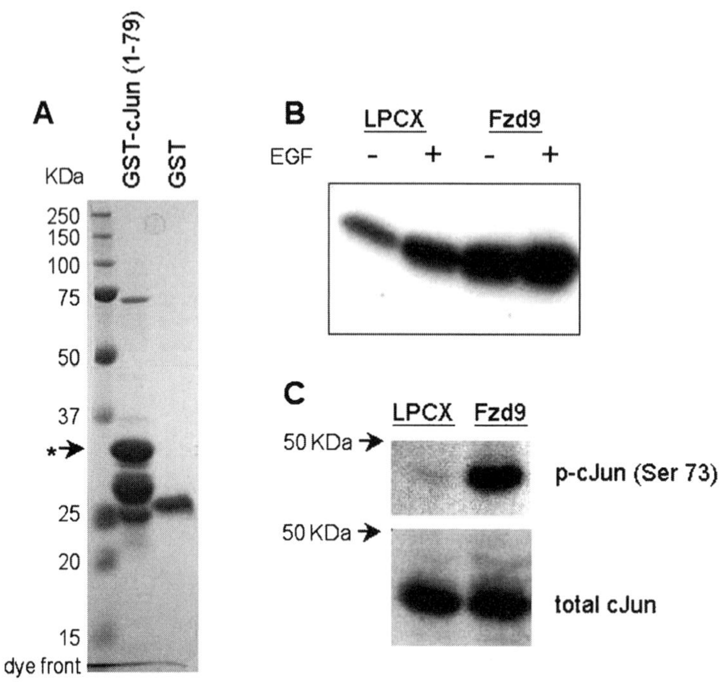

Fig. 14.1. A representative GST-cJun (1–79) preparation, JNK assay, and phospho-cJun (Ser73) immunoblot. A 10 μL of a typical GST-cJun (1–79) preparation was resolved by SDS-PAGE and stained with Coomassie R-250. A preparation of GST alone is shown for reference. The asterisk indicates the full-length GST-cJun (1–79) polypeptide that becomes phosphorylated in the JNK assay. B H2122-LPCX and H2122-Fzd9 cells were incubated for 10 min with or without 10 ng/mL epidermal growth factor (EGF) and extracts were prepared and assayed for JNK activity. The autoradiograph of the resulting experiment is shown. The excised bands were submitted to scintillation counting and yielded 13,734 cpm, 33,246 cpm, 64,828 cpm, and 87,347 cpm for H2122-LPCX, H2122-LPCX + EGF, H2122-Fzd9, and H2122-Fzd9 + EGF, respectively; the assay blank (see Section 3.4, step 1) was 288 cpm. C Cell extracts were prepared from confluent dishes of H2122-LPCX and H2122-Fzd9 cells and submitted to immunoblot analysis of phospho-cJun (Ser73). The filter was stripped and reprobed for total cJun.

this solution onto the pre-poured separating gel with inserted comb and allow the solution to polymerize for 30 min.

3.4. Assay of JNK Activity

1. In 1.5-mL microfuge tubes, bring portions of the cell extracts containing 200 μg protein to a total volume of 400 μL with MKLB and then add 100-μL aliquots of a 10% suspension of GST-cJun (1–79) immobilized to GSH–agarose (*see* **Note 2**). Incubate the tubes for 2 h at 4°C with rocking or rotation. For estimation of the assay blank, incubate 100 μL of GST-cJun (1–79) beads with 400 μL of MKLB and process as described in steps 2–5.
2. Wash the GST-cJun (1–79) beads with bound JNK proteins three times by repetitive centrifugation (3 min; 1,000×*g*) with MKLB in a refrigerated microcentrifuge.
3. Following aspiration of the final wash, suspend the GST-cJun (1–79)/JNK complexes in 40 μL of JNK assay buffer

and incubate for 20 min at 30°C. Mix the assay components by frequent vortexing (*see* **Note 3**).

4. The reactions are terminated by addition of 10 μL of 5× sample buffer, boiled and microcentrifuged for 10 sec at 1,000×*g*.
5. Load the supernatants (50 μL) (*see* **Note 4**) into wells of a 10% polyacrylamide gel and submit to electrophoresis. When the bromphenol blue dye front is 2 to 3 cm from the bottom of the gel, terminate electrophoresis, remove the gel from the glass plates, and trim the bottom portion of the gel, including the dye front, with a razor blade and discard in a suitable radioactive waste container (*see* **Note 5**).
6. Incubate the remaining upper portion of the SDS polyacrylamide gel at room temperature for 5 min in Coomassie stain solution and then destain by three to four washes (~30 min per wash) in 200 mL destain solution.
7. When the GST-cJun (1–79) polypeptides (~35 kDa) can be easily identified in the destained gel, position the gel on Whatmann 3MM blotting paper and dry with a vacuum gel drier.
8. Wrap the Whatman paper with the dried gel in a single layer of plastic wrap and submit to autoradiography with enhancer screens for 1 to 16 h, depending on the level of JNK activity. A representative autoradiograph of a JNK assay is shown in **Fig. 14.1b**.
9. To quantitate the ^{32}P-GST-cJun (1–79) in the dried gel, excise the uppermost GST-cJun (1–79) band (including 5 mm of the gel above the band) with scissors and count with commercial scintillation cocktails in a scintillation counter (*see* **Note 6**).

3.5. Immunoblot Analysis for Phospho-cJun (Ser 73)

1. In 1.5-mL microfuge tubes, bring portions of the extracts containing 100 μg protein to a total volume of 100 μL with MKLB and add 25 μL of 5× sample buffer.
2. Boil the samples for 3 min, cool to room temperature, and load into wells of a 10% polyacrylamide SDS gel. Load prestained markers into an adjacent well and load unused wells with equal volumes of 1× sample buffer.
3. Assemble the loaded polyacrylamide gel into the electrophoresis device and fill the upper and lower chambers with 1× electrophoresis buffer. Connect the assembly to a power supply and run at constant current (~7 mA per gel) for 16 h or until the bromphenol blue dye front runs off the bottom of the gel.
4. Transfer the resolved proteins in the SDS gel to NitroPure membranes (400 mA, 2 h) using a tank transfer apparatus with transfer buffer.
5. Seal the nitrocellulose membrane with transferred proteins in a plastic pouch with 10 mL of blocking/antibody

dilution buffer (*see* **Note 7**) and rock for 1 h at room temperature.

6. Open the pouch at a corner and pipette 10 μL of anti-phospho-cJun (Ser73) rabbit polyclonal antibody into the blocking solution. Incubate overnight at 4°C with constant rocking.
7. Remove the nitrocellulose membrane from the pouch and wash the membrane (200 mL/wash) four times in TTBS (5 min per wash) at room temperature.
8. Seal the filter in a new pouch with 10 mL blocking/antibody dilution buffer containing 10 μL of goat anti-rabbit alkaline phosphatase-conjugated antibody and rock for 1 h at room temperature.
9. Wash the membrane in TTBS as in **Section 3.5**, step 7 for 5 min per wash at room temperature.
10. Incubate the washed nitrocellulose membrane for 3 min in 10 mL Lumi-Phos WB, wrap in plastic wrap, and submit the membrane to autoradiography with Kodak Biomax MR film, using exposure times of 30 sec to 5 min.
11. Overlay the developed films on the membrane and note the position of the molecular weight markers on the film with a pen or marker. A representative phospho-cJun (Ser73) immunoblot is shown in **Fig. 14.1c**.
12. Remove the nitrocellulose membrane from the plastic wrap and place it in 200 mL TTBS prior to stripping and reprobing.

3.6. Stripping and Reprobing Blots for Total cJun Protein

1. Incubate the nitrocellulose membrane in 100 mL of stripping buffer at 50°C for 30 min within a hybridization bottle rotated in a hybridization oven.
2. Wash the membrane extensively in 200-mL washes of TTBS (15 min per wash).
3. Seal the membrane in a plastic pouch with 10 mL blocking/antibody dilution buffer (3% nonfat dry milk in TTBS) and incubate for 1 h at room temperature with rocking.
4. Open the pouch, add 50 μL anti-cJun rabbit polyclonal antibody, reseal the pouch, and continue incubation for 16 h with rocking at 4°C.
5. Following four washes with TTBS, incubate the membrane in 10 mL blocking/antibody dilution buffer containing 10 μL goat anti-rabbit alkaline phosphatase-conjugated antibody for 1 h at room temperature.
6. Wash the membrane with four changes of TTBS, incubated for 3 min with 10 mL of Lumi-Phos WB and submit to autoradiography as described in **Section 3.5**.

4. Notes

1. A typical GST-cJun (1–79) preparation resolved on a 10% polyacrylamide SDS gel and stained with Coomassie R-250 is shown in Fig. 14.1a. The upper band resolving at 35 KDa (indicated by an asterisk) represents the full-length GST-cJun (1–79) species and is the only recombinant protein that becomes phosphorylated in the JNK assay.
2. To ensure accurate delivery of the GST-cJun (1–79) beads into the JNK binding assay, snip the ends of pipette tips to prevent plugging by the beads.
3. The GSH–agarose beads with immobilized GST-cJun (1–79) protein rapidly settle during the kinase assay. Thus, mix by vortexing briefly (2–3 sec) every 3 min.
4. While we avoid intentional transfer of GSH–agarose beads, there is no apparent problem with aspirating and transferring a small amount of the beads.
5. Only a small fraction of ^{32}P becomes incorporated into GST-cJun (1–79) and the bulk of the radioactivity remains associated with [γ-^{32}P]ATP, which resolves at or near the dye front following SDS-PAGE. Contamination of the electrophoresis buffer is minimized if electrophoresis is terminated when the dye front is 2–3 cm from the bottom of the gel, and greater than 95% of the radioactive waste can be isolated in the trimmed gel and easily collected in a suitable radioactive waste container.
6. The GST-cJun (1–79) polypeptides exhibit retarded mobility in SDS-PAGE relative to nonphosphorylated cJun. Thus, cut at least 5 mm above the GST-cJun (1–79) species indicated with an asterisk in **Fig. 14.1a** to ensure isolation of the ^{32}P-GST-cJun (1–79) protein.
7. While Cell Signaling Technology, Inc. recommends the use of nonfat dry milk in blocking and antibody dilution buffers for the anti-phospho-cJun (Ser 73) antibody, we observe greatly increased sensitivity in buffers containing 3% Cohn-crystallized bovine serum albumin.

Acknowledgments

This work was supported by the National Institutes of Health (NIH) grants RO1 CA116527, PO1 CA58187, and K22 CA113700, and a VA Merit Award to RAW.

References

1. Bogoyevitch, M.A., Boehm, I., Oakley, A., et al. (2004) Targeting the JNK MAPK cascade for inhibition: basic science and therapeutic potential. *Biochim Biophys Acta* **1697**, 89–101.
2. Xia, Y. and Karin, M. (2004) The control of cell motility and epithelial morphogenesis by Jun kinases. *Trends Cell Biol* **14**, 94–101.
3. Kennedy, N.J. and Davis, R.J. (2003) Role of JNK in tumor development. *Cell Cycle* **2**, 199–201.
4. Heasley, L.E. and Han, S.Y. (2006) JNK regulation of oncogenesis. *Mol Cells* **21**, 167–173.
5. Dérijard, B., Hibi, M., Wu, I.-H., Barrett, T., Su, B., Deng, T., Karin, M., and Davis, R.J. (1994) JNK1: a protein kinase stimulated by UV light and Ha-Ras that binds and phosphorylates the c-Jun activation domain. *Cell* **76**, 1025–1037.
6. Kyriakis, J.M., Banerjee, P., Nikolakaki, E., Dai, T., Rubie, E.A., Ahmad, M.F., Avruch, J., and Woodgett, J.R. (1994) The stress-activated protein kinase subfamily of c-Jun kinases. *Nature* **369**, 156–160.
7. May, G.H., Allen, K.E, Clark, W., et al. (1998) Analysis of the interaction between c-Jun and c-Jun N-terminal kinase in vivo. *J Biol Chem* **273**, 33429–33435.
8. Hibi, M., Lin, A., Smeal, T., Minden, A., and Karin, M. (1993) Identification of an oncoprotein- and UV-responsive protein kinase that binds and potentiates the c-Jun activation domain. *Genes Dev* **7**, 2135–2148.
9. Boutros, M., Paricio, N., Strutt, D.I., and Mlodzik, M. (1998) Dishevelled activates JNK and discriminates between JNK pathways in planar polarity and wingless signaling. *Cell* **94**, 109–118.
10. Yamanaka, H., Moriguchi, T., Masuyama, N., et al. (2002) JNK functions in the non-canonical Wnt pathway to regulate convergent extension movements in vertebrates. *EMBO Rep* **3**, 69–75.
11. Wallingford, J.B., Fraser, S.E., and Harland, R.M. (2002) Convergent extension: the molecular control of polarized cell movement during embryonic development. *Dev Cell* **2**, 695–706.
12. Winn, R.A., Marek, L., Han, S.Y., et al. (2005) Restoration of Wnt-7a expression reverses non-small cell lung cancer cellular transformation through frizzled-9-mediated growth inhibition and promotion of cell differentiation. *J Biol Chem* **280**, 19625–19634.
13. Butterfield, L., Storey, B., Maas, L., and Heasley, L.E. (1997) cJun NH_2-terminal kinase-regulation of the apoptotic response of small cell lung cancer cells to ultraviolet radiation. *J Biol Chem* **272**, 10110–10116.

Chapter 15

ROCK Enzymatic Assay

John D. Doran and Marc D. Jacobs

Abstract

We describe the protocols for measuring Rho-associated coiled-coil-containing kinase (ROCK) activity *in vitro*. A His-tagged, constitutively active form of the protein (lacking C-terminal inhibitory domains) is expressed in baculovirus. The protein is purified by a combination of metal affinity, ion exchange, and size exclusion chromatography. Enzymatic activity is measured spectrophotometrically in a coupled assay format wherein a molecule of NADH is oxidized to NAD+ each time a phosphate is transferred by ROCK.

Key words: Kinase, Phosphorylation, Coupled-assay, Baculovirus, Dimer, Rho, ROCK.

1. Introduction

One role of Wnt is to establish long-range cellular polarity within the plane of a tissue (planar cell polarity [PCP]) *(1, 2)*. This higher-order organization is regulated in part through the Wnt non-canonical signaling pathway, which includes RhoA and Rho-associated coiled-coil-containing kinase (ROCK) *(3, 4)*. Herein we present protocols for the production of constitutively active ROCK and for assaying ROCK activity *in vitro*.

1.1. ROCK Protein

ROCK is a serine/threonine kinase that acts as an effector of Rho-dependent signaling. It is involved in cell contractility, motility, and cell adhesion, smooth muscle contraction, and actin-cytoskeleton organization *(5)*. ROCK is a potential therapeutic target in cardiovascular indications such as hypertension *(6–9)*, pulmonary indications including asthma *(10)*, and immunosup-

Elizabeth Vincan (ed.), *Wnt Signaling, Volume I: Pathway Methods and Mammalian Models, vol. 468*

Book doi: 10.1007/978-1-59745-249-6

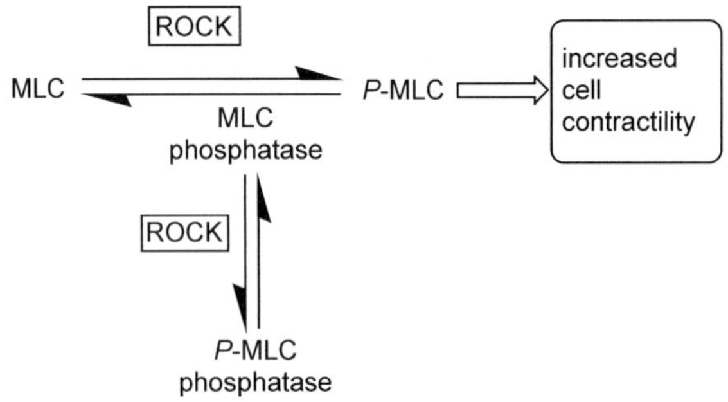

Fig. 15.1. ROCK regulates the activity of muscle myosin regulatory light chain (*MLC*) proteins both by direct phosphorylation, and also by phosphorylation and inhibition of the myosin phosphatase. This leads to an increase in the amount of phosphorylated MLC, upregulating cell contractility and mobility.

pression and inflammation *(11, 12)*. Substrates of ROCK include the LIM kinases, myosin regulatory light chain (MLC) and MLC phosphatase (MCLP) *(5)*. Increased actomyosin contractility in nonmuscle cells results from elevated phosphorylation of MLC, either directly by ROCK or indirectly through inhibition of MLCP *(13–15)* (Fig. 15.1).

There are two isoforms of ROCK, known as ROCK I and II, or Rho-kinase β and α, respectively *(16)*. ROCK I is composed of a dimerization domain (residues 1–72), a catalytic kinase domain (residues 73–405), a coiled-coil region (residues 425–1100), and a C-terminal pleckstrin homology (PH) domain (residues 1103–1230) *(17)*. This domain organization is shared with three closely related kinases: myotonic dystrophy kinase (DMPK) *(18)*, myotonic dystrophy kinase-related Cdc42-binding kinase (MRCK) *(19)*, and citron kinase *(20)*.

1.2. ROCK Activity

Full-length ROCK, in the absence of RhoA-GTP, exists as an autoinhibited structure. The C-terminal region of ROCK (coiled-coil and PH domains) has been shown to partially inhibit the kinase catalytic activity by binding directly to the kinase domain *(21)*. When GTP-bound RhoA binds to the Rho-binding region of the coiled-coil domain, the interaction with the catalytic kinase domain is disrupted, partially increasing kinase activity *(22)*. Cleavage of the C-terminal inhibitory domain from the catalytic domain by caspase-3 also activates the enzyme *(23, 24)*.

The effect of RhoA binding on enzymatic activity is modest. The addition of RhoA/GTP to full-length ROCK in vitro decreases the K_m for MLC and increases K_{cat} approximately two-fold *(25–27)*. For enzymatic assays, a constitutively active, truncated form of ROCK was produced that lacks the C-terminal autoinhibitory domains.

Phosphorylation of the activation loop does not appear to be necessary for ROCK activity. In many kinases, phosphorylation

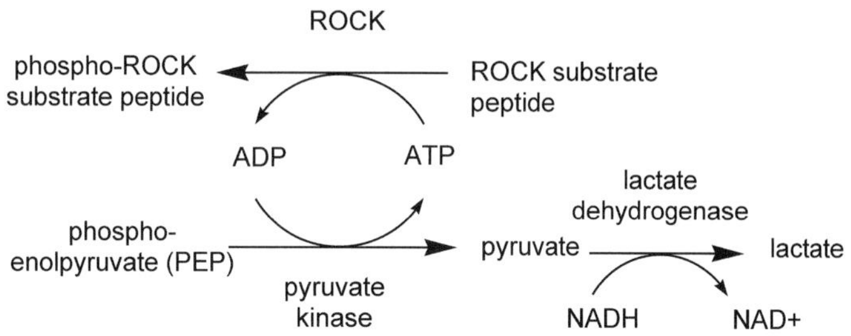

Fig. 15.2. ROCK enzymatic activity is measured using a coupled assay. As ROCK transfers a phosphate from ATP to the protein substrate, pyruvate kinase regenerates the ATP and converts phosphoenolpyruvate to pyruvate. As pyruvate is converted to lactate by lactate dehydrogenase, the change in NADH concentration is monitored spectrophotometrically.

of a residue in the activation loop stabilizes a conformation that is compatible with substrate binding (reviewed in ref. *(28)*). However, the crystal structures of both ROCK I and II reveal that the unphosphorylated protein exists in an active conformation *(17, 29)*. Therefore, for the assays described herein, the protein is not activated by phosphorylation prior to use.

1.3. Coupled Assay

Protein kinase activity can be assayed either by monitoring the conversion of ATP to ADP or by the accumulation of phosphorylated substrate. Here we describe a regenerative assay that couples ATP dephosphorylation with NADH oxidation *(30–32)* (**Fig. 15.2**). Each time phosphoryl transfer from an ATP molecule is catalyzed by the kinase, the product ADP is subsequently phosphorylated by pyruvate kinase using phosphoenolpyruvate (PEP) as a substrate. The pyruvate is then converted to lactate by lactate dehydrogenase (LDH) using NADH. NADH has a high optical absorbance at 340 nm, while NAD+ does not, providing a convenient and sensitive spectrophotometric signal. Alternatively, an assay measuring the transfer of radiolabeled phosphate from ATP to the substrate may also be used *(33, 34)*.

2. Materials

2.1. ROCK Expression

1. ROCK I expression vector: ROCK I was isolated from a human leukocyte complementary DNA (cDNA) library *(25)*. A truncated version of ROCK (residues 6–553) containing an N-terminal hexa-histidine tag was cloned into a baculoviral transfer vector, pBEV10, and expressed in insect cells *(35, 36)*.
2. Insect cell lines: *Trichoplusia ni* (High-5) and *Spodoptera frugiperda* (*Sf*12 and *Sf*9) were used for the expression of recombinant

protein. The latter was also used for the generation of recombinant viral stocks. *Sf*9 and *Sf*21 cells were obtained from American Type Culture Collection (ATCC, Manassas, VA). High-5 cells are available from Invitrogen, Carlsbad, CA.

3. EX-Cell 405 media with glutamine (JRH Bioscience, Lenexa, KS) was used for expression in High-5.
4. ESF921 media (Expression Systems, Woodland, CA) was used for expression in *Sf*9 and *Sf*21 cells.
5. Wave Bioreactor System 20/50EH (Wave Biotech, Bridgewater, NJ).
6. CedexAS20 analysis system (Innovartis, Bielefeld, Germany).

2.2. ROCK Protein Purification

2.2.1. Chromatography and Concentration

1. His-Select nickel affinity resin (Sigma-Aldrich, St. Louis, MO).
2. 50-mL Amicon stirred ultrafiltration cell (Millipore, Billerica, MA).
3. Amicon YM-10 Ultracel membrane (10-kDa molecular weight cutoff).
4. Mono-Q HR 5/5 or 10/10 anion-exchange column (GE Healthcare, Piscataway, NJ).
5. Sephadex-200 16/60 size-exclusion chromatography column (GE Healthcare).
6. Tris-glycine 4–20% gradient polyacrylamide gels (Invitrogen).

2.2.2. Buffers

1. Lysis buffer: 50 mM HEPES, 300 mM NaCl, 10% glycerol, 5 mM 2-mercaptoethanol, and 5 mM imidazole (final pH 7.8).
2. Nickel elution buffer: Lysis buffer + 500 mM imidazole (final pH 7.8).
3. Mono-Q buffers: Buffer A: 20 mM HEPES, pH 7.8; and Buffer B: 20 mM HEPES and 1 M NaCl (final pH 7.8).
4. Size-exclusion buffer: 50 mM HEPES, 300 mM NaCl, 10% glycerol, 5 mM 2-mercaptoethanol, 5 mM imidazole (final pH 7.8).

2.3. ROCK Enzymatic Assay

1. Phosphoenolpyruvate (PEP) (Roche Diagnostics, Basel, Switzerland): 100 mM stock in 100 mM HEPES, pH 7.6.
2. NADH (Roche Diagnostics): 20 mM stock in 100 mM HEPES, pH 7.6.
3. Lactate dehydrogenase (LDH) (Roche Diagnostics).
4. Pyruvate kinase (PK) (Roche Diagnostics).
5. ATP (Sigma-Aldrich): 10 mM stock in 100 mM HEPES, pH 7.6.
6. Dithiothreitol (DTT): 1.0 M stock in water.

7. 1 M HEPES, pH 7.6.
8. 1 M $MgCl_2$.
9. ROCK Peptide Substrate: KKRNRTLSV (American Peptide Company, Sunnyvale, CA) 10 mM stock in 100 mM HEPES, pH 7.0.
10. 96-Well flat-bottom assay plate.
11. SpectraMax spectrophotometer (Molecular Devices, Sunnyvale, CA).

3. Methods

3.1. ROCK Baculovirus Expression

1. Grow insect cells used for expression to a density of 2×10^6 cells/mL in a Fernbach flask. Transfer 1 L of insect cells directly into a Cellbag 20 (WaveBiotech).
2. Dilute cells to a density of 3.5×10^5 cells/mL with media (~6–7.5 L total volume). Grow at 27°C for 72 hours with a rocking speed of 16 per minute and infused with 0.1 L/min of air.
3. Adjust cell density to 2×10^6 cells/mL with fresh media (final volume, 6–10 L). Infect insect cells at a multiplicity of infection (MOI) of 2.5 with clonal, high-titer baculovirus.
4. Incubate insect cells at 27°C with a rocking speed of 20–22.5 per minute and infused with 0.15–0.2 L/min of air.
5. Harvest insect cells after a further 72 hours growth, usually at 65–75% viability as measured using a Cedex AS20. Cells are harvested by centrifugation at $6{,}000\times g$ for 10 min, prior to flash freezing and storage at –70°C.

3.2. ROCK Protein Purification

3.2.1. Lysis and Nickel Affinity Purification (See Note 1)

1. Add 500 mL of lysis buffer for every 50 g ROCK insect cell paste.
2. Stir at room temperature until completely dispersed, typically 30 min.
3. Mechanically disrupt cells by passing once through a microfluidizer (Microfluidics Corp., Newton, MA).
4. Centrifuge the cell lysate at 54,000×g for 1 hour at 4°C.
5. Incubate the supernatant with nickel resin (preequilibrated with lysis buffer) overnight at 4°C. Use 1 mL of resin for every 50 g of cell paste.
6. Wash the resin with 200 mL of lysis buffer, and elute protein with 4–5 column volumes of nickel elution buffer.
7. Analyze eluant with sodium dodecyl sulfate (SDS)-polyacrylamide gel electrophoresis (PAGE).

3.2.2. Anion-Exchange Chromatography

1. Dilute the eluant 10-fold using Buffer A. The sample is then concentrated to approximately 50 mL using an Amicon stirred-cell concentrator.
2. Load the sample onto the Mono-Q column and wash with 10 column volumes of Buffer A.
3. Elute the protein with a 10–40% Buffer B gradient over 20 column volumes at room temperature.
4. Analyze fractions by SDS-PAGE.

3.2.3. Size-Exclusion Chromatography

1. Concentrate the protein sample to 2–4 mL using an Amicon stirred-cell concentrator. Since the solubility of ROCK protein is high, concentrations at or above 20 mg/mL may be used.
2. Load protein onto a Superdex-200 16/60 column equilibrated with size-exclusion buffer.
3. Analyze fractions by SDS-PAGE.
4. Pool desired fractions and concentrate to 1–2 mg/mL. The protein concentration may be determined spectrophotometrically using an extinction coefficient of 66,710 mol^{-1} cm^{-1} at 280 nm *(37)*.

3.3. ROCK Enzymatic Assay (See Note 2)

3.3.1. Preparation of Reaction Mix

A reaction mix is made with all components of the coupled assay except for ATP, which is added to initiate the assay. The final volume of each reaction will be 100 μL, of which 10 μL will be the ATP stock solution. The reagents should be mixed such that they are at the final concentrations listed below after the addition of ATP:

0.1 M HEPES, pH 7.6

10 mM $MgCl_2$

2.5 mM PEP

0.2 mM NADH

0.03 mg/mL PK

0.01 mg/mL LDH

2 mM DTT

100 nM ROCK protein

100 μM ROCK peptide

3.3.2. Initiation of the Enzyme Assay

Ninety microliters of the reaction mix is added to each well in a 96-well plate. The reaction is then initiated by the addition of 10 μL ATP, giving a final volume of 100 μL. Mix the reagents well using a plate shaker for 1 min. The ROCK K_m for ATP is 30 μM when using the ROCK peptide, so a final ATP concentration of 300 μM yields a high signal-to-noise ratio.

3.3.3. Monitoring the Reaction

The progress of the reaction is monitored (using the SpectraMax spectrophotometer) as a function of time by the decrease in absorbance at 340 nm due to the oxidation of NADH. During the reaction, the plate temperature is maintained at 25°C. Typically, data points can be collected at 10-second intervals for a total of 5 min. Data can be analyzed using the method of initial rates or by nonlinear fitting of the progress curve *(38)*.

4. Notes

1. Early in the protein purification, the cell lysis and the nickel resin binding should be performed at 4°C or on ice. However, the nickel column elution and subsequent steps can be performed at room temperature if desired, due to the high stability of the ROCK protein.
2. For the coupled reaction, the NADH and PEP stock solutions should be prepared just prior to use. The DTT and ATP stock solutions can be stored at –20°C and thawed when needed.

For more accurate rate measurements, background control reaction rates measured in the absence of ROCK enzyme should be determined.

Full-length ROCK protein can also be used in this assay providing that RhoA and GTP-γS are included *(25)*. We have also used MLC protein and several other peptides as ROCK substrates *(25, 38)*.

Acknowledgments

We thank Stephen Chambers, Douglas Austen, Scott Raybuck, and Edward Fox for helpful discussions and critical comments on this manuscript.

References

1. Jenny, A. and Mlodzik, M. (2006) Planar cell polarity signaling: a common mechanism for cellular polarization. *Mt Sinai J Med* 73, 738–750.
2. Fanto, M. and McNeill, H. (2004) Planar polarity from flies to vertebrates. *J Cell Sci* **117**, 527–533.
3. Masckauchan, T. N. and Kitajewski, J. (2006) Wnt/Frizzled signaling in the vasculature: new angiogenic factors in sight. *Physiology (Bethesda)* **21**, 181–188.
4. Huelsken, J. and Behrens, J. (2002) The Wnt signalling pathway. *J Cell Sci* **115**, 3977–3978.

5. Riento, K. and Ridley, A. J. (2003) Rocks: multifunctional kinases in cell behaviour. *Nat Rev Mol Cell Biol* **4**, 446–456.
6. Loirand, G., Guerin, P., and Pacaud, P. (2006) Rho kinases in cardiovascular physiology and pathophysiology. *Circ Res* **98**, 322–334.
7. Noma, K., Oyama, N., and Liao, J. K. (2006) Physiological role of ROCKs in the cardiovascular system. *Am J Physiol Cell Physiol* **290**, C661–C668.
8. Mukai, Y., Shimokawa, H., Matoba, T., Kandabashi, T., Satoh, S., Hiroki, J., Kaibuchi, K., and Takeshita, A. (2001) Involvement of Rho-kinase in hypertensive vascular disease: a novel therapeutic target in hypertension. *FASEB J.* **15**, 1062–1064.
9. Shimokawa, H. (2002) Rho-kinase as a novel therapeutic target in treatment of cardiovascular diseases. *J Cardiovasc Pharmacol* **39**, 319–327.
10. Gosens, R., Schaafsma, D., Nelemans, S. A., and Halayko, A. J. (2006) Rho-kinase as a drug target for the treatment of airway hyperrespon-siveness in asthma. *Mini Rev Med Chem* **6**, 339–348.
11. Ohki, S., Iizuka, K., Ishikawa, S., Kano, M., Dobashi, K., Yoshii, A., Shimizu, Y., Mori, M., and Morishita, Y. (2001) A highly selective inhibitor of Rho-associated coiled-coil forming protein kinase, Y-27632, prolongs cardiac allograft survival of the BALB/c-to-C3H/He mouse model. *J Heart Lung Transplant* **20**, 956–963.
12. Zhao, D. and Pothoulakis, C. (2003) Rho GTPases as therapeutic targets for the treatment of inflammatory diseases. *Expert Opin Ther Targets* **7**, 583–592.
13. Kureishi, Y., Kobayashi, S., Amano, M., Kimura, K., Kanaide, H., Nakano, T., Kaibuchi, K., and Ito, M. (1997) Rho-associated kinase directly induces smooth muscle contraction through myosin light chain phosphorylation. *J Biol Chem* **272**, 12257–12260.
14. Amano, M., Ito, M., Kimura, K., Fukata, Y., Chihara, K., Nakano, T., Matsuura, Y., and Kaibuchi, K. (1996) Phosphorylation and activation of myosin by Rho-associated kinase (Rho-kinase). *J Biol Chem* **271**, 20246–20249.
15. Kawano, Y., Fukata, Y., Oshiro, N., Amano, M., Nakamura, T., Ito, M., Matsumura, F., Inagaki, M., and Kaibuchi, K. (1999) Phosphorylation of myosin-binding subunit (MBS) of myosin phosphatase by Rho-kinase in vivo. *J Cell Biol* **147**, 1023–1038.
16. Leung, T., Chen, X. Q., Manser, E., and Lim, L. (1996) The p160 RhoA-binding kinase ROK alpha is a member of a kinase family and is involved in the reorganization of the cytoskeleton. *Mol Cell Biol* **16**, 5313–5327.
17. Jacobs, M., Hayakawa, K., Swenson, L., Bellon, S., Fleming, M., Taslimi, P., and Doran, J. (2006) The structure of dimeric ROCK I reveals the mechanism for ligand selectivity. *J Biol Chem* **281**, 260–268.
18. Bush, E. W., Helmke, S. M., Birnbaum, R. A., and Perryman, M. B. (2000) Myotonic dystrophy protein kinase domains mediate localization, oligomerization, novel catalytic activity, and autoinhibition. *Biochemistry (Mosc)* **39**, 8480–8490.
19. Tan, I., Seow, K. T., Lim, L., and Leung, T. (2001) Intermolecular and intramolecular interactions regulate catalytic activity of myotonic dystrophy kinase-related Cdc42-binding kinase alpha. *Mol Cell Biol* **21**, 2767–2778.
20. Di Cunto, F., Calautti, E., Hsiao, J., Ong, L., Topley, G., Turco, E., and Dotto, G. P. (1998) Citron rho-interacting kinase, a novel tissue-specific ser/thr kinase encompassing the Rho-Rac-binding protein Citron. *J Biol Chem* **273**, 29706–29711.
21. Amano, M., Chihara, K., Nakamura, N., Kaneko, T., Matsuura, Y., and Kaibuchi, K. (1999) The COOH terminus of Rho-kinase negatively regulates rho-kinase activity. *J Biol Chem* **274**, 32418–32424.
22. Ishizaki, T., Maekawa, M., Fujisawa, K., Okawa, K., Iwamatsu, A., Fujita, A., Watanabe, N., Saito, Y., Kakizuka, A., Morii, N., and Narumiya, S. (1996) The small GTP-binding protein Rho binds to and activates a 160 kDa Ser/Thr protein kinase homologous to myotonic dystrophy kinase. *EMBO J* **15**, 1885–1893.
23. Coleman, M. L., Sahai, E. A., Yeo, M., Bosch, M., Dewar, A., and Olson, M. F. (2001) Membrane blebbing during apoptosis results from caspase-mediated activation of ROCK I. *Nat Cell Biol* **3**, 339–345.
24. Sebbagh, M., Renvoize, C., Hamelin, J., Riche, N., Bertoglio, J., and Breard, J. (2001) Caspase-3-mediated cleavage of ROCK I induces MLC phosphorylation and apoptotic membrane blebbing. *Nat Cell Biol* **3**, 346–352.
25. Doran, J. D., Liu, X., Taslimi, P., Saadat, A., and Fox, T. (2004) New insights into the structure-function relationships of Rho-associated kinase: a thermodynamic and hydrodynamic study of the dimer-to-monomer transition and its kinetic implications. *Biochem J* **384**, 255–262.

26. Hammond, S. M., Jenco, J. M., Nakashima, S., Cadwallader, K., Gu, Q., Cook, S., Nozawa, Y., Prestwich, G. D., Frohman, M. A., and Morris, A. J. (1997) Characterization of two alternately spliced forms of phospholipase D1. Activation of the purified enzymes by phosphatidylinositol 4,5-bisphosphate, ADP-ribosylation factor, and Rho family monomeric GTP-binding proteins and protein kinase C-alpha. *J Biol Chem* **272**, 3860–3868.
27. Hathaway, D. R. and Haeberle, J. R. (1983) Selective purification of the 20,000-Da light chains of smooth muscle myosin. *Anal Biochem* **135**, 37–43.
28. Nolen, B., Taylor, S., and Ghosh, G. (2004) Regulation of protein kinases; controlling activity through activation segment conformation. *Mol Cell* **15**, 661–675.
29. Yamaguchi, H., Kasa, M., Amano, M., Kaibuchi, K., and Hakoshima, T. (2006) Molecular mechanism for the regulation of rho-kinase by dimerization and its inhibition by fasudil. *Structure* **14**, 589–600.
30. Kiianitsa, K., Solinger, J. A., and Heyer, W. D. (2003) NADH-coupled microplate photometric assay for kinetic studies of ATP-hydrolyzing enzymes with low and high specific activities. *Anal Biochem* **321**, 266–271.
31. Chen, G., Porter, M. D., Bristol, J. R., Fitzgibbon, M. J., and Pazhanisamy, S. (2000) Kinetic mechanism of the p38-alpha MAP kinase: phosphoryl transfer to synthetic peptides. *Biochemistry* **39**, 2079–2087.
32. Fox, T., Coll, J. T., Xie, X., Ford, P. J., Germann, U. A., Porter, M. D., Pazhanisamy, S., Fleming, M. A., Galullo, V., Su, M. S., and Wilson, K. P. (1998) A single amino acid substitution makes ERK2 susceptible to pyridinyl imidazole inhibitors of p38 MAP kinase. *Protein Sci* **7**, 2249–2255.
33. Trauger, J. W., Lin, F. F., Turner, M. S., Stephens, J., and LoGrasso, P. V. (2002) Kinetic mechanism for human Rho-Kinase II (ROCK-II). *Biochemistry* **41**, 8948–8953.
34. Turner, M. S., Fen Fen, L., Trauger, J. W., Stephens, J., and LoGrasso, P. (2002) Characterization and purification of truncated human Rho-kinase II expressed in Sf-21 cells. *Arch Biochem Biophys* **405**, 13–20.
35. Chambers, S. P., Austen, D. A., Fulghum, J. R., and Kim, W. M. (2004) High-throughput screening for soluble recombinant expressed kinases in Escherichia coli and insect cells. *Protein Expr Purif* **36**, 40–47.
36. Chambers, S. P. (2002) High-throughput protein expression for the post-genomic era. *Drug Discov Today* **7**, 759–765.
37. Gasteiger, E., Gattiker, A., Hoogland, C., Ivanyi, I., Appel, R. D., and Bairoch, A. (2003) ExPASy: The proteomics server for in-depth protein knowledge and analysis. *Nucleic Acids Res* **31**, 3784–3788.
38. Futer, O., Saadat, A. R., Doran, J. D., Raybuck, S. A., and Pazhanisamy, S. (2006) Phosphoryl transfer is not rate-limiting for the ROCK I-catalyzed kinase reaction. *Biochemistry* **45**, 7913–7923.

Chapter 16

Detection of Planar Polarity Proteins in Mammalian Cochlea

Mireille Montcouquiol, Jennifer M. Jones, and Nathalie Sans

Abstract

The "core genes" were identified as a group of genes believed to function as a conserved signaling cassette for the specification of planar polarity in *Drosophila Melanogaster*, and includes *frizzled* (*fz*), *van gogh* (*vang*) or *strabismus* (*stbm*), *prickle* (*Pk*), *dishevelled* (*dsh*), *flamingo* (*fmi*), and *diego*. The mutation of each of these genes not only causes the disruption of planar polarity within the wing or the eye of the animal, but also affects the localization of all the other protein members of the core group. These properties emphasize the importance of the interrelations between the proteins of this group. All of these core genes have homologs in vertebrates. Studies in *Danio Rerio* (zebrafish) and *Xenopus laevis* (frog) have uncovered other roles for some of these molecules in gastrulation and neurulation, during which the shape of a given tissue will undergo major transformation through cell movements. A disruption in these processes can lead to severe neural tube defects in diverse organisms, including humans. In fact, a large body of evidence suggests that planar polarity proteins are not involved in one specific cascade but in many different ones and many different mechanisms such as, but not limited to, hair or cilia orientation, asymmetric division, cellular movements, or neuronal migration.

In mice cochleae, mutations in planar polarity genes lead to defects in the orientation of the stereociliary bundles at the apex of each hair cell. This phenotype established the cochlea as one of the clearest examples of planar polarity in mammals. Although significant progress has been made toward understanding the molecular basis required for the development of planar polarity in invertebrates, similar advances in vertebrates are more recent and rely mainly on the identification of a group of mammalian mutants that affect hair cell stereociliary bundle orientation. These include mutation of *vangl2*, *scrb1*, *celsr1*, *PTK-7*, *dvl1-2*, and more recently *fz3* and *fz6* (1). In this chapter, we describe how to use the mammalian cochlea, which represents one of the best systems to study planar polarity in mammals, to identify planar polarity mutants, study protein distribution, do in vitro analysis, and perform Western blots to analyze putative planar polarity proteins.

Key words: Vangl2, Planar polarity, Asymmetry, Culture, Cochlea, Immunofluorescence.

Elizabeth Vincan (ed.), *Wnt Signaling, Volume I: Pathway Methods and Mammalian Models, vol. 468*

Book doi: 10.1007/978-1-59745-249-6

1. Introduction

Before 2003, little was known about planar polarity genes, proteins, or molecular cascades in mammals. However, as early as 1997, a review from Suzan Eaton compared the onset and polarization of the actin-rich single hair at the apex of each *Drosophila* epithelial wing cell with those of the actin-rich hair bundle at the apex of each hair cell in the mammalian inner ear epithelia *(2)*. The comparison clearly emphasized similarities between the two structures and possibly between mechanisms. In 2003, we identified *vangl2* and *scribble1* as the first two genes involved in planar polarity in the mouse, and demonstrated at the same time that the inner ear was one of the clearest systems to study planar polarity in mammals *(3)*. Homozygote mutations for *vangl2* or *scribble1* strongly disrupted the orientation of the hair bundle, which is uniformly polarized in a wild-type animal. Since then, a few other genes have been identified, based on a similar phenotype in the inner ear. In 2006, the group of Jeremy Nathans and our group developed specific antibodies and showed that the asymmetric distribution of proteins such as Frizzled3 and Frizzled6 or Vangl2 exist in mammalian cochleae *(4, 5)*. Even though we still do not fully understand the mechanisms that set up this asymmetry or the reasons for it, it is a hallmark of planar polarity proteins in *Drosophila*, and the conservation of this asymmetry in mammals seems to be an important characteristic of mammalian planar polarity proteins. Conducting studies at the protein level is therefore very important if we are to advance our understanding of the molecular mechanisms of planar polarity in mammals, and the inner ear system is a fantastic tool for this analysis. We describe various techniques that allow the analysis of proteins in the mammalian cochlea.

2. Materials

2.1. Animals

Sprague Dawley rats or CD-1 mice at various gestational days (Janvier, Le Genest St. Isle, France).

2.2. Dissection

1. Kit of Sylgard 184 from Dow Corning (World Precision Instruments, Stevenage, UK; #SYLG184).
2. Kimax glass petri dishes with covers (Fisher Bioblock Scientific, Illkirch, France; #08746A).
3. Charcoal ("Norit A" Activated Charcoal Powder).
4. Forceps (Dumont 5; World Precision Instruments).

5. Minutien pins (World Precision Instruments).
6. Micro dissecting curette (World Precision Instruments).

2.3. Cell Culture

1. MatTek glass-bottom dishes: MatTek Corp. P35G-0-10CM (Ashland, MA).
2. Matrigel, growth factor-reduced, without phenol red (BD Bioscience, Pont de Claix, France).
3. HBSS (10×), without phenol red (Fisher Bioblock Scientific).
4. 1 M HEPES (Fisher Bioblock Scientific).
5. Dulbecco's Modified Eagle's Medium DMEM (1×), liquid (high glucose), with l-glutamine, 4,500 mg/L d-glucose, and 25 mM HEPES, without sodium pyruvate (Invitrogen, Fisher Bioblock Scientific).
6. Certified fetal bovine serum, USA origin (Invitrogen, Fisher Bioblock Scientific).
7. N2 Supplement (100×) (Invitrogen, Fisher Bioblock Scientific).
8. 500-mL, 0.22-µm filter (Fisher Bioblock Scientific).

2.4. Electroporation

1. Maxi prep kit (Qiagen, Courtaboeuf-France) resuspended in water.
2. Electroporator, BTX830 (BTX Instrument Division, Harvard Apparatus, Holliston, MA).
3. BTX Genepaddles (Harvard Apparatus, Holliston, MA).

2.5. Immunofluore scence

1. Microscope cover slips (No. 0) (VWR-International, Fontenay-sous-Bois).
2. Phosphate-buffered saline (PBS): Prepare 10× stock with 1.37 M NaCl, 27 mM KCl, 100 mM Na_2HPO_4, and 18 mM KH_2PO_4 (adjust to pH 7.4 with HCl if necessary) and autoclave before storage at room temperature. Prepare working solution by diluting one part with nine parts water.
3. Stock solution 16% (w/v) paraformaldehyde (PFA; Electron Microscopy Sciences, Hatfield, PA): Prepare a 2% (w/v) solution in PBS fresh for each experiment.
4. Permeabilization solution: 0.5% (v/v) Triton X-100 in PBS.
5. Antibody dilution buffer: 2% (w/v) Normal goat serum in PBS (Sigma-Aldrich, St. Louis, MO).
6. Secondary antibody: Alexa Fluor® 488, 546, or 647 goat anti-rabbit IgG (H+L), Alexa Fluor® 488, 546, or 647 goat anti-mouse (Molecular Probes, Invitrogen, Carlsbad, CA).
7. Alexa Fluor® 488, 546, or 647 phalloidin (Molecular Probes, Invitrogen).

8. Mounting medium: Antifade (Molecular Probes, Invitrogen).
9. OS30 solvent (Dow Corning).

2.6. SDS-Polyacrylamide Gel Electrophoresis (SDS-PAGE) and Western Blotting

2.6.1. Samples

1. Modified Laemmli buffer: 50 mM Tris-HCl, pH 6.8, 4% (w/v) sodium dodecyl sulfate (SDS), 20% (v/v) glycerol, 0.025% (v/v) β-mercaptoethanol, and 0.005% (v/v) bromophenol blue. Store aliquots at –20°C.
2. Sonicator with microprobe (VWR-International).

2.6.2. Gels

1. Separating buffer (4×): 1.5 M Tris-HCl, pH 8.8, and 0.1% (w/v) SDS. Store at 4°C.
2. Stacking buffer (4×): 0.5 M Tris-HCl, pH 6.8, and 0.1% (w/v) SDS. Store at 4°C.
3. 30% acrylamide/bis solution (37.5:1 with 2.6% C) (this is a neurotoxin when unpolymerized and so care should be taken not to receive exposure—*see* **Note 1**) and *N,N,N,N′*-tetramethyl-ethylenediamine (TEMED; Bio-Rad, Hercules, CA).
4. Ammonium persulfate: prepare 40% (w/v) solution in water and store at 4°C for a week at most.
5. Water-saturated isobutanol: Shake equal volumes of water and isobutanol in a glass bottle and allow separation of the two phases, use the top layer. Store at room temperature.
6. Running buffer (1×): 25 mM Tris pH 8.3, 192 mM glycine, and 0.1% (w/v) SDS. Store at room temperature (Bio-Rad).
7. Prestained molecular weight markers: BenchMark Pre-Stained Protein Ladder or SeeBlue Pre-Stained Standard (Invitrogen).

2.6.3. Transfer

1. Transfer buffer (1×): 25 mM Tris (do not adjust pH), 192 mM glycine, and 20% (v/v) methanol. Store at 4°C.
2. Polyvinyl difluoride (PVDF) membrane from Millipore, 0.45-μm pores (Sigma-Aldrich), and Gel Blot Paper from Schleicher & Schuell (Bassel, Germany).
3. Filter paper: Whatman 3MM or, preferably, blotting paper.
4. Tris-buffered saline (TBS) with Tween (TBS-T): Prepare 10× stock with 150 mM NaCl, 100 mM Tris-HCl, pH 7.5, and 0.05% (v/v) Tween-20. Dilute 100 mL with 900 mL distilled water for use.
5. Blocking buffer: 5% (w/v) nonfat dry milk in TBS-T.
6. Primary antibody dilution buffer: TBS-T supplemented with 0.02% (v/v) sodium azide.
7. Secondary antibody: Horseradish peroxidase (HRP)-conjugated anti-mouse IgG (GE Healthcare Europe GmbH, Zurich, Switzerland).

8. Enhanced chemiluminescent (ECL) reagents from Pierce or Amersham and Hyperfilm ECL (GE Healthcare Europe GmbH).
9. Stripping buffer: 62.5 mM Tris-HCl, pH 6.8, and 2% (w/v) SDS. Store at room temperature. Warm to working temperature of 70°C and add 100 mM β-mercaptoethanol.

3. Methods

3.1. Coating Dishes with Matrigel

1. Dilute matrigel by adding 300 μL growth factor-reduced matrigel to 2,700 μL of cold DMEM. Add 200 μL of this mix to the glass bottom of each MatTek dish in order to cover it entirely (*see* **Note 2**).
2. Place the MatTek dishes in a sterile large culture dish (Corning 150×25 mm) and transfer to the tissue culture incubator at 37°C for at least 1 h so that the matrigel can transform into a gel.

3.2. Preparation of Organotypic Cultures of the Organ of Corti

The entire dissection is done in chilled HBSS: 100 mL of 10× HBSS is diluted with 895 mL of ultrapure water, and 5 mL of a solution of 1 M HEPES is added to the mix. The pH is adjusted at approximately 7.2 and sterilized using two 500-mL filter units.

1. Euthanize a pregnant rat or mouse on gestational day 16.5 (rat) or 13.5 (mouse), and collect the litters in cold HBSS. Embryos are individually staged based on the developmental series presented in Kaufman *(6)*. Clean the embryos from placental tissue and quickly separate their head from the body with forceps.
2. Cut open the top of the cranium of each embryo with forceps so that the entire brain tissue can be removed; simply flip the brain over toward the spinal chord. Care should be taken when disconnecting the nerve fibers that connect the central nervous system to the inner ear embedded in the temporal bone. At this stage, the inner ear is easy to see because it is outlined by a vein running along the semicircular canals.
3. Place the forceps between the inner ear and the skull, if possible, along the line of the blood vessel surrounding the semicircular canals, and use the forceps as scissors to delicately disconnect the inner ear from the skull (*see* **Fig. 16.1**).
4. Transfer the inner ear to a sylgard dish filled with chilled HBSS. Sylgard dishes are 60 or 100-mm glass Petri dishes filled with a mix of sylgard and charcoal polymerized according to the manufacturer's instructions. These dishes allow the use

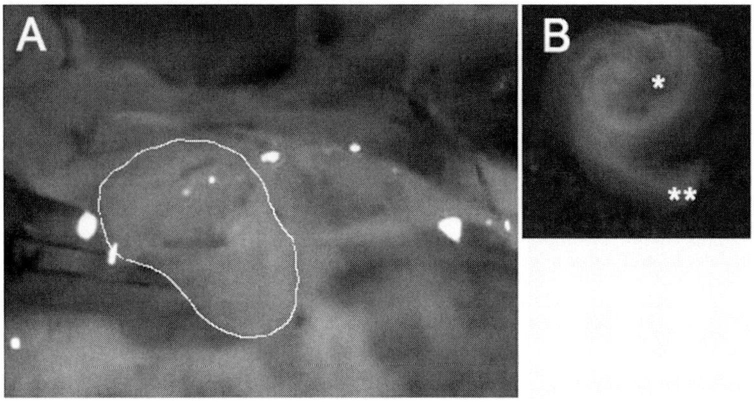

Fig. 16.1. Dissection of mouse inner ear. **A** The inner ear of an E18.5 mouse is carefully detached from the temporal bone (cartilage at this stage). The forceps can be seen on the left of the photo intercalated between the cartilage and the inner ear. The inner ear is outlined by a *white line*. **B** Illustration of the cochlear organ ready to be placed in culture after full dissection. At this late stage, the cochlea forms a spiral of approximately a turn and a half (base indicated with *two asterisks*, apex indicated with *one asterisk*).

of minutia, tiny pins that we use to pin the cochlea for the fine dissection. The inner ear is pinned through the vestibular portion, with the oval window facing up.

5. To remove the cartilage covering the cochlea, chip the cartilage with the forceps beginning close to the oval window. Carefully separate the cochlea from the cartilage with the forceps.
6. To open the cochlear duct, it is easier to pinch the anlagae of the Reissner's membrane at the base of the cochlear duct—where the duct is larger—with the forceps, and to pull it, exposing the developing sensory epithelium.
7. The cochlea can now be disconnected from the vestibular portion of the inner ear with the forceps, and the last step consists of cleaning the sensory epithelium from the underlying connective tissue, so that the base of the cochlea is flat enough to allow adequate sticking to the dish.
8. At this stage, the matrigel is removed from one MatTek dish and replaced with culture medium made of DMEM supplemented with 10% fetal bovine serum and N2 supplement.
9. The cochlea can then be transferred with a curette tool in that medium.
10. Using forceps to maintain the cochlea in the center of the dish, the entire culture medium is aspirated slowly to flatten out the tissue, and then replaced slowly to cover the tissue up again.

11. The MatTek dish is placed carefully in the incubator and more medium is added about 1 h later to fill up the dish (approximately 150 μL).
12. The cultures are incubated for 6 days, to reach a stage equivalent to P0.
13. After 6–7 days in vitro, the culture is fixed and processed for immunocytochemistry, as indicated below (*see* **Fig. 16.2**).

This culturing method allows one to 1) analyze the cochlear phenotype of mutants that do not survive past E13.5 (i.e., development is extended by culturing the cochlea for 6 days in vitro), 2) test the effect of an inhibitor on the orientation of the hair bundle simply by adding this inhibitor in the culture medium.

3.3. Electroporation

1. Transfer the fully dissected cochlea (**Section 3.2**, item 7) in a 10 μL complementary DNA (cDNA) drop with the curette tool.
2. Orient the cochlea in the drop as perpendicular to the plane of the dish as possible, so that the cochlear epithelium faces the cathode of the gold-plated probe, while the mesenchyme at the base of the explant faces the anode. This idea is that the DNA will enter first and mostly through the epithelium.

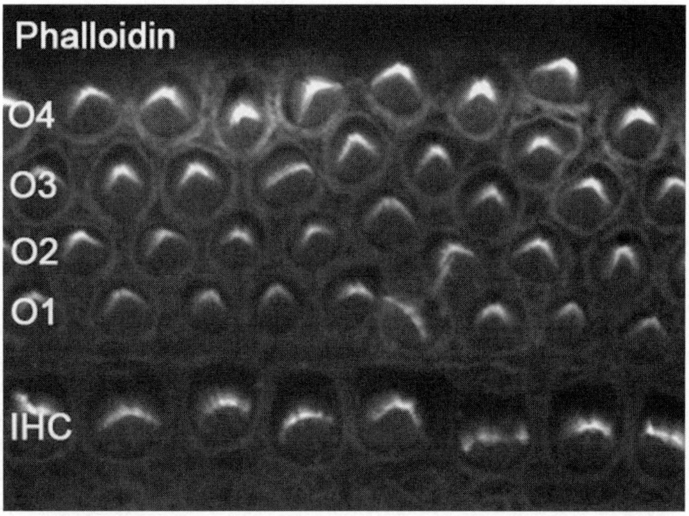

Fig. 16.2. Rat cochlear culture. The cochlea of an E16.5 rat was harvested and placed in culture for 6 days. The cochlea was then fixed and processed for immunocytochemistry. In this example, the tissue was labeled with phalloidin Alexa-546, which allows us to see the hair bundles, which are strongly stained and pointing roughly all in the same direction (*white arrow*). We can observe one row of inner hair cells (*IHC*), and four rows of outer hair cells (*01–04*). The additional row of OHC is not unusual in culture. The F actin is also apparent at the border of each epithelial cell.

3. The parameters for the eletroporator should be 25 V, seven to nine pulses for 25 msec with 500-msec intervals, but this can be species and stage dependent. We were not able to get cDNA into the sensory epithelium of rats past E17 or mice past E15. However, at these stages, the inner or outer sulcus can be positively electroporated (see Note 3).
4. Leave the cochlea in the cDNA for 2–3 min and then transfer them to the MatTek well filled with culture medium. Force the cochlea to adhere as indicated previously (Section 3.2, item 10).
5. The tissue is maintained for 6–7 days in vitro, then fixed and processed for immunocytochemistry (*see* **Fig. 16.3**).

3.4. Immunocytochemistry and Immunohistochemistry

For immunocytochemistry, the inner ears are harvested as detailed in **Section 3.2**, except that cold PBS is used instead of HBSS.

1. Drop the inner ears into freshly prepared 2% PFA and fix the tissues at 4°C with slow agitation for 45 min.
2. The PFA is then replaced by PBS and the tissue is rinsed three times for 15 min with PBS.
3. Dissect the cochlea as in **Section 3.2**, item 7, except that when the dissection is done, the tissue is transferred to a 1.5-mL Eppendorf tube, where it will be processed for immunocytochemistry.
4. Typically, the cochlea is permeabilized with a solution of PBS with 0.5% Tween-20 (PBS-T), before blocking with 10% goat serum in PBS-T for 1 h at room temperature.

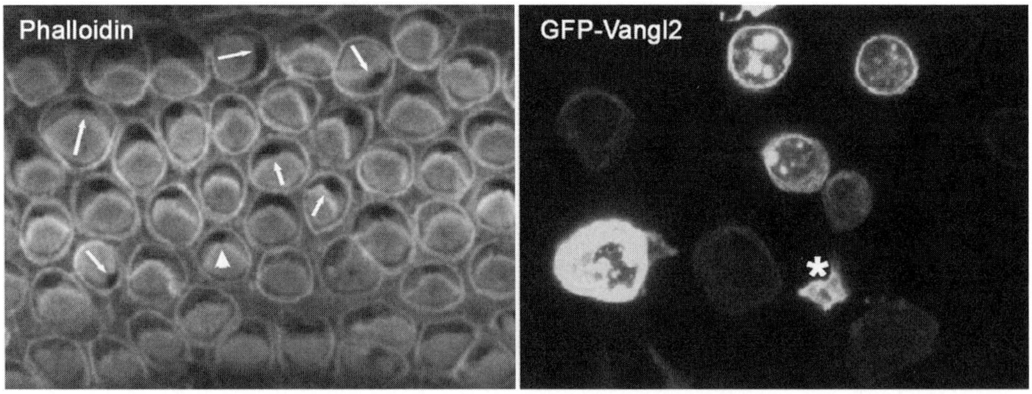

Fig. 16.3. Rat cochlea electroporated with a green fluorescent protein (*GFP*)–*Vangl2* construct. The cochlea of an E16.5 rat was harvested and electroporated with a GFP-Vangl2 plasmid, and then placed in culture and maintained for 6 days in vitro. The cochlea was then labeled with phalloidin Alexa 546. We can easily differentiate between the GFP expression in the hair cells transfected (*round shapes* in the *right panel*, *arrows* in the *left panel*), versus supporting cells (*asterisk*). The hair bundle orientation is also evident, and seems to be affected by *vangl2* overexpression (*arrows* in the *left panel*). The *arrowhead* indicates the normal hair bundle planar polarity.

5. Our primary antibody for Vangl2 was raised in rabbit, and is a polyclonal antibody that we use typically at 1:500 for immunocytochemistry (2 mg/mL) in PBS-T with 2% goat serum, overnight at 4°C with gentle agitation.
6. The next day, rinse the cochlea three times for 30 min each, then add an anti-rabbit Alexa 488 (1:1,000) for 45 min in the presence of phalloidin 546 (1:200).
7. Rinse the tissue three times, 30 min in PBS before mounting.
8. For mounting, place a drop of antifade mounting medium on a clean slide, the cochlea is then transferred in the drop with the curette tool, and orientated correctly with the forceps so that the epithelium is face-up (*see* **Note 4**).
9. The drop of medium containing the cochlea is then gently and slowly covered by a coverslip.
10. If the immunocytochemistry is done in a MatTek dish, the coverslip can be detached from the dish using OS30 solvent. Simply add 200 μL of OS30 in the cover of the MatTek dish positioned upside-down, and place the corresponding dish inside the cover. Let it sit at room temperature for 1–2 h, then discard the excess solvent, and carefully detach the coverslip from the dish with a razor blade. You can now mount the coverslip on a slide and observe under a fluorescent microscope (*see* **Fig. 16.4**).

This immunocytochemistry will allow you to not only localize a protein within your cochlea, but will also let you know whether the planar polarity of the bundles or a mutant is affected.

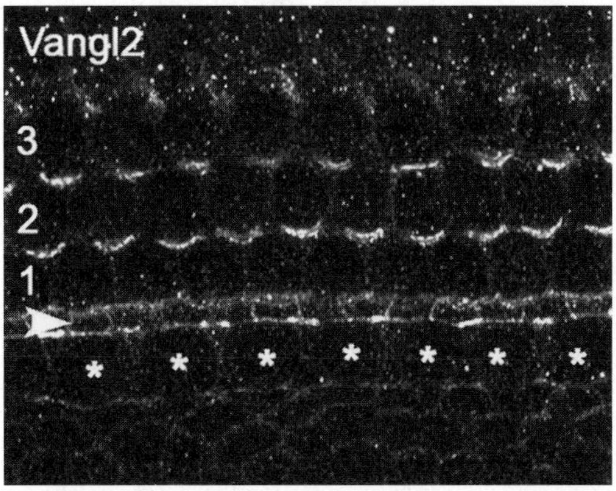

Fig. 16.4. Vangl2 immunocytochemistry on a rat cochlea. A rat cochlea was harvested and fixed at birth (P0), dissected, and processed for immunocytochemistry with a custom-made polyclonal anti-Vangl2 antibody. Vangl2 expression is very strong on the proximal side of the OHC numbered 1–3, as well as at the borders of the pillar cells (*arrowhead*), which are supporting cells separating the IHC (*asterisks*) and the OHC.

3.5. Preparation of Cochlea Samples for Immunoblotting

1. For immunoblotting, harvest the cochleae in cold PBS (**Section 3.4**) and freeze immediately on dry ice.
2. Frozen cochleae are homogenized in cold PBS plus protease inhibitors. Typically, four cochleae can be diluted in 50 μL of the mix.
3. Add 2× SDS sample buffer to the tube and then heat at 95°C for 5 min (50 μL for 50 μL of the PBS/protease mix above).
4. After cooling at room temperature, the lysates are ready for separation by SDS-PAGE.

3.6. SDS-PAGE and Western Blots

We use a Hoeffer Mighty Gel system. Glass plates for the gels are scrubbed clean with ethanol after use and rinsed extensively with distilled water.

1. Prepare a 1-mm-thick, 10% gel by mixing 1.75 mL of Tris-HCl, pH 8.8, with 2.33 mL of 30% acrylamide/bis solution, 35 μL of 20% SDS solution, 0.8 mL of 66% (v/v) glycerol solution, 2.07 mL of water, 11.6 μL of 40% ammonium persulfate solution, and 4.7 μL of TEMED.
2. Pour the gel, leaving space for a stacking gel, and overlay with water. The gel should polymerize in about 30 min.
3. Pour off the water and rinse the top of the gel twice with water. Carefully remove all of the water.
4. Prepare the stacking gel by mixing 1.25 mL of Tris-HCl, pH 6.8, with 0.83 mL of 30% acrylamide/bis solution, 25 μL of 20% SDS solution, 1.12 mL of 66% glycerol solution, 1.66 mL of water, 6.9 μL of 40% ammonium persulfate solution, and 5 μL TEMED.
5. Use about 0.5 mL of water to quickly rinse the top of the gel, and then pour the stacking gel and insert the comb. The stacking gel should polymerize within 30 min.
6. Prepare the running buffer by diluting 100 mL of the 10× running buffer with 900 mL of water in a measuring cylinder. Cover the measuring cylinder with Parafilm and invert to mix.
7. Once the stacking gel has set, carefully remove the comb and use a 3-mL syringe fitted with a 22-gauge needle to wash the wells with running buffer.
8. Add the running buffer to the upper and lower chambers of the gel unit and load 20 μL (or less) of each sample in a well. Include one well for prestained molecular weight markers.
9. Complete the assembly of the gel unit and connect to a power supply. The gel can be run at 125 V (60 mA). The dye fronts (blue and pink) can be run off the gel if desired.
10. Following SDS-PAGE, the gel is dismantled from the electrophoresis unit and submerged in cold transfer buffer.

11. For transfer, we use PVDF membrane. Check the instructions, but usually immerse the PVDF membrane in 100% methanol for 15 sec, pure water for 2 min, then at least 5 min in transfer buffer. Do not let the membrane dry after immersion.
12. Prepare a tray (that is large enough to lay out a transfer cassette) for transfer. Immerse the cassette with its pieces of foam in transfer buffer. The two sheets of 3MM paper are submerged in the transfer tank most of the time. Lay out the foam, the paper, and the sheet of the PVDF cut just larger than the size of the separating gel, and lay on the gel. Add an additional sheet of paper and foam. Be sure that no bubbles are trapped in the resulting sandwich, and close the cassette.
13. The cassette is placed into the transfer tank such that the PVDF membrane is between the gel and the anode. Of course, this orientation is very important to ensure that proteins will be transferred to the membrane and not lost from the gel into the buffer. Transfer for 2.5 h at 60 V in cold buffer (and in a cold room) in a Hoefer transfer tank system.
14. After 2.5 h, the transfer is complete and the cassette is taken out of the tank and carefully disassembled in the inverse order that it was made. The excess membrane is cut using a razor blade. The colored molecular weight markers should be clearly visible on the membrane and not on the gel. The membrane is then washed with TBS buffer at room temperature on a rocking platform.
15. For the blocking step, we use 5% nonfat dried milk in TBST overnight.
16. Rinse the milk off the membrane prior to addition of the antibody in TBST for 1 h at room temperature on a rocking platform. The optimal primary antibody concentration is determined experimentally. Primary antibodies can be reused many times. Store primary antibodies at 4 °C with sodium azide.
17. Wash the membrane in a large volume of TBST three to four times for a total of about 45 min.
18. Just before use, prepare the secondary antibody in TBST with 1% milk, 1:5,000 dilution for the Amersham secondaries.
19. Rinse the membrane three to four times with TBST for total of 30 min and then with TBS 3× 10 min.
20. We use both Amersham and Pierce ECL reagents according to the manufacturer's instructions. They both need to be warmed up to room temperature. One point to keep in mind is that you do not want to dry the membrane for the exposure. We use Parafilm to do the incubation and then pour off

the liquid, but not let the membrane dry. We then cover he membrane with Saran wrap for the ECL development.

21. The wrapped membrane is then placed in an X-ray film cassette with film for a suitable exposure time, typically a few minutes.
22. We usually warm stripping buffer (50 mL per blot) to 60–70°C before adding the β-mercaptoethanol and then the blot. The blot is incubated three times for 10 min each with occasional agitation. Once the blot is stripped, it is extensively washed in TBS buffer and then blocked again in blocking buffer (*see* **item 14**).

4. Notes

1. Acrylamide/bis solution is a neurotoxin when unpolymerized, so care should be taken to avoid exposure.
2. The DMEM must be cold because the matrigel is only liquid at 4°C and becomes a gel at higher temperatures.
3. Not all vectors will successfully get cDNA expressed within the sensory epithelium of the cochlea, which is different from the vestibular system. We found that pCLIG is a vector that works in the cochlea *(7)*.
4. To avoid the cochlea curling, manipulating the cochlea at the edge of the drop helps because of the surface tension.

Acknowledgments

We thank F. Loll for technical assistance, Elodie Richard for proofreading, and Dr. Ronna Hertzano for the photographs in Fig. 16.1. This work was supported by grants from Institut National de la Santé et de la Recherche Médicale, Fondation pour la Recherche Médicale, and Région Aquitaine (MM and NS).

References

1. Montcouquiol, M., Crenshaw, E. B., and Kelley, M.W. (2006) Non canonical Wnt signalling and neural polarity. *Annu Rev Neurosci.* **29**, 363–386.
2. Eaton, S. (1997) Planar polarization of Drosophila and vertebrate epithelia. *Curr. Opin. Cell Biol.* **9**, 860–866.
3. Montcouquiol, M., Rachel, R. A., Lanford P. J., Copeland, N. G., Jenkins, N, A., and Kelley, M.W. (2003) Identification of Vangl2 and Scrb1 as planar polarity genes in mammals. *Nature* **423**, 173–177.
4., Wang, Y., Guo, N., and Nathans, J. (2006) The role of Frizzled3 and Frizzled6 in neural tube

closure and in the planar polarity of inner-ear sensory hair cells. *J. Neurosci.* **26**, 2147–2156.

5. Montcouquiol, M., Sans, N., Huss, D., Kach, J., Dickman, J.D., Forge, A., Rachel, R.A., Copeland, N. G., Jenkins, N. A., Bogani, D., Murdoch, J., Warchol, M. E., Wenthold, R.J., and Kelley, M. W. (2006) Asymmetric localization of Vangl2 and Fz3 indicate novel mechanisms for planar cell polarity in mammals. *J. Neurosci.* **26**, 5265–5275.

6., Kaufman, M. H. (1992) The atlas of mouse development. Academic Press, San Diego, California. 386 pp.

7. Jones, J. M., Montcouquiol, M., Dabdoub, A., Woods, C., and Kelley, M. W. (2006) Inhibitors of differentiation and DNA binding (Ids) regulate Math1 and hair cell formation during the development of the organ of Corti. *J. Neurosci.* **26**, 550–558.

Part II

Mammalian Model Systems for Wnt/FZD Function

Chapter 17

Proteomic Analyses of Protein Complexes in the Wnt Pathway

Stephane Angers

Abstract

Multiple screens performed in *Drosophila*, *Caenorhabditis elegans*, *Xenopus*, and zebrafish have identified dozens of proteins participating in Wnt signal transduction. Epistasis experiments, enhancer and suppressor screens, and protein–protein interaction techniques have also been efficient at finding new pathway members, connecting proteins together, and establishing the architectural framework of how the Wnt signaling pathway functions. In the last few years, spectacular technological breakthroughs in the field of mass spectrometry have allowed the study of proteins and peptides with unprecedented sensitivity and accuracy. Recently, we have developed methods to study the Wnt pathway using mass spectrometry by studying the composition of protein complexes isolated from mammalian cells. In addition to identifying novel proteins acting in this pathway, this approach is providing information about the supramolecular organization of the protein complexes in the pathway and how the individual proteins are activated and regulated. This chapter details the experimental procedure that we developed to study mammalian protein complexes using a gel-free mass spectrometry approach.

Key words: Proteomics, Mass spectrometry, Tandem-affinity purification, Protein complexes, Wnt.

1. Introduction

The isolation of proteins using a traditional immunopurification approach with antibodies directed against a specific protein or to a heterologous epitope tag (i.e., FLAG or HA) are generally well suited for the study of protein–protein interactions *(1)*. They are, however, generally considered crude methods for the purification of proteins since abundant proteins and intrinsically "sticky" proteins are often found nonspecifically in the immunoprecipitates. When mass spectrometry is used to study the composition of protein complexes, the presence of nonspecific proteins in samples complicates the differentiation of "bona fide" associated proteins present in these complexes from contaminating proteins. To circumvent this

Elizabeth Vincan (ed.), *Wnt Signaling, Volume I: Pathway Methods and Mammalian Models, vol. 468*

Book doi: 10.1007/978-1-59745-249-6

problem, the laboratory of Séraphin has pioneered the development of the tandem-affinity purification (TAP) methodology *(2)*. In the original TAP version, the protein A IgG-binding motifs and the calmodulin-binding peptide (CBP) affinity tags are genetically fused in frame with a protein of interest. Two rounds of purification using protein A and calmodulin matrices are then performed to isolate the tagged protein along with its interacting proteins. The added stringency provided by the second round of purification minimizes the carryover of unwanted contaminating proteins. Because large amount of cells are needed to isolate sufficient protein complexes amenable for mass spectrometry, these TAP experiments have generally been conducted in yeast cells. In the last few years, we have developed a variation of the original TAP method and optimized protocols for the efficient isolation of protein complexes from mammalian cells *(3, 4)* in order to study the Wnt signaling pathway.

2. Materials

2.1 Reagents and Buffers

1. Fast-Flow Streptavidin Sepharose Amersham (GE Healthcare, Piscatouvay; NJ; #17-513-01).
2. Calmodulin-Sepharose 4B (GE Healthcare; #17-0529-01).
3. Protease inhibitor complete mini (Roche, Nutley, NJ; #11 836 170 001).
4. Phosphatase inhibitors (Calbiochem, Gibbstown, NJ; #524628).
5. Chromatography mini spin columns (Bio-Rad, Hercules, CA; #732-6204).
6. Lysis buffer: 10% (v/v) glycerol, 50 mM HEPES-NaOH, pH 8.0, 150 mM NaCl, 2 mM EDTA, 0.1% (v/v) Igepal CA-650, 2 mM DTT, complete mini protease inhibitors cocktail, 10 mM NaF, and phosphatase inhibitors cocktail (*see* **Notes 1** to **4**).
7. Calmodulin-binding buffer (CBB): 10 mM β-mercaptoethanol, 50 mM HEPES-NaOH, pH 8.0, 150 mM NaCl, 1 mM MgOAc, 1 mM imidazole, 0.1% (v/v) Igepal CA-650, and 2 mM $CaCl_2$.
8. Streptavidin-elution buffer: CBB supplemented with 10 mM D-biotin (*see* **Note 5**).
9. Calmodulin-rinsing buffer: 50 mM ammonium bicarbonate, pH 8.0, 75 mM NaCl, 1 mM MgOAc, 1 mM imidazole, and 2 mM $CaCl_2$.
10. Calmodulin-elution buffer: 50 mM ammonium bicarbonate, pH 8.0, and 25 mM EGTA (*see* **Note 6**).

11. HPLC buffer A: 95% (v/v) water, 5% (v/v) acetonitrile (Burdick & Jackson, Muskegon, MI), and 0.1% (v/v) formic acid (JT Baker, Phillipsburg NJ) (*see* **Note 7**).
12. HPLC buffer B: 20% (v/v) water, 80% (v/v) acetonitrile (Burdick & Jackson), and 0.1% (v/v) formic acid (JT Baker).

2.2. Mammalian Expression Vectors

We have developed two expression vectors containing a dual-affinity tag composed of a streptavidin-binding peptide (SBP) and a calmodulin-binding peptide (CBP) to allow for the TAP purification and a HA epitope to monitor the level of expression and efficiency of purification when an antibody for the desired bait protein is not available. The pGlue-N and pGlue-C vectors are used when the affinity cassette is placed at the N terminus and C terminus of the protein of interest, respectively (**Fig. 17.1**). We use the SBP because of its relatively small size compared with the IgG domain used in the original TAP version, which we find often negatively impairs the functional activity of many proteins. Using polymerase chain reaction (PCR), the gene of interest is amplified using primers harboring restriction sites compatible for its insertion in-frame with the TAP cassette present in the desired pGlue vector. Both of these vectors also contain an internal ribosome entry site (IRES) driving the expression of a puromycin-resistance gene needed for the establishment of cell lines stably expressing the engineered fusion proteins.

2.3. Mammalian Stable Cell Lines

Once the sequence integrity of the gene of interest inserted in the pGlue vector is confirmed, we normally test for protein expression using transient transfection of mammalian cells and Western blotting. To do so, 2 μg of pGlue plasmid containing the protein can be transfected along with 8 μg of carrier plasmid DNA into a 10-cm dish containing cells at 60–75% confluency. Although potentially any mammalian cells could be used, we routinely use HEK293 cells

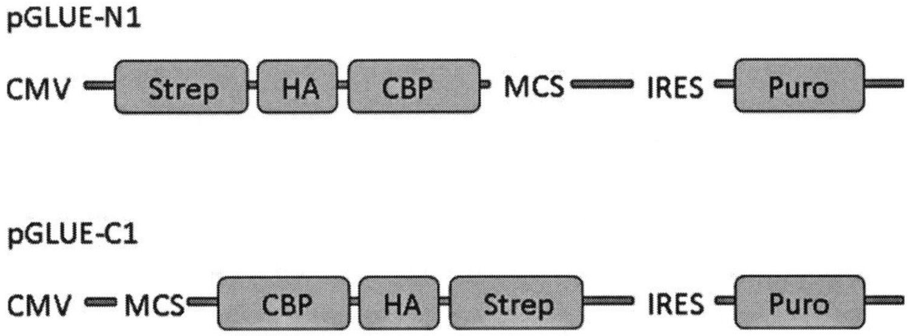

Fig. 17.1. Mammalian expression vectors used for the generation of dual-affinity tagged proteins. pGLUE-N1 and pGLUE-C1 vectors are respectively designed for linking the affinity cassette to the N terminus or C terminus of the protein of interest. A puromycin (*Puro*) selection gene driven by an IRES is present downstream of the fusion protein to allow for the selection of cells that have stably integrated the construct. An HA epitope is also present on both vectors. CMV = promoter; MCS = multiple cloning site; CBP = calmodulin binding peptide; Strep = streptavidin binding peptide.

because of their rapid growth and good transfection efficiency. Using $CaPO_4$, 70–80% of HEK293 cells can be transfected. Using standard Western blotting techniques, cell extracts can then be probed for the expression of the fusion proteins using anti-HA antibodies.

If functional assays are available to test the activity of the fusion protein, we recommend testing the fusion protein compared with the wild-type counterpart. When we studied the Dishevelled protein complex (3), for instance, we found that a Dsh3 construct fused to the original TAP version with CBP and IgG could only activate TOPFlash (a luciferase reporter assay monitoring the activation state of the Wnt/β-catenin signaling pathway) at 5% of the wild-type level when overexpressed in HEK293 cells. In contrast, the pGlue–Dsh3 (tagged with SBP and CBP) was indistinguishable from the wild-type protein in this assay. When we isolated the IgG–CBP–Dsh3 protein complex, all that was identified using mass spectrometry were tubulin and chaperone proteins, whereas most of the known Dishevelled-associated proteins could be identified using the pGlue–Dsh3 construct.

Although some preliminary experiments can be performed using transient transfection experiments, the best results are obtained when cell lines stably expressing the fusion protein are derived. The high and variable levels of expression obtained in transient transfection lead to protein aggregation and prolonged association with cellular chaperones. Because the bait proteins become "coated" with chaperones, the binding sites for the other proteins are masked and most of the proteins subsequently identified in these conditions are chaperones.

To establish a cell line stably expressing the desired affinity-tagged protein, we normally transfect 5 μg of the appropriate pGlue plasmid DNA in a 10-cm dish containing HEK293 cells at 60–70% confluency using the $CaPO_4$ transfection procedure. Forty-eight hours posttransfection, the cells are then rinsed 1× in PBS and dissociated using 1 mL of trypsin-EDTA. Cells are resuspended in 10 mL of Dulbecco's modified Eagle's medium (DMEM)-fetal bovine serum (FBS) and 3 mL of the cell suspension is passaged into a 15-cm dish containing 20 mL of DMEM-FBS supplemented with 2 μg/mL of puromycin (effective concentration for HEK293 cells; independent kill curves should be performed for other cell lines to find the effective drug concentration). Stable integrants are then isolated by replacing the selective media every 2 to 3 days. A polyclonal stable cell line is normally obtained after 2 weeks of selection. Since the gene driving the resistance to puromycin is driven by an IRES present on the same messenger RNA (mRNA) as the fusion protein, all the cells selected also contain the protein of interest. Soon thereafter we normally freeze a subsequent passage of cells for future use. It is also a good idea to reassess the level of expression of the fusion protein in the stable cell line before performing a large-scale purification.

3. Methods

3.1. Amplification of Cells

In order to obtain sufficient amounts of protein complexes amenable for mass spectrometry, a relatively large amount of starting material is needed. To begin the expansion of the desired cell line stably expressing the protein of interest, we normally split a confluent 10-cm dish into two 15-cm dishes. After 2 to 3 days, the 15-cm dishes reach confluency, and are at that point split into 10–15 (more if needed) 15-cm dishes.

3.2. Preparation of Cell Extract

1. Different procedures may be use to rinse and harvest the cells, however, for HEK293 cells, remove the media from the cells and add 10 mL of PBS to each 15-cm dish.
2. After 5–10 minutes, the cells can be easily squirted out of the dish into a 50-mL conical tube. The cells are then collected by centrifugation at 800 × *g* for 5 minutes, combined and washed in 50 mL of PBS.
3. After pelleting the cells for 5 minutes at 800×*g*, the cells are lysed in 10 mL of TAP lysis buffer supplemented with the protease (Roche) and phosphatase inhibitors (Calbiochem). We normally lyse the cells in a 15-mL conical tube at 4°C with gentle rocking for 15 minutes.
4. To ensure complete cell lysis, proceed with two freeze–thaw cycles by immersing the tube in liquid nitrogen (*see* **Note 8**).
5. Thaw the lysate, aliquot into 10 Eppendorf tubes, and clear by centrifugation at ≥15,000 × *g* for 15 minutes in a microcentrifuge. We recommend keeping an aliquot of the pelleted (nonsoluble) material and a fraction (40 μL) of the cleared lysate (input). Western blotting with HA antibodies could then be performed to monitor the efficiency of the lysate preparation. If the majority of your protein of interest is in the pellet, the solubilization conditions or alternative lysate preparation need to be optimized.

3.3. Streptavidin Affinity Chromatography

All spins and incubation are performed at 4°C and all buffers are prechilled on ice.

1. 100 μL of packed Sepharose–streptavidin beads are first equilibrated using two 800 μL washes with TAP lysis buffer (protease and phosphatase inhibitors are not necessary at this point). The beads are sedimented by centrifugation at 800 × *g* for 1 minute using a microcentrifuge. We use a 26-gauge needle attached to a vacuum pump to remove the buffer during the washes. Between 20 and 50 μL of buffer is left on the beads to prevent the beads from drying.
2. Transfer the beads to a 15-mL conical tube to which the cleared lysates from the 10 Eppendorf tubes is added.

Although an incubation of 2 hours at 4°C with gentle rocking is sufficient to isolate the majority of the proteins, we normally prepare the lysate in the afternoon of day 1 and leave the lysate on the streptavidin beads overnight.

3. After the incubation, the beads are sedimented by centrifugation at 800 × *g* and the supernatant is discarded (keep an aliquot of the discarded supernatant to evaluate proteins that did not bind). At that point, the beads are transferred to an Eppendorf tube to perform the washes.
4. Perform two washes with 800 μL of TAP lysis buffer and three washes with 800 μL calmodulin-binding buffer (CBB) (at the last wash when the beads are resuspended in 800 μL of CBB, keep a 20-μL aliquot to evaluate material bound to streptavidin beads).
5. Elute the protein complexes from the streptavidin beads using two consecutive elutions with 200 μL of biotin elution buffer (CBB supplemented with 10 mM D-biotin). Because of the very high affinity of biotin for streptavidin, the elution of the proteins from the beads is immediate.
6. The 400 μL of eluted material is supplemented with 400 μL of CBB (800 μL total volume) and 5 μL of 1 M $CaCl_2$ and applied to calmodulin–Sepharose beads (*see* **Section 3.4**). Aliquots of eluted material (10 μL of 800 μL) and of streptavidin beads post-elution (resuspend beads in 100 μL CBB and save 20 μL) can then be saved for troubleshooting.

3.4. Calmodulin Affinity Chromatography

1. Equilibrate 100 μL of packed calmodulin–Sepharose beads with two washes with 800 μL CBB.
2. The streptavidin elution from **Section 3.3** is then transferred to the beads and incubated for 2 hours with gentle agitation at 4°C.
3. After the incubation, the calmodulin beads are centrifuged and the supernatant is removed (save a 20-μL aliquot to analyze what fraction did not bind to the calmodulin beads). The beads are washed twice with 800 μL of CBB and three times with calmodulin rinsing buffer. The beads are then resuspended in 800 μL of calmodulin-rinsing buffer (save a 20-μL aliquot to analyze the material bound to beads before elution) and transferred to an empty chromatography mini spin-column (Bio-Rad).
4. The beads are then allowed to sediment while the calmodulin rinsing buffer flows through the column (*see* **Note 9**).
5. For the elution, the column is placed in an Eppendorf tube and the protein complexes are eluted twice using 100 μL of calmodulin-elution buffer directly applied to the beads. A 10-μL aliquot of the final elution and a 20-μL aliquot of

beads resuspended in 200 μL of calmodulin rinsing buffer after the elution can then be saved to assess the efficiency of elution (*see* **Note 10**).

3.5. Trypsin Digestion of Protein Complexes

Once eluted, the sample is then directly processed for trypsin digestion.

1. The sample is reduced with the addition of 5 μL of 1 M DTT (25 mM final) and heated at 50°C for 20 minutes.
2. The free sulfhydryl groups are then alkylated by adding 40 μL of freshly prepared 500 mM iodoacetamide (100 mM final concentration) in the dark for 20 minutes.
3. The sample is then digested overnight at 37°C by adding 1 μg of sequence-grade trypsin.

3.6. Preparation of Sample for Liquid Chromatography and Tandem Mass Spectrometry Analysis

To maximize the sensitivity for the mass spectrometry analysis, we directly load the sample on an analytical column using a pressure valve. The analytical columns are made of 75-μm inner diameter (ID) fused silica and the tip is pulled either manually or using a laser puller (Sutter Instruments, Novato, CA). We normally use 30-cm-long columns that are packed with 20- to 25-cm of reverse-phase material (Jupiter 4μ Proteo 90A; Phenomenex, Inc., Torrance, CA) using a pressure valve. To minimize sample loading time, the volume of the sample is reduced to 50 μL using a Speed-Vac. Half of the sample (25 μL) is then manually loaded on the column. In our setup, the analytical column is then placed online with the LTQ linear ion-trap mass spectrometer and the peptides are eluted using a 2-hour gradient method where the aqueous buffer A is progressively mixed with higher proportion of the organic buffer B by the HPLC. To reach nanoflow capabilities, a flow-split system is used to reduce the flow of 150 nL/min coming out of the HPLC to 20–50 nL/min on the analytical column. Peptide ions are dynamically selected for fragmentation using data-dependent acquisition by the operating software (the five more intense precursor ions of each mass spectrometry (MS) scan are selected for subsequent MS/MS).

4. Notes

1. All chemicals are molecular biology grade from Sigma, St. Louis, MO or other vendors.
2. A 100× phosphatase inhibitor cocktail is available from Calbiochem.
3. The protease and phosphatase inhibitors in the lysis buffer can be left out during the washes.

4. We generally freeze the lysis buffer with all the inhibitors in 10-mL aliquots and thaw one tube per experiment for the lysis and keep 100 mL of lysis buffer without inhibitors at 4°C for the washes.
5. If the biotin does not go into solution, verify that the pH is 8.0.
6. Elution buffer can be supplemented with 0.1% (w/v) Rapi-Gest (Waters, Milford, MA) if elution is problematic.
7. We always use the same glass cylinder to make up the HPLC buffer and only rinse it with pure water. Never use soap because this will be a source of contamination in the mass spectrometer.
8. This could represent a good stopping point as the cells can be stored in liquid nitrogen or at –80°C for several weeks without any decrease in protein complex isolation.
9. If the buffer does not flow through, it can gently be forced out of the column using the top of a Pasteur pipette rubber bulb.
10. Although on average 70–90% of the bait protein is eluted, for some bait proteins, this step can be very inefficient. The addition of 0.1% (w/v) of the acid-cleavable detergent Rapi-Gest to the calmodulin-elution buffer can improve the elution in these cases. In this case, the RapiGest needs to be cleaved using 1 M HCl before LC-MS/MS analysis.

Acknowledgements

I am in debt to Dr. Anne-Claude Gingras for her insights and expertise during the optimization of the affinity purification method and Dr. Michael Maccoss for teaching me mass spectrometry.

References

1. Masters, S. C. (2004) Co-immunoprecipitation from transfected cells. *Methods Mol Biol* **261**, 337–350.
2. Rigaut, G., Shevchenko, A., Rutz, B., Wilm, M., Mann, M., and Seraphin, B. (1999) A generic protein purification method for protein complex characterization and proteome exploration. *Nat Biotechnol* **17**, 1030–1032.
3. Angers, S., Thorpe, C. J., Biechele, T. L., Goldenberg, S. J., Zheng, N., MacCoss, M. J., and Moon, R. T. (2006) The KLHL12-Cullin-3 ubiquitin ligase negatively regulates the Wnt-beta-catenin pathway by targeting Dishevelled for degradation. *Nat Cell Biol* **8**, 348–357.
4. Angers, S., Li, T., Yi, X., MacCoss, M. J., Moon, R. T., and Zheng, N. (2006) Molecular architecture and assembly of the DDB1-CUL4A ubiquitin ligase machinery. *Nature* **443**, 590–593.

Chapter 18

In Situ Hybridization to Evaluate the Expression of Wnt and Frizzled Genes in Mammalian Tissues

Kestutis Planutis, Marina Planutiene, and Randall F. Holcombe

Abstract

In situ hybridization can be utilized to specifically define the expression of genes and to determine their localization in complex mammalian tissues *(1)*. The expression of specific members of the Wnt ligand and frizzled receptor families of molecules can be defined using an antisense RNA probe that will specifically hybridize with messenger RNA (mRNA) in the tissue to form a double-stranded product. The double-stranded product can then be detected microscopically by identifying digoxigenin groups that are attached to the probe during its synthesis. Probe sequence selection is critical to ensure specificity among different Wnt or frizzled family members. Controls are needed at every step in the technique to confirm appropriate quality of the tissue sections, quality of the prepared probe, and specificity of the hybridization reactions. If performed properly, in situ hybridization can be utilized to define gene expression and specific localization of RNA in human and other mammalian tissues, and can be utilized in previously fixed and paraffin-embedded tissue samples.

Key words: In situ hybridization, Gene expression, RNA probes, Mammalian tissues, Wnt, Frizzled.

1. Introduction

In this chapter, we describe methodologies for in situ hybridization to define the expression of genes of the Wnt ligand and frizzled receptor families in mammalian tissues. In situ hybridization on human mammalian tissues can be problematic as the samples available are often fixed and paraffin embedded. Thus, specific techniques are required to obtain successful hybridizations yielding meaningful results. We describe methods and reagents necessary for tissue/slide preparation, probe preparation, hybridization, and visualization (**Fig. 18.1**). It is critical for this technique that adequate controls are utilized to ensure

Elizabeth Vincan (ed.), *Wnt Signaling, Volume I: Pathway Methods and Mammalian Models, vol. 468*

Book doi: 10.1007/978-1-59745-249-6

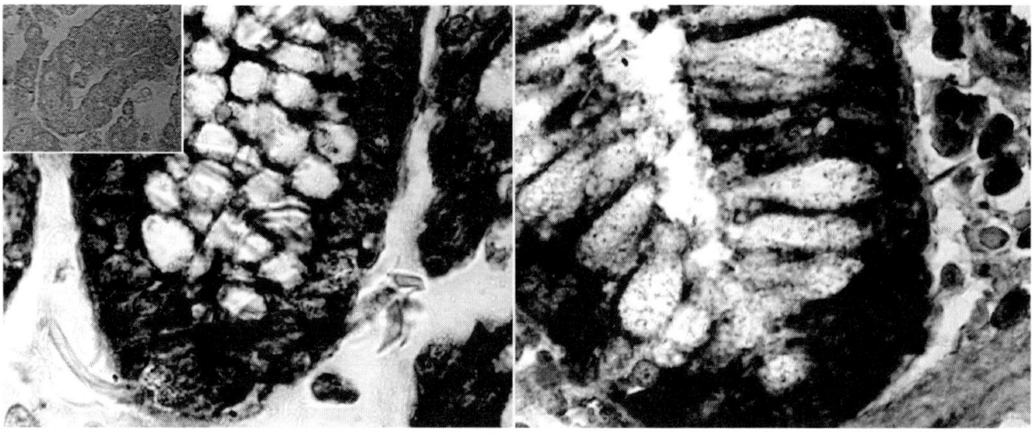

Fig. 18.1. Expression of Wnt7b and Wnt10B genes in colon epithelium. An example of in situ hybridization with an antisense probe directed against the 3′ end of the coding sequence for Wnt7B gene, and at the center of coding sequence of Wnt10B gene. The insert demonstrates negative staining with a sense probe. The tissue is human colon that was obtained from a fixed, paraffin-embedded tissue sample.

the quality of the tissues, probes, and hybridization. The steps involved are 1) deparaffinization and rehydration of sections, 2) treatment with proteinase K, 3) fixation with formaldehyde, 4) dehydration of sections, 5) probe synthesis and hybridization, and 6) probe detection and counterstaining.

2. Materials

2.1. Stock Solutions

1. Diethyl pyrocarbonate (DEPC). All buffers and solutions used for in situ hybridization must be DEPC treated. DEPC treatment of water and solutions is done to remove RNAses and prevent RNA degradation. Mix 2 mL DEPC for each 1,000 mL of solution or water to be treated. This mixture needs to be shaken vigorously since droplets of DEPC float in the water. Let the mixture stand in a fume hood overnight then autoclave for 70 min to remove the remainder of DEPC, because traces of the DEPC might inhibit enzymes used in later steps. The autoclaving time should be increased proportionally if the volume of the treated water increases. Adjust the pH of any buffers with 1 M NaOH since DEPC may cause a fall in pH upon degradation. Alternatively, molecular-grade water (distilled, deionized, DNases, RNAses, and proteases tested) from commercial sources could be used.
2. 1 M NaOH. Store at room temperature.

3. 4 M LiCl. Mix 1.7 g lithium chloride and bring up to 10 mL with DEPC-treated water. Aliquot in 1-mL portions and store at –20°C.
4. Glycogen solution: To prepare glycogen solution, mix 10 mg glycogen and bring up to 1 mL with DEPC water. Store at –20°C.
5. 10 × phosphate-buffered saline (PBS): 25 mM sodium phosphate buffer with 150 mM NaCl. Mix 37.99 g of NaCl, 4.97 g of Na_2 HPO_4 anhydrous, and 1.8 g of NaH_2 PO_4 anhydrous. Bring up to a total volume of 500 mL with DEPC-treated water. Autoclave and keep the solution at room temperature as it will precipitate if it is stored at 4°C.
6. 1 × PBS with Tween-20 (PBT). To prepare 1 × PBT, mix the following in a DEPC-treated glass bottle using a DEPC-treated stir bar: 40 mL of DEPC-treated 10 × PBS, 360 mL of DEPC-treated water, and 0.4 mL Tween-20 (final concentration, 0.1%). Store at 4°C.
7. 20 × SSC. To prepare 20 × SSC, mix 175 g NaCl (final concentration, 3 M), 88 g Na_3 citrate·2 H_2O (final concentration, 0.3 M), and bring up to 1 L with DEPC-treated water. Adjust the pH to 7.0 with 1 M HCl. Store at 4°C.
8. 5 M NaCl. Store at room temperature.
9. 1 M $MgCl_2$. Store at room temperature.
10. 1 M Tris-HCl, pH 9.5. Store at 4°C.
11. Proteinase K solution. Prepare the proteinase K solution by adding 0.8 μL of 5 mg/mL proteinase K to 0.8 mL of PBT (working concentration is 5 μg/mL). Store at –80°C.

2.2. Buffers

1. Hybridization buffer: Denhardt's solution with 10% (w/v) dextran sulfate, 40% (v/v) formamide, 10 mM DTT, 4 × SSC, 1 mg/mL yeast transfer RNA (tRNA), and 1 mg/mL salmon sperm DNA. In an RNAse-free tube, mix 0.125 g dextran sulfate and 0.5 mL formamide. Then add 25 μL of 50 × Denhardt's solution, 250 μL of 20 × SSC, 12.5 μL of 1 M DTT, 280 μL of 4.5 mg/mL yeast tRNA, and 125 μL of 10 mg/mL salmon sperm DNA. Mix well again. For an experiment with 10 slides, prepare two tubes of 1.25 mL of hybridization buffer. Keep one tube on ice, store another tube at –80°C.
2. Prehybridization buffer: 50% (v/v) formamide and 4 × SSC. In an RNAse-free tube, mix 0.375 mL of DEPC-treated water, 0.625 mL of formamide, and 0.25 mL of 20 × SSC. Prepare two tubes of 1.25 mL prehybridization buffer for one experiment with 10 slides. Keep both tubes of the prehybridization buffer on ice.

3. Alkaline wash buffer: 100 mM Tris-HCl, pH 9.5, 100 mM NaCl, 50 mM $MgCl_2$, and 0.1% (v/v) Tween-20. To prepare alkaline wash buffer, mix in a DEPC-treated glass bottle, 99.48 mL of DEPC water, 2.4 mL of 5 M NaCl, 6.0 mL of 1 M $MgCl_2$, 12 mL of 1 M Tris-HCl, pH 9.5, and 120 µL of Tween-20. Keep the solution on ice.
4. Fixation solution: 4% (w/v) formaldehyde in PBT. To prepare fixation solution, rinse a 50-mL cylinder capped with aluminum foil twice with RNAseZap and twice with DEPC-treated water. Keep the cylinder at 4°C and use it to measure all PBT and ethanol solutions. In the hood, into an RNAse-free Coplin staining jar, mix 44.6 mL of PBT and 5.4 mL of 37% (w/v) formaldehyde. Keep the solution on ice.
5. 0.2 M sodium acetate solution. To prepare the sodium acetate solution, rinse a glass bottle or suitable container once with RNAseZap and twice with DEPC-treated water. Add 0.82 g anhydrous sodium acetate and bring up to 50 mL with DEPC-treated water. Mix to dissolve and keep on ice.
6. Bicarbonate buffer: 40 mM $NaHCO_3$ and 60 mM Na_2CO_3. To prepare bicarbonate buffer, rinse a bottle once with RNAseZap and then twice with DEPC-treated water. Add 0.17 g $NaHCO_3$ and 0.32 g Na_2CO_3 to DEPC-treated water. Adjust the pH to 10.2 and make up to 50 mL with DEPC-treated water. Keep buffer on ice.

2.3. Probe Detection

1. Anti-digoxigenin (DIG) antibody that is conjugated with alkaline phosphatase (AP). The antibody is specifically directed against the DIG epitope incorporated into the probe. Prepare 200 µL of 1:200 diluted anti-digoxigenin–AP Fab fragments for each slide. The final concentration of antibody is 150 U/200 µL.
2. Substrate for alkaline phosphatase, BCIP/NBT. Dilute substrate 1:50 in alkaline wash buffer, utilizing 200 µL for each slide. To 1 mL buffer, add 20 µL substrate or similar proportions.
3. Counterstain (0.01% Neutral Red solution) to highlight the structures of the tissue. Dilute the 1% Neutral Red solution 1:100 by adding 10 µL of dye to 1 mL of DEPC-treated water.
4. Water-based media to mount the slides (Crystal/Mount or similar product).

Use only RNAse-free tubes and pipette tips. To remove RNAse, tubes can be treated with RNAseZap (Sigma; R2020 or similar product) and rinsed twice with DEPC-treated water. RNAse-free plastic from commercial sources is also suitable for this procedure.

3. Methods

3.1. Generation of Probes

Generation of a good quality RNA probe is essential for successful in situ hybridization (*see* **Note 1**).

3.1.1. Probe Synthesis

1. Mix the following to a 20-μL reaction volume:

   12 μL of DEPC water
   1.0 μL (1.0 μg) of plasmid DNA, linearized with appropriate restriction enzyme, representing the gene for which expression is to be defined
   2.0 μL of 10× dig-dNTP RNA labeling mix, kept on ice
   2.0 μL of 5× transcription buffer
   0.5 μL of 40 U/μL RNasin RNAse inhibitor, keep enzyme on ice
   2.0 μL of 20 U/μL T7 RNA polymerase (or other polymerases applicable for the probe synthesis with a given vector), keep enzyme on ice

2. Incubate for 2 h at 37°C.
3. Run a small amount (1.5 μL) on a 1% agarose gel to check for correct size and presence of the probe (*see* **Note 2**).

3.1.2. Probe Precipitation

The probe needs to be precipitated prior to hybridization to remove other contaminants.

1. Mix 18.5 μL RNA probe, 2.5 μL of 4 M LiCl (final concentration, 103 mM), and 75 μL of 95% (v/v) ethanol (final concentration, 73%).
2. Incubate overnight at -80°C.
3. The next day, warm to room temperature and spin at 20,000 ×*g* for 30 minutes.
4. Wash with 0.8 mL of 70% (v/v) ethanol in DEPC-treated water.
5. Dry at room temperature for ~30 min.
6. Resuspend pellet to 100 μL in DEPC-treated water by pipetting.
7. Divide into five aliquots of 20 μL each and store aliquots at –80°C until needed.

3.1.3. Probe Hydrolysis

The probe is hydrolyzed into short segments prior to hybridization. This facilitates hybridization to mRNA present within the tissue. To hydrolyze probe, mix a 20-μL aliquot of probe with 20 μL DEPC-treated water and 40 μL bicarbonate buffer. Incubate for 45 minutes at 60°C.

3.1.4. Probe Re-precipitation

1. Mix 0.4 mL of 0.2 M sodium acetate (final concentration, 0.16 M), 80 μL of 4 M LiCl (final concentration, 0.64 M), and 20 μL of 10 mg/mL glycogen (final concentration, 0.4 mg/mL). Keep at 4°C.
2. Add hydrolyzed probe to 100 μL of the mixture.
3. Add 480 μL of 95% ethanol in DEPC water (final concentration, 79%) to precipitate the probe (total volume is 580 μL).
4. Keep on dry ice for 30 minutes or –80°C overnight (*see* **Note 3**).
5. Spin probe at 20,000 ×*g* at 4°C for 30 minutes.
6. Wash with 0.8 mL of 70% (v/v) ethanol in DEPC-treated water.
7. Dry at room temperature.
8. Resuspend the pellet by pipetting in 200 μL of hybridization buffer.
9. Dilute the probe 1:5 with hybridization buffer by adding 25 μL of the probe plus 125 μL of the hybridization buffer for one experimental slide (150 μL of the final probe solution per slide). Prepare as much as needed for the number of slides (including controls) to be hybridized.
10. Denature for 10 min at 80°C and then cool on ice until use.

3.2. Preparation of the Slides

For most human tissue work and some animal work, tissue sections on microscope slides will be prepared from tissues that have been fixed and paraffin embedded. Frozen tissues can also be utilized for in situ hybridization. In contrast to fixed and paraffin-embedded tissue, frozen tissue does not require rehydration. The integrity of mRNA is generally better in frozen tissues (*see* **Note 4**). Preparation of the slides can begin concurrently with probe preparation.

1. Label every slide in pencil with the source of the tissue, the hybridization planned for the tissue, and the date. Any tools utilized for handling slides, such as forceps, need to be treated with RNAseZap.
2. Rehydrate slides: Treat each slide sequentially through a rehydration series (first xylene then 100%, 95%, 70%, 50% [v/v] ethanol, and finally DEPC-treated water). This also serves to abolish any RNAse activity in the tissue on the slide. Allow 5–10 minutes in each solution for equilibration. All containers should be rinsed once in RNAseZap and twice in DEPC water.
3. Treat slides with proteinase K: Proteinase K treatment is a critical factor in this method, giving the probes an access to the tissue mRNAs. Prepare a humid chamber, which will be used for this step as well as prehybridization and

hybridization buffers (*see* **Note 5**). Cover each slide with 200 μL of 5 μg/mL proteinase K solution (*see* **Note 6**). Incubate in the humid chamber for 15 minutes at 37°C. Wash 2 × 5 minutes with PBT (*see* **Note 7**).

4. Fix with formaldehyde: Proteinase K partly reversed the effects of tissue fixation, so the tissues must be fixed again prior to hybridization, to secure nucleic acids within the cells. In a hood, coat each slide for 20 min at room temperature with the fixation solution and then wash twice (5 min each) with PBT. Dispose of all rinses with formaldehyde properly, utilizing the hood.
5. Dehydrate slides: Sequentially immerse slides for 2 min each in a graded alcohol series, 50%, 70%, and 100% ethanol, and then air-dry the slides in the *unsealed* humid chamber.

3.3. Probe Hybridization

1. Prehybridization: Coat each slide with 200 μL of the prehybridization buffer and incubate for 1 hour at 37°C in the humid chamber.
2. Hybridization: Cover each slide with the specimens with 150 μL of probe in hybridization buffer. Cover each slide with a coverslip and seal each coverslip with rubber cement. First, add a thick layer at the edges of the slide, after that, close the lateral surfaces more precisely. Observe for leaks and, if present, seal the leaks. Hybridize the slides for 72 hours at 37°C in the humid chamber.
3. Posthybridization washes: This step reduces nonspecific binding of the probe. If the background is too high, the concentration of washing solutions, the time, and the temperature of washing should be adjusted first. Remove the coverslips and wash the slides with the following solutions—ensure that the slides are completely covered for each wash:

   1 × 20 minutes 2 × SSC/50% formamide at 37°C in a shaking water bath
   1 × 15 minutes 2 × SSC, room temperature, rocking platform
   1 × 15 minutes 1 × SSC, room temperature, rocking platform.

3.4. Probe Detection

To detect digoxigenin-labeled probe, incubate with an anti-digoxigenin antibody conjugated with alkaline phosphatase.

1. Cover with 200 μL of 1:200 diluted anti-digoxigenin–AP Fab fragments for each slide.
2. Hybridize for 1 hour at 37°C in the humid chamber or overnight in the humid chamber at 4°C.
3. Wash 5 × 5 minutes in PBT. *Do not forget these washes*!
4. Alkaline wash: This step sets the pH in the optimal range of the alkaline phosphatase enzyme. Wash 2 × 5 minutes with

alkaline wash buffer. Do not leave the samples in this buffer for a long time because alkaline pH destabilizes proteins. This buffer should be cold (4°C).

5. Incubation with substrate: Cover the slides with substrate for alkaline phosphatase, BCIP/NBT diluted 1:50 in alkaline wash buffer, utilizing 200 μL for each slide. Place slides in the dark in the humid chamber at room temperature for 20 minutes to 2 hours (*see* **Note 8**). Wash 2 × 5 minutes with PBT.
6. Counterstain using 0.01% Neutral Red solution to highlight the structures of the tissue. Quickly cover the slide with this solution for 10 seconds and rinse twice for 1 minute with DEPC-treated water.
7. Mount the slides with water-based media (Crystal/Mount or similar product). Avoid bubbles! Bubbles will make it difficult to interpret the in situ hybridization and will also dry out the slides. Do not stack the slides before the mounting medium is completely dry or the slides will stick one to another.

3.5. Controls To Be Processed in Parallel

3.5.1. Control for the Integrity of the Tissues

It is crucial to determine if any viable RNA is present within the tissue sample, an issue most important for fixed, paraffin-embedded, older tissues. This is best determined by utilizing a probe to a gene known to be expressed, such as actin or tubulin. If the tissues do not stain positively for actin or tubulin, it can be assumed that the RNA in the samples is substantially degraded. Tissues from this source should not be utilized further for in situ hybridization.

3.5.2. Positive Controls for the Integrity of the Probe

It is important to have a reliable positive control for each probe. We utilize cell lines with known expression patterns for the various members of the Wnt ligand and frizzled receptor families. Typically, this information has been obtained via Northern blot or reverse transcriptase (RT) polymerase chain reaction (PCR). Cells can be fixed, pelleted, embedded in paraffin, and sectioned on a microtome to mimic the tissue slides. Hybridization with these cell line-derived "tissue" slides should be done concurrently. If no hybridization is seen, the probe integrity needs to be questioned and results from the in situ hybridization on the tissue samples cannot be considered reliable.

3.5.3. Controls for the Integrity of the Hybridization

The in situ hybridization utilizes an antisense probe to bind to the sense mRNA, forming a double-stranded complex that can then be detected. Because nonspecific binding and background staining can be a significant problem, all experiments should be conducted concurrently with a sense probe. The sense probe should not bind to the native mRNA molecules and thus serve as an excellent negative control for the hybridization process.

If the sense strand hybridization produces what appears to be a positive result, indicating detection of messenger RNA, then it is likely that all of the positive staining seen is nonspecific. Attempts to reduce nonspecific hybridization, such as additional blocking steps, more efficient proteinase K treatment, or shorter incubation times with substrate should be considered.

4. Notes

1. Probes must have minimal homology to other members of either the Wnt or frizzled family. In general, it is best to utilize a 3′ or 5′ untranslated sequences for generating the probe. These tend to have minimal homology across family members, although a unique stretch of the coding sequence can also be utilized. However, a full sequence database search should be performed prior to settling on a specific sequence to avoid cross-reactivity with other genes. Longer than 120-base pair (bp) intervals of sequence with more than 90% homology with other RNA species from the same organism should be avoided in the sequence chosen for the probe *(2)*. Optimal probe length is between 100 and 500 bp. Longer probes are subject to higher rates of degradation and have increased cross-reactivity. Shorter probes are easily lost during the preparation process. The cloning vector should contain T7 and T3 (or other suitable polymerases) promoters either side of the multiple cloning site so that the inserted complementary DNA (cDNA) fragment will be flanked by two different polymerase promoters. This means that the same vector can be utilized to produce both antisense and sense probes simply by choosing the appropriate polymerase.
2. Occasionally, no probe is seen following gel electrophoresis. This could be because no probe was generated or because the amount is insufficient to permit visualization in the 1.5-μL aliquot. If you are a novice at generating probes, run more on the gel to ensure that you have synthesized the probe correctly. If you use up too much however, you might have to start the synthesis over to generate more for hybridization. If you are experienced with the technique, it is likely that sufficient probe for hybridization is present even if none is seen on the gel electrophoresis. Often, you can proceed with the hybridization and obtain an excellent in situ result even if no band is seen on the gel. You must see a band for your sense probe since this represents your negative control

and you must ensure that you generated the probe prior to hybridization.

3. This is the best place to take a break in the procedure if needed. The probe can be stored at –80°C at this step for up to 7 days prior to use. After 7 days, the integrity of the probe may be compromised.
4. Fixed tissues may have poor-quality mRNA, making the in situ hybridization difficult. You will be testing the quality of the mRNA directly (*see* **Section 3.5**). To save time, however, any tissue sections obtained for use should be as new as possible. For human tissues, however, RNA degradation can occur in the operating room and in the Pathology department prior to fixation. You have no control over these variables and, indeed, have limited if any ability to gather any information related to them. RNA degradation can occur during storage, and in this case, is time dependent. The more recent the samples, the more likely they will be useful for the in situ hybridization.
5. The humid chamber is a box on which the slides can rest horizontally while they are being incubated with proteinase K solution, prehybridization buffer, or probe/hybridization buffer. We utilize a plastic pipette box lid, approximately 10×15 cm. The lid is inverted and on the bottom is placed blotting paper soaked with DEPC-treated 1× PBS. Along the longer length, on each side, is a plastic pipette cut to exactly the length of the lid and treated with RNAseZap and DEPC-treated water. The slides rest horizontally on top of the pipettes. The lid can be covered with plastic wrap to retain humidity.
6. Proteinase K can be kept as 5-μL aliquots of 20 mg/mL proteinase K stored at –20°C. One aliquot is thawed and added to 95 μL PBT. Five microliters of this solution is added to 1 mL of PBT to obtain a solution used to cover the slides (final concentration, 5 μg/mL).
7. Each lot of proteinase K will need to be titrated. Generally, it is best to titrate the proteinase K using slides with cell lines (*see* **Section 3.5**) and a probe identifying a gene known to be expressed. A sense probe will be utilized that should yield no positive signal. If insufficient proteinase K activity is utilized, background levels will be high, and the specific signal will be low. The proteinase K step is the first candidate for any in situ protocol optimization if optimization is required. The best optimization criteria are morphology of the tissue and intensity of the specific staining; too prolonged proteinase treatment or too high proteinase activity lead to the degradation of tissue structures. Too weak treatment gives

suboptimal access of the probe to the mRNA, and low specific signal-to-background ratio.

8. Incubation with substrate may take as little as 20 minutes if you have excellent-quality RNA in the tissues, excellent probe quality, and the message is abundant. However, if the RNA in the tissues is partially degraded, your technique in making the probe is not yet optimal, or you are looking for a gene expressed at low levels, then incubation may require 2 hours. Longer than 2 hours is associated with unacceptable background staining. The shorter the incubation you can get away with, the better. You can "take a peak" at the slides with an inverted microscope periodically to gauge the progress.

Acknowledgments

The authors appreciate the suggestions of Dr. Yi Guo.

References

1. Milovanovic, T., Planutis, K., Nguyen, A., Marsh JL, Lin, F., Hope, C., Holcombe, R.F. (2004) Expression of Wnt genes and Frizzled 1 & 2 receptors in normal breast epithelium and infiltrating breast carcinoma. *Int. J. Oncol.* **25**, 1337–1342.

2. Data Production Processes, Allen Institute for Brain Science, Allen Brain Atlas. http://www.brain-map.org/pdf/ABADataProductionProcesses.pdf.

Chapter 19

Assaying Wnt5A-Mediated Invasion in Melanoma Cells

Michael P. O'Connell, Amanda D. French, Poloko D. Leotlela, and Ashani T. Weeraratna

Abstract

Wnt5A has been implicated in melanoma metastasis, and the progression of other cancers including pancreatic, gastric, prostate, and lung cancers. Assays to test motility and invasion include both in vivo assays and in vitro assays. The in vivo assays include the use of tail vein or footpad injections of metastatic cells, and are often laborious and expensive. In vitro invasion assays provide quick readouts that can help to establish conditions that either activate or inhibit melanoma cell motility, and to assess whether the conditions in question are worth translating into an in vivo model. Here we describe two standard methods for assaying motility and invasion in vitro including wound healing assays and Matrigel invasion assays (Boyden chamber assays). In addition, we and several other laboratories have previously shown that melanoma cells require matrix metalloproteinase (MMP)-2 for their invasion, and have recently shown that Wnt5A treatment can increase the levels of this enzyme in melanoma cells, as demonstrated by gelatin zymography. The use of these techniques can help to assess the migratory capacity of melanoma cells in response to Wnt treatment.

Key words: Melanoma, Wnt5A, Matrigel, Boyden chamber, Invasion, Wound-healing assays.

1. Introduction

Tumor cell invasion occurs via three steps: attachment of cells to the basement membrane, proteolytic dissolution of the basement membrane, and finally movement through the basement membrane into the bloodstream (intravasation). Extravasation at the site of distant metastasis occurs via the homing of tumor cells due to the expression of molecules such as CD44 on both tumor and endothelial cells, and the release of proteolytic enzymes such as the matrix metalloproteinases (MMPs), which allow the tumor cells to digest their way out of the blood vessels into distant

Elizabeth Vincan (ed.), *Wnt Signaling, Volume I: Pathway Methods and Mammalian Models, vol. 468*

Book doi: 10.1007/978-1-59745-249-6

organs *(1)*. This complex process is governed by a delicate balance between angiogenic or pro-metastatic factors such as vascular endothelial growth factor (VEGF) and CD44 *(2, 3)*, and metastasis suppressors such as Kiss-1 and nm-23 *(4)*, and between matrix metalloproteinases and inhibitors of these enzymes *(5, 6)*, or tissue inhibitors of metalloproteinases (TIMPs).

We have previously demonstrated that Wnt5A signaling can increase the invasive ability of melanoma cells, and the increased expression of Wnt5A correlates to increased malignancy in melanoma patients. A recent study from our laboratory has indicated that Wnt5A is able to mediate these effects via the increase of tumor-associated antigens such as CD44, and the decrease of metastasis suppressors such as Kiss-1 *(7)*. Using recombinant Wnt5A we are able to increase the motility of noninvasive melanoma cells, upregulate the expression of CD44, and downregulate the expression of Kiss-1. Conversely, we can inhibit the invasion of highly invasive melanoma cells using small interfering RNA (siRNA) against Wnt5A (ref. *(7)* and **Fig. 19.1a**), downregulate the expression of CD44, and upregulate the expression of Kiss-1, indicating that Wnt5A is able to regulate the metastatic phenotype of melanoma cells. These effects can be mimicked by regulating protein kinase C (PKC), where, for example, inhibiting PKC in highly metastatic cells can inhibit motility in a wound-healing assay (**Fig. 19.1b**). Other laboratories have confirmed the importance of Wnt5A in melanoma progression *(8–10)*, and Wnt5A has also been implicated in the invasion of several other

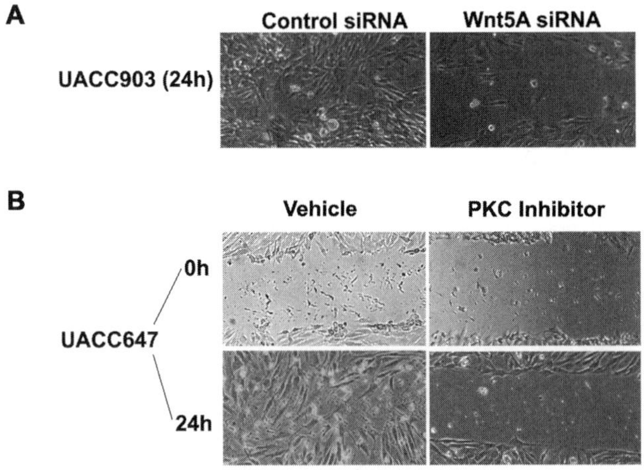

Fig. 19.1. Inhibiting PKC and Wnt5A in melanoma cells results in an inhibition of melanoma cell motility in a wound-healing assay. **A** UACC647 melanoma cells (highly invasive, high Wnt5A) were treated with either a vehicle control or a PKC inhibitor, and subjected to a wound-healing assay. 24 hours post-treatment, vehicle controls had healed the wound, while PKC-inhibited cells could not. **B** Treating highly invasive Wnt5A-high UACC903 melanoma cells with an siRNA against Wnt5A results in a decrease in the ability of these cells to close a wound.

cancers, including pancreatic cancer *(11)*, gastric cancer *(12)*, prostate cancer *(13)*, and non-small cell lung cancer *(14)*. Interestingly, in some cancers such as breast cancer and colon cancer, increased Wnt5A expression signifies a positive outcome for patients *(15, 16)*, indicating that Wnt5A may be acting as tumor suppressor in these patients. Here, we provide detailed protocols for assaying the invasion potential of cancer cells, including wound-healing assays, Matrigel (Boyden Chamber) invasion assays, and zymography.

2. Materials

2.1. Scratch Assay

1. Collagen IV- and fibronectin-coated 24-well plates or slide chambers (BD Biosciences, San Jose CA), *see* **Note 1**.
2. 200-μL yellow p200 pipet tips.
3. Light microscope with imaging capabilities.

2.2. Matrigel Invasion Assay

1. Fluoroblok HTS transwell filters, 12 mm, either 3 μm or 8 μm, available from Fisher Scientific (Rochester, NY), depending on the size of the cell. Nuclear diameter is the most critical measurement and must be smaller than the pore size selected.
2. Reconstituted basement membrane (Matrigel ®), from BD Biosciences.
3. Phosphate-buffered saline (PBS), pH 7.4.
4. 24-well tissue culture plates.
5. Calcein-AM (Molecular Probes, Salem, OR).

2.3. Gelatin Zymography

1. 10% Tris-glycine gel with 0.1% gelatin (Invitrogen, Carlsbad, CA).
2. 4× Zymogram Sample Buffer: To make 50 mL, add 6.25 mL of 2 M Tris-HCl, pH 6.8, 25 mL of glycerol, 5 g of sodium dodecyl sulfate (SDS), 2.5 mL of beta-mercaptoethanol, and 5 mg of bromophenol blue.
3. 10× Zymogram Developing Buffer (BioRad, Hercules, CA).
4. 10× Zymogram Renaturing Buffer (BioRad).
5. Coomassie Blue Stain: 10% (v/v) glacial acetic acid, 30% (v/v) methanol, and 0.25% (w/v) Coomassie brilliant blue.
6. Destaining solution: 10% (v/v) glacial acetic acid and 10% (v/v) methanol.
7. Recombinant MMP-2 (R&D Systems, Minneapolis, MN).

3. Methods

3.1. Wound-Healing Assays

1. Transfect or treat cells as described in **Chapter 12 (Volume 1), Section 3.1**, item 2. Briefly, seed melanoma cells at 0.5×10^5 in fibronectin- or collagen IV-treated 24-well plates and grow cells to 60% confluence for transfection or 80% confluence for treatment (*see* **Note 2**).
2. Once the culture has reached complete confluency, seal a 200-μL plastic pipette tip using a blue flame (*see* **Note 3**). Take the cooled pipette tip and make two scratches in each well, one horizontal and one vertical (*see* **Notes 4** and **5**).
3. Allow the cells to sit for 30 minutes in the incubator and then begin imaging, treating the first image as 0 h (**Fig. 19.1b**). Wounds can heal in as little as 12–24 hours for highly metastatic cells, or may take up to 72 hours for less metastatic cells (*see* **Note 6**).
4. Using the intersection of the two lines, select a spot to follow. Take images of the same field at 0, 12, 24, 36, 48, and 72 hours, using phase-contrast light microscopy. It is advisable to select one spot to follow, but to also assess the overall rate of closure of the entire wound. To do this we image at both ×2.5 and ×10 magnifications.

3.2. Matrigel Invasion Assays

These in vitro invasion assays, also known as Boyden chamber assays, provide an assessment of the three basic steps of tumor cell metastasis—attachment to a basement membrane, proteolytic dissolution of this membrane, and migration through the membrane. Older versions of this assay required that researchers remove the filter from the chamber, scrape off the Matrigel, stain the filter, and assess the number of cells left on the underside of the filter. This method was highly prone to human error. The advent of new filters (Fluorblok ®) that have patented membranes that do not permit fluorescence to pass through them allows for a much more accurate assessment of this process, as cells can be stained with fluorescent markers such as Calcein-AM. The amount of fluorescence that accumulates in the bottom chamber due to migrating cells can then be assessed using a fluorescent plate reader and assays can be followed in real time, in a more quantitative fashion. An added benefit of using Calcein-AM is that it requires a cell to be alive and actively cleaving the ester (AM) in order to produce fluorescence.

1. Two days prior to the assay, place filters, Matrigel, PBS pH 7.4, and pipette tips at 4°C to chill.
2. The day before the assay, take the filters out of 4°C, and place them immediately on ice—be sure to keep them either

in their plastic container, or place them immediately in a 24-well dish that you will use for the assay. Place pipette tips on Parafilm on ice (*see* **Note 7**).

3. Dilute the Matrigel to 80 µg/mL in ice-cold PBS. Use ice-cold tips. Pipette in the Matrigel first, then use a small volume (200 µL or less) to really resuspend the Matrigel before adding the rest of the PBS—make sure the Matrigel is completely dissolved.
4. Pipette 150 µL of 80 µg/mL Matrigel onto each filter (be careful not to touch the tip to the filter), get rid of air bubbles in the solution by lightly running the tip over the solution (*see* **Note 8**).
5. Place filters at 37°C for 2 hours, then allow to dry overnight in the tissue culture hood. Examine filters for even dispersion of the Matrigel. Dispersion is even if the surface looks smooth and slightly cloudy, with no dark holes or gaps in the cloudy layer. Filters can be stored at 4°C for up to 1 week after this point. Make enough filters for the assay, with a few extra in case of uneven coating (*see* **Note 9**).
6. Place cells in serum-free or low-serum media the day before the assay (so that the cells have been exposed to low serum for 16 hours by the time of assay). This allows the cells to respond better to the chemoattractant and also allows them to cycle together. Treat or transfect the cells such that the ideal time point for the treatment coincides with the start of the assay.
7. Two hours prior to commencing the assay, place cells in Calcein-AM. Determine to which concentration of Calcein-AM the cells in question best respond. Typically, it is a good idea to test a range of Calcein-AM from 2 to 10 µM and time ranging from 45 minutes to 2 hours. A typical concentration is 5 µM for 1 hour for melanoma cells (*see* **Note 10**).
8. Place 800 µL of chemoattractant in the bottom of a 24-well plate—this chemoattractant can vary, e.g., conditioned media from 3T3 cells, media containing chemokines, etc.; for basic assessments of invasion, we use standard culture media with 20% (v/v) fetal bovine serum.
9. Gently place the transwell chamber with the filter on top of this solution (in the well)—again make sure there are no air bubbles. Overlay the filter with 150 µL of serum-free or low-serum medium—the same solution that the cells were placed in the night before. Place cells in the 37°C incubator until you are ready to add cells to the filter. At least 30 minutes is required for efficient reconstitution of the Matrigel.

10. After incubation in Calcein-AM, count the cells, and seed 50,000 cells per filter. Bring the volume of cells in the transwell chamber up to 800 µL using no- or low-serum medium (this takes into account the 150 µL that is already in the chamber from the Matrigel reconstitution).
11. Allow to sit for 15 minutes at 37°C, then take the first reading. We use a Cytofluor-4000 from Perkin-Elmer (Waltham, MA), but any bottom-reading fluorescence reader will do. Use an excitation/emission of 480/530 and a gain of 50, using bottom-read fluorescence only (*see* **Note 11**).
12. Take readings every hour for 4 to 6 hours. Invasion can be assessed by subtracting the average of the first read for each condition from the last read. Background fluorescence can be assessed and subtracted from these measurements using a well with only the chemoattractant media in it. Results can be measured in one of three ways: 1) as a fold increase over time zero for each condition; 2) as a fold increase over a control sample (e.g., the least invasive cell line); or 3) the amount of fluorescence/cell can be quantified by plating a serial dilution of Calcein-AM-labeled cells directly into the bottom well, and calculating fluorescence units per cell. However, due to the fact that the levels of Calcein-AM appear to change over the 6-hour period, perhaps due to breakdown within the cells, we find method 2) to be the most informative when assessing changes in invasion. A bar graph is usually the best way to represent the results.

3.3. Zymography

We and others have demonstrated the importance of MMP-2 in melanoma cell motility *(17, 18)*. We have also shown that Wnt5A treatment increases MMP-2 secretion in melanoma cells with low motility, and also increases their invasion in a scratch assay (ref. *(7)* and **Fig. 19.2**). Assaying increases in MMP-2 secretion, or inhibition of MMP-2 is another method to assess changes in the invasive ability of melanoma cells and is assayed using zymography. Zymography can be used to detect and characterize metalloproteinases, collagenases, and various other proteases.

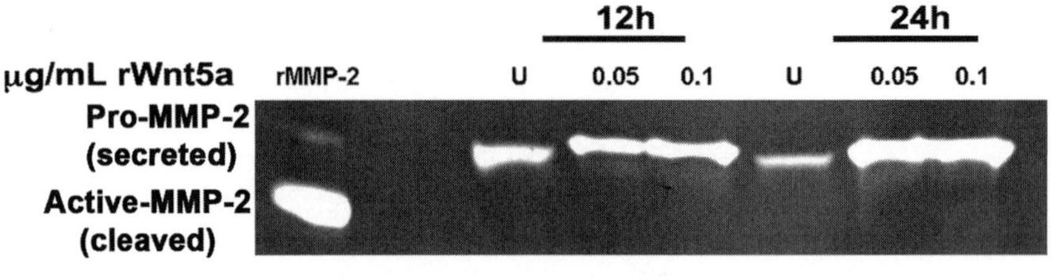

Fig. 19.2. Treatment of melanoma cells with recombinant Wnt5A increases MMP-2 secretion as determined by gelatin zymography. Wnt5A-low UACC1273EV and G361 cells were treated with recombinant Wnt5A at different concentrations, and subjected to gelatin zymography. Wnt5A treatment increases the secretion of MMP-2. *U*, untreated.

The type and quantity of metalloproteinases expressed by tumor cells may allow determination of the metastatic potential of the tumor. Areas of proteolytic activity appear as clear bands against a dark blue background due to proteolytic digestion of the gelatin protein (*see* **Note 12**).

1. Mix one part sample (conditioned medium from cells—*see* **Notes 13** and **14**) with three parts Zymogram Sample Buffer (4×) and let stand 10 minutes at room temperature. Do not heat.
2. Apply samples (typically 10–25 μL containing 30–50 μg) and run the gel with 1× Tris-glycine SDS Running Buffer according to the following running conditions: voltage: 125 V constant, run time: approximately 90 minutes (*see* **Note 15**).
3. After running, dilute the Zymogram Renaturing Buffer (10×) 1:9 and incubate the gel in the buffer (100 mL for two gels) with gentle agitation for 30 minutes at room temperature.
4. Decant the Zymogram Renaturing Buffer and replace with 1× Zymogram Developing Buffer (100 mL for two gels). Equilibrate the gel for 30 minutes at room temperature with gentle agitation then replace with fresh 1× Zymogram Developing Buffer and incubate at 37°C for at least 4 hours (*see* **Note 16**). The optimal result can be determined empirically by varying the sample load or incubation time.
5. Stain with Coomassie Blue solution for 30 minutes to 1 hour. Gels should be destained with Destaining solution. Change destain every 30 minutes for the first 2 hours. Areas of protease activity will show up as clear bands. Stained gels can be wrapped in plastic and stored at 4°C for several months.

4. Notes

1. The type of plate used will affect the motile and invasive capability of the cells. We use plates that have some sort of a matrix on them, usually Collagen IV or fibronectin. The researcher should be aware that the same cell line may exhibit marked differences between the different matrices, depending most likely upon their integrin profile. Matrigel plates cannot be used because a scratch cannot be performed without disrupting the gel. Further, some cells are not suitable for the scratch assay because, once the scratch has been made, the entire layer of cells peels off. Other cells are not suitable because they cannot form a confluent monolayer.

Obviously, this assay can only be used for adherent cells. Therefore, when performing a scratch assay, a range of plates need to be tested for the cell line being used.

2. Because cells have to be completely confluent for the assay, plan transfections and experiments very carefully around the doubling time of each specific cell line. We transfect siRNAs at a slightly higher cell confluency because transfection is still efficient at higher confluencies. Wounds can then be inflicted within 24 hours onto a confluent monolayer. Transfecting DNA is slightly more tricky, as a lower confluency is required for efficient transfection but a confluent monolayer is required for an efficient assay, so these conditions should be optimized for each cell line.
3. The sealing of the tip needs to be complete and clean. When sealing the tip, it needs to be placed very briefly in the blue part of the flame, and a complete seal has occurred once the end of the tip is smooth and round. The exposure to the flame must be short but efficient—an overly long exposure will cause bubbling of the tip, resulting in an overly thick tip, which will inflict too wide a scratch. Overexposure to the flame, or exposure to the yellow part of the flame will result in charring (evident by a smoky residue on the tips), which we find causes cell death around the scratch, perhaps due to toxins that are released when the plastic burns. It is imperative to attempt to just lightly seal the tip, as that will keep the size of the tips consistent, such that scratches do not vary widely in width. A tip that is not completely sealed will result in cells left behind in the center of the scratch. Although images can still be taken whilst the scratch is open, once it begins to close an inaccurate result may develop as the scratch may appear to be closed due to the cells left behind in the center of the scratch.
4. The pressure applied to make the scratch must be firm and even across the entire well. A scratch that has been made using too little pressure results in live cells left in the center of the scratch and leads to the problems associated with this, as mentioned in **Note 3**. A scratch made using too much pressure will scrape the bottom of the well and leave grooves in the matrix. This can be visualized under the microscope as large black lines seen in the middle of the scratch. Often when this occurs, the cells take longer to migrate across these grooves, causing an inaccurate result, although highly metastatic cells may remain unaffected by this.
5. It is important to make two scratches in each well, one horizontal and one vertical. This results in a cross that can be seen under the microscope and helps orient the investigator to the same field. The same field must be used when taking

images as the size of the scratch will vary slightly across the well. Therefore, choosing a similar size scratch across samples and using the same field for imaging is very important. Often cells may die around the scratch, leaving some areas less confluent, which will affect the ability of the cells to invade. Thus, even though one area is followed, it is important to assess invasion across the entire scratch, to get an idea of whether that portion of the scratch is truly representative of the whole field.

6. The media level used must be sufficient to last the entire duration of the wound-healing assay. Changes of media can disrupt the cells at the edges of the scratch leaving debris and causing an inaccurate measure of invasiveness. Therefore, if using a 24-well plate, add at least 1 mL of medium.
7. It is crucial to keep everything ice-cold for the Matrigel invasion assay. We suggest prechilling tips, filters etc, and using Parafilm laid over ice in an ice bucket in the hood as a surface on which to place filters, tips, etc.
8. A significant source of air bubbles can be avoided by not ejecting the solution out all the way, just to the end of the first ejection.
9. For each condition you are testing, it is advisable to do triplicates in each experiment, and repeat each invasion assay three times, as the standard error bars can be quite large.
10. It is advisable to do a trypan blue viability assay to ensure that Calcein-AM is not affecting the viability of the cells.
11. If you do not have a bottom-reading fluorescent reader, it is possible to get data by removing the transwell and assaying only the plate. This will count the fluorescent cells that have migrated all the way through, although not the ones that have migrated through the filter, but are still attached to the underside. Cell counts will be smaller but still representative. This can also be done if Fluorblok filters are not readily available and regular transwell migration chambers are used.
12. As an added confirmation to zymography, we also stain for cell-associated MMP-2, using goat-anti-MMP-2 (R&D Systems), and assess membrane-associated MMP-2 using confocal microscopy. For examples, see refs. *(7, 18)*.
13. It is advisable to use serum-free medium where possible, and to concentrate the medium.
14. Phenol red-free medium should be used because phenol red can interfere with colorimetric quantitation in a BCA assay.
15. Because the gels used are so small, make sure to run them for an extended period of time to fully separate the secreted and active band for better visualization.

16. Many protocols state that zymograms can be developed for 4 hours: we find this to be far too short a period of time for most cells examined.

Acknowledgments

We thank Dr. Michel Bernier and Dr. Paritosh Ghosh for helpful comments on this manuscript. Any data represented in this chapter was generated with the support of funds from the Intramural Research Program of the National Institute on Aging, Baltimore, MD.

References

1. Fidler, I. J. (2002) Critical determinants of metastasis. *Semin Cancer Biol* **12,** 89–96.
2. Bourguignon, L. Y. (2001) CD44-mediated oncogenic signaling and cytoskeleton activation during mammary tumor progression. *J Mammary Gland Biol Neoplasia* **6,** 287–297.
3. Agnantis, N. J., Goussia, A. C., Batistatou, A., and Stefanou, D. (2004) Tumor markers in cancer patients. An update of their prognostic significance. Part II. *In Vivo* **18,** 481–48.
4. Steeg, P. S., Ouatas, T., Halverson, D., Palmieri, D., and Salerno, M. (2003) Metastasis suppressor genes: basic biology and potential clinical use. *Clin Breast Cancer* **4,** 51–62.
5. Hofmann, U. B., Westphal, J. R., Van Muijen, G. N., and Ruiter, D. J. (2000) Matrix metalloproteinases in human melanoma. *J Invest Dermatol* **115,** 337–344.
6. Hofmann, U. B., Houben, R., Brocker, E. B., and Becker, J. C. (2005) Role of matrix metalloproteinases in melanoma cell invasion. *Biochimie* **87,** 307–314.
7. Dissanayake, S. K., Wade, M., Johnson, C. E., O'Connell, M. P., Leotlela, P. D., French A. D., et al. (2007) The Wnt5a/Pkc pathway mediates motility in melanoma cells via the inhibition of metastasis suppressors, and initiation of an epithelial to mesenchymal transition. *J Biol Chem* **282,** 17259–17271.
8. Lewis, T. B., Robison, J. E., Bastien, R., Milash, B., Boucher, K., Samlowski, W. E., et al. (2005) Molecular classification of melanoma using real-time quantitative reverse transcriptase-polymerase chain reaction. *Cancer* **104,** 1678–1686.
9. Bachmann, I. M., Straume, O., Puntervoll, H. E., Kalvenes, M. B., and Akslen, L. A. (2005) Importance of P-cadherin, beta-catenin, and Wnt5a/frizzled for progression of melanocytic tumors and prognosis in cutaneous melanoma. *Clin Cancer Res* **11,** 8606–8614.
10. Rodolfo, M., Daniotti, M., and Vallacchi, V. (2004) Genetic progression of metastatic melanoma. *Cancer Lett* **214,** 133–147.
11. Ripka, S., Konig, A., Buchholz, M., Wagner, M., Sipos, B., Kloppel, G., et al. (2007) WNT5A—target of CUTL1 and potent modulator of tumor cell migration and invasion in pancreatic cancer. *Carcinogenesis* **28,** 1178–1187.
12. Kurayoshi, M., Oue, N., Yamamoto, H., Kishida, M., Inoue, A., Asahara, T., et al. (2006) Expression of Wnt-5a is correlated with aggressiveness of gastric cancer by stimulating cell migration and invasion. *Cancer Res* **66,** 10439–10448.
13. Stanbrough, M., Bubley, G. J., Ross, K., Golub, T. R., Rubin, M. A., Penning, T. M., et al. (2006) Increased expression of genes converting adrenal androgens to testosterone in androgen-independent prostate cancer. *Cancer Res* **66,** 2815–2825.
14. Huang, C. L., Liu, D., Nakano, J., Ishikawa, S., Kontani, K., Yokomise, H., et al. (2005) Wnt5a expression is associated with the tumor proliferation and the stromal vascular endothelial growth factor—an

expression in non-small-cell lung cancer. *J Clin Oncol* **23,** 8765–8773.

15. Dejmek, J., Dejmek, A., Safholm, A., Sjolander, A., and Andersson, T. (2005) Wnt-5a protein expression in primary dukes B colon cancers identifies a subgroup of patients with good prognosis. *Cancer Res* **65,** 9142–9146.
16. Jonsson, M., Dejmek, J., Bendahl, P. O., and Andersson, T. (2002) Loss of Wnt-5a protein is associated with early relapse in invasive ductal breast carcinomas. *Cancer Res* **62,** 409–416.
17. Bartolome, R. A., Molina-Ortiz, I., Samaniego, R., Sanchez-Mateos, P., Bustelo, X. R., and Teixido, J. (2006) Activation of Vav/Rho GTPase signaling by CXCL12 controls membrane-type matrix metalloproteinase-dependent melanoma cell invasion. *Cancer Res* **66,** 248–258.
18. Leotlela, P. D., Wade, M. S., Duray, P. H., Rhode, M. J., Brown, H. F., Rosenthal, D. T., et al. (2006) Claudin-1 overexpression in melanoma is regulated by PKC and contributes to melanoma cell motility. *Oncogene* **26,** 3846–3856.

Chapter 20

Coculture Methodologies for the Study of Wnt Signals

Kestutis Planutis, Marina Planutiene, and Randall F. Holcombe

Abstract

In vivo, responses to extracellular Wnt ligands are context dependent; the temporal characteristics and intensity of the signal are critical in determining the target cell response. In general, Wnt ligand-induced differentiation in mammalian cells requires several days of exposure. In order to better characterize Wnt-induced signaling in vitro, side-by-side and partitioned cocultures can be utilized. These methodologies closely mimic how Wnt signals are transmitted in the tumor microenvironment.

Key words: Coculture, Tumor microenvironment, Temporal response.

1. Introduction

Wnt ligands are morphogens; that is, substances produced by a specific set of cells that are present in a concentration gradient that defines the fate of each cell along this gradient *(1)*. In development, the timing and strength of Wnt signals are critical in defining the response within a target cell population in vivo. Given this, it is important to develop defined in vitro methods that incorporate these characteristics.

Recently, the study of the response to Wnt ligand-induced stimulation in vitro has been facilitated by the utilization of Wnt-containing conditioned medium as described by Willert et al. *(2, 3)*. This approach provides sufficient Wnt ligand to induce a response in a target cell population. Purified Wnts help to analyze functions of separate Wnt ligands *(2, 3)*. Coculture adds the temporal and spatial intricacies that define Wnt responses in vivo. In order to develop an in vitro system that more closely mimicked physiologic conditions, especially those in the tumor

Elizabeth Vincan (ed.), *Wnt Signaling, Volume I: Pathway Methods and Mammalian Models, vol. 468*

Book doi: 10.1007/978-1-59745-249-6

microenvironment, we utilize two different coculture methodologies originally introduced for the study of Wnt signaling over a decade ago *(4)*. In this chapter, we provide detailed descriptions of the methodology and information regarding comparison of responses obtained via side-by-side and partitioned coculture techniques as compared with responses seen following the addition of conditioned medium.

2. Materials

2.1. Materials Required for Partitioned Coculture (Pcc)

1. RPMI-1640 media with 10% (v/v) fetal bovine serum (FBS) or other suitable medium to culture the cells.
2. Cell culture grade 0.2 g/L EDTA per liter solution in PBS at 37°C to remove adherent cells from plastic after transfection.
3. Cell culture inserts, 0.4-μm pore size polyethylene terephthalate (PET) tracketched membrane, 6-well format (Becton Dickinson Labware, Franklin Lakes, NJ; catalog No. 35-3090) to plate the target cells on. Sterile forceps for work with the inserts.
4. RKO cells previously transfected with vector only or Wnt expression constructs, or LWnt3a cells (ATCC, Manassas, VA; #CRL-2647) and control L cells.

2.2. Materials for Side-by-Side (SbS) Coculture

1. SuperTOPflash or control reporters, Wnt expression constructs or vector only to transfect the cells. RPMI-1640 media with 10% FBS for recovery. Cell culture-grade 0.2 g/L EDTA per liter solution in PBS to remove the cells from the plastic.
2. Celltracker Orange CMRA (Molecular Probes, Invitrogen, Carlsbad, CA; catalog No. C-34551; absorption at 548 nm, emission at 576 nm) to stain the suspension of detached transfected cells. Cell culture-grade DMSO to dissolve the Celltracker. Serum-free medium to dilute the stock solution.
3. Lab-Tek II Chamber Slides (4-well, glass, Nalge Nunc International, Rochester, NY) for 24–48 hours for the coculture.

2.3. Materials for Transfection of Cells with Expression Plasmids

1. Six-well dishes or cell culture inserts, to grow the cells.
2. Wnt expression constructs or vector only control. Opti-MEM I Reduced Serum Medium (Invitrogen; catalog No. 11058-021). Lipofectamine 2000 (Invitrogen; catalog No. 11668-019).
3. Minimum essential medium with Earle's salts to rinse the wells, removing antibiotics. Dulbecco's modified Eagle's medium, containing non-essential amino acids and 10% FBS without antibiotics.

4. A benchtop shaker.
5. Sterile PBS to rinse the cells.

3. Methods

3.1. Partitioned Coculture (Pcc)

1. If necessary, transfect the target cells with SuperTOPflash or control reporters prior to plating on culture inserts (**Fig. 20.1**). Culture for 6–12 hours in RPMI-1640 media with 10% (v/v) FBS as a brief recovery period. This step is not necessary if Wnt throughput, as measured by SuperTOPflash activity, is not the desired readout.
2. Remove adherent cells from plastic with cell culture-grade 0.2 g/L EDTA per liter solution in PBS at 37°C.
3. Plate the target cells on cell culture inserts, and grow overnight until 5–10% confluence. Use sterile forceps for work with the inserts (*see* **Note 1**).
4. Place inserts with target cells into wells containing RKO cells previously transfected with vector only or Wnt expression constructs, or LWnt3a cells (ATCC; #CRL-2647) and control L cells.
5. Maintain the coculture for 17 to 69 hours. At the end of the incubation, harvest cells grown on the inserts for immunohistochemistry (**Fig. 20.2**), RNA analysis, or luciferase assays (**Fig. 20.3**).

3.2. Side-by-Side (SbS) Coculture

1. Transfect cells with Wnt expression constructs or vector only and then culture for 6–12 hours in RPMI-1640 media with 10% FBS as a brief recovery period. For luciferase assays, transfect target cells with SuperTOPflash or control reporters prior to plating in 6-well plates, and let them also recover after removal of DNA for 6–12 hours.

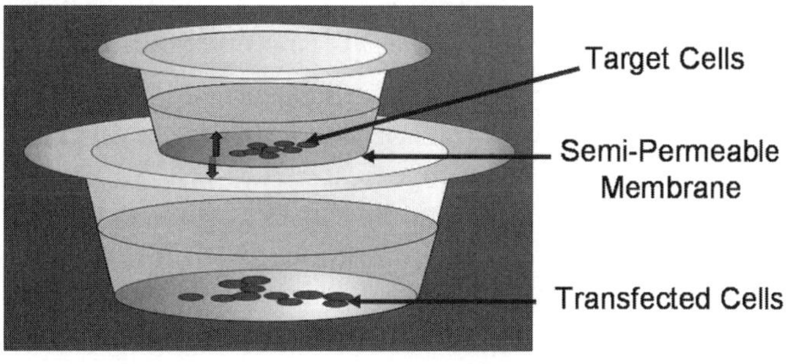

Fig. 20.1. A diagram of the partitioned coculture apparatus.

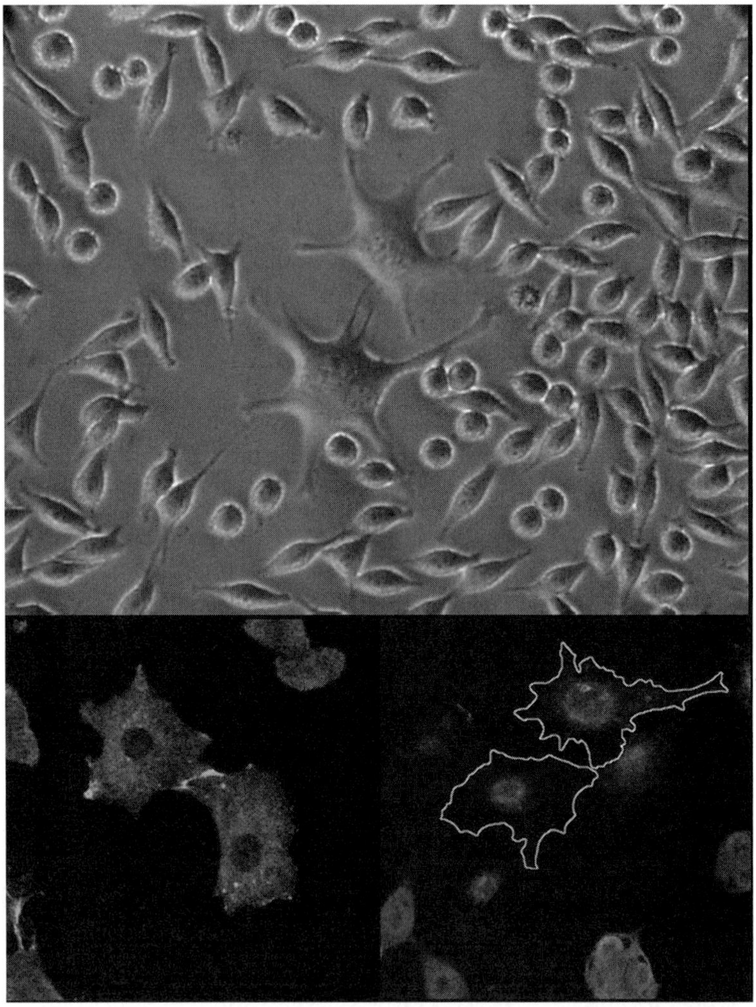

Fig. 20.2. Schematic of the side-by-side coculture (SbS). *Top panel* depicts large endothelial cells cocultured with Wnt-transfected RKO cells. Bottom panels depict β-catenin staining in endothelial cells in SbS coculture with vector-only transfected (*left*) or Wnt2 expression construct-transfected (*right*) RKo cells.

2. Remove adherent cells from plastic with cell culture-grade 0.2 g/L EDTA per liter solution in PBS at 37°C.
3. Stain the suspension of detached transfected cells with Celltracker. Warm up the vial with lyophilized Celltracker to room temperature. Add cell culture grade DMSO to final concentration 10 mM. Dilute the stock solution to a final working concentration of 25 μM in serum-free medium. Warm the working solution to 37°C (*see* **Note 2**).
4. Resuspend the cells and mix with target cells for the coculture in Chamber Slides. Incubate at 37°C for 24–48 hours.

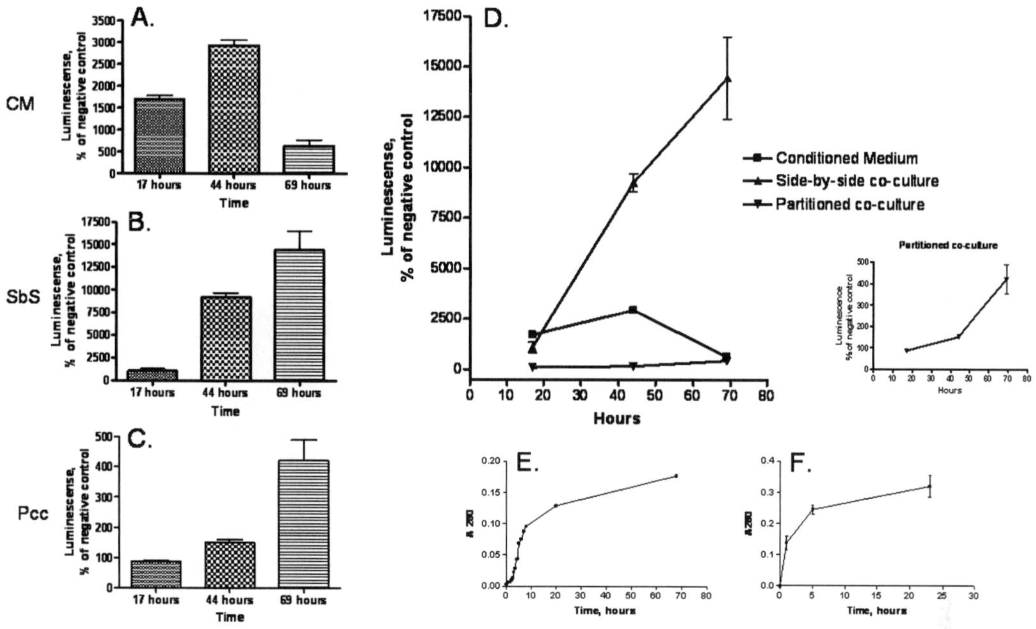

Fig. 20.3. Coculture and conditioned medium treatment have a distinct time course. Comparison of Wnt3a effects on canonical Wnt pathway throughput, measured by the SuperTOPflash reporter construct in RKO cells after treatment with conditioned medium (*CM*), produced by Wnt3a L-cell stables (**A**), side-by-side (*SbS*) coculture (**B**), and partitioned coculture (*Pcc*) (**C**). **D** Wnt throughput responses of each of the methodologies on the same scale (*inset* depicts Pcc at a scale so that the pattern of Wnt throughput can be appreciated). SbS and Pcc have a similar pattern of Wnt throughput stimulation, distinct from the pattern seen with CM. SbS coculture produces the greatest magnitude of response and Pcc, the lowest. The diffusion of BSA through two different types of insert membranes utilized in Pcc is represented in (**E**) (0.4-μm PET track-etched membrane) and (**F**) (0.2-μm Anopore membrane).

5. After incubation, fix the cells for immunohistochemistry (**Fig. 20.2**) or for luciferase assays (**Fig. 20.3**) if target cells have been previously transfected with SuperTOPflash (*see* **Note 3**).

3.3. Transfection of Cells with Expression Plasmids

1. Plate the cells in 6-well dishes or cell culture inserts; grow them until they are ~60% confluent.
2. Transfect cells with Wnt expression constructs or vector only control. For one well, add 2 μg of DNA to 100 μL of Opti-MEM I. Separately, to another 100 μL of the Opti-MEM, add 6 μL of Lipofectamine 2000.
3. Rinse the wells with 1 mL of minimum essential medium to remove antibiotics and then add 0.8 mL Dulbecco's modified Eagle's medium, containing non-essential amino acids and 10% (v/v) FBS without antibiotics.
4. After incubation for not more than 5 min at room temperature, combine both portions of Opti-MEM based reagents, mix on a benchtop shaker at room temperature for 20 min, and layer over the cells and incubate at 37°C for 24 hours.

5. Rinse the cells with sterile PBS, and use for various coculture experiments.
6. At the end of the experimental incubation period, quantify luciferase activity as described in Chapter 9 in this volume.

4. Notes

1. Two types of membranes are available, a 0.4-μm PET track-etched membrane (Falcon, 35-3090; Becton-Dickenson) in 6-well plates and a 0.2-μm Anopore membrane (Nalge Nunc) in 96-well plates. To define the stimulation occurring in Pcc, we have evaluated the transfer of bovine serum albumin (BSA) through both types of semi-permeable membranes to the upper chamber (**Fig. 20.3**). BSA was placed in phosphate-buffered saline (PBS) at a concentration of 20 mg/mL and volume of 2 mL for 6-well plates and 150 μL for 96-well plates in the lower wells prior to placement of the semipermeable membranes in the inserts. The volume of PBS in the inserts was 1 mL for the 6-well plates and 50 μL for the 96-well plates. The BSA concentration in the upper well was measured at 280 nm absorbance at various time points (**Fig. 20.2**). The 0.2-μm membrane permitted the most rapid transfer of BSA, achieving nearly complete transfer by 5 hours. In experiments with the 0.4-μm membrane, nearly complete transfer of BSA was achieved by 10 hours. The maximal concentration of BSA was greatest with the 0.2-μm membrane, approximately 50% greater than with the 0.4-μm membrane. Thus, transfer of BSA was most efficient with the 0.2-μm Anopore membrane. However, the transfer of Wnt3a, as measured by

Table 20.1
Comparison of conditioned medium (CM) with two coculture methodologies

	Advantages	Disadvantages
CM	Fast, strong, easy analysis	Least physiologic
SbS	Strongest, allows cell: cell contact	Requires labeling of cells
Pcc	Easy separation of target cells	Stimulation is slow and weak

SbS, side-by-side coculture; Pcc, partitioned coculture

SuperTOPflash activity of target cells across the membranes, was seen only with the 0.4-μm PET track-etched membrane. These differences are likely due to the intrinsic characteristics of the membranes, particularly their ability to permit passage of highly hydrophobic molecules such as Wnt ligands.

2. In SbS coculture, Wnt producing cells are prestained with orange dye Celltracker Orange CMRA to distinguish them from the cocultured target cells on the same slide.

3. We demonstrate here that both SbS and Pcc can be utilized in studying Wnt ligand-induced responses (**Table 20.1**). Both methodologies appear to have a more physiologic time course than the addition of CM and both may more closely represent the transmission of signals by Wnt ligands in the tumor microenvironment. SbS coculture permits cell:cell interaction and provides the strongest Wnt response. Pcc provides weak and slow stimulation but facilitates the isolation of target cells for molecular studies. In cell culture models for Wnt-induced differentiation and studies aimed at defining interactions between cancer cells and nonmalignant cells as would be found in the tumor microenvironment, the prolonged stimulation obtainable by using coculture methodologies may be superior to other methodologies.

References

1. Mehlen, P., Mille, F., Thibert, C. (2005) Morphogens and cell survival during development. *J. Neurobiol.* 64, 357–366.
2. Willert, K., Brown, J. D., Danenberg, E., Duncan, A. W., Weissman, I. L., Reya, T., Yates, J. R. 3rd, Nusse, R. (2003) Wnt proteins are lipid-modified and can act as stem cell growth factors. *Nature* **423,** 448–452.
3. Willert, K. (2003) *Protocol for the purification of Wnt proteins.* http: //www.stanford.edu/~rnusse/assays/W3aPurif.htm.
4. Jue, S. F., Bradley, R. S., Rudnicki, J. A., Varmus, H. E., Brown, A. M. (1992) The mouse Wnt-1 gene can act via a paracrine mechanism in the transformation of mammary epithelial cells. *Mol. Cell. Biol.* **12,** 321–328.

Chapter 21

Analysis of Wnt/FZD-Mediated Signalling in a Cell Line Model of Colorectal Cancer Morphogenesis

Elizabeth Vincan, Robert H. Whitehead, and Maree C. Faux

Abstract

Cells in tissues do not exist as isolated entities but are part of the three-dimensional tissue architecture. Consequently, some aspects of cell behaviour cannot be mimicked by simple in vitro monolayer culture systems. Moreover, cell shape and behaviour is not rigid but is dynamic and can be regulated by intrinsic and extrinsic factors. For example, tumour cells in epithelium-derived cancer such as colorectal cancer often retain significant features of the colonic mucosa. However, as the tumour progresses, the morphology of the tumour cells often undergoes a transition from an epithelial morphology to a mesenchymal morphology. This transition is important as it signifies a change in the tumour phenotype to a more aggressive, invasive, and eventually metastatic phenotype. In vitro models that allow the study of this transition are needed. One such model is the LIM1863 colon carcinoma cells that normally grow as organoids but can be adapted to efficiently undergo an epithelial to mesenchymal transition that can be reversed. This system has allowed the study of the genes such as Frizzled 7 that are involved in this dynamic and reversible epithelial to mesenchymal transition.

Key words: Frizzled7, Wnt signalling, Colorectal cancer, Tumour morphogenesis.

1. Introduction

The unequivocal link between β-catenin-dependent Wnt signalling and the genesis and progression of colorectal cancer has been made through studies with the intracellular (downstream) components of the pathway *(1, 2)*. Intriguingly, additional roles have recently been uncovered for the upstream components of the pathway (i.e. the ligand and receptor, Wnt and FZD, respectively) *(3, 4)* and the molecules that regulate them (e.g. sFRP *(5, 6)* and DKK *(7)*). However, assigning causative roles to

Elizabeth Vincan (ed.), *Wnt Signaling, Volume I: Pathway Methods and Mammalian Models, vol. 468*

Book doi: 10.1007/978-1-59745-249-6

specific molecules is hampered by the large number of family members and potential functional redundancy; there are 10 FZD, 19 Wnt, 5 SFRP, and 4 DKK genes in the mammalian genome (see the Wnt home page for more information *(8)*). Moreover, the functional outcome of activating a specific Wnt or FZD is dependent on the cellular and tissue context *(9)*. This means that studies in simple cell culture systems may not adequately reflect the in vivo situation. On the other hand, generating genetically modified mice, for example, for in vivo mammalian studies, is time consuming and expensive. These limitations of in vitro and in vivo studies have brought three-dimensional (3D) *(10)* and ex vivo *(11)* mammalian culture systems into the forefront.

Here we describe a dynamic human organoid in vitro culture system that mimics many aspects of colorectal cancer morphogenesis *(4, 12)*. Most importantly, it mimics the dynamic and reversible transition of tumour cells to a mesenchymal and invasive cell phenotype; a transition that is considered to be the initial step in tumour spread to other organs in the body (metastasis). The model system was established by adapting the human colon cancer cell line LIM1863, which contains several epithelial cell types, including goblet and absorptive columnar cells, and grows as organoids in culture *(13)*. The adapted cell line is referred to as LIM1863-*Mph,* for morphogenetic, to distinguish it from the parental cell line and other derivations of LIM1863. Importantly, LIM1863-*Mph* cells are susceptible to retroviral-mediated gene transfer and thus can be manipulated to either increase or decrease gene expression as a means to study function. This dynamic model system facilitates the study of genes, for example FZD7 *(4, 12)*, that are involved in the epithelial plasticity that underscores colorectal cancer morphogenesis. In this chapter, we use an enhanced green fluorescence protein (EGFP)-encoding retrovirus to exemplify the methodology.

2. Materials

2.1. Preparing Retrovirus for Gene Transfer

There are several different methods for preparing replication incompetent retroviruses (*see* **Note 1**). We routinely prepare culture supernatant fluids containing retroviral particles by transiently transfecting HEK293T cells with "helper" plasmids (encode the viral envelope, and the GAG [structural] and POL [polymerase] proteins) and the appropriate retroviral plasmids (encode the gene to be expressed, for example EGFP) *(3, 4, 14)*. Although this is a little labour intensive (*see* **Note 2**), it allows more versatility with choice of retroviral *env* and *gag* products, and the host cell line used for virus production.

1. Phosphate-buffered saline (PBS), pH 7.4.
2. 2× HEPES (*N*-2-hydroxyl piperazine-*N'*-2-ethane sulphonic acid)-buffered saline (HEBS): 280 mM NaCl, 1.5 mM Na_2HPO_4, and 0.05 M HEPES, pH 7.05. Dissolve 4.92 g NaCl, 3.57 g HEPES acid, and 0.063 g Na_2HPO_4 in 250 mL of ddH_2O, adjust pH to 7.05 with 0.5 M NaOH, and adjust volume to 300 mL with ddH_2O. The pH of the solution is critical. Sterilize the solution by passage through a 0.22-µm filter and store as 20- to 25-mL aliquots at –20°C.
3. 2 M $CaCl_2$ solution. Store as 1-mL aliquots at -20°C.
4. HEK293T tissue culture medium (DF-10): DMEM supplemented with 10% (v/v) heat inactivated foetal calf serum (FCS), 20 mM HEPES, and antibiotics (100 U/mL penicillin and 100 µg/mL streptomycin).
5. Trypsin/EDTA solution: 0.05% (w/v) Trypsin 1:250 and 0.02% (w/v) EDTA, pH 7.0.
6. Retroviral and helper plasmids: For example, we have used a plasmid (pRDF) that encodes the envelope protein of the feline endogenous virus, RD114, and a second plasmid (pEQPAM3-E) encoding the GAG and POL proteins of murine leukaemia virus to introduce diverse retroviral plasmids into human colon cancer cell lines *(3, 4)*. To exemplify the methods, we will use a retroviral plasmid (MSCV-EGFP) that encodes the enhanced green fluorescence protein (EGFP). In this plasmid, the reading frames of the inserted gene of interest and EGFP are separated by an internal ribosomal entry site and are transcribed into a bicistronic transcript under the control of the mouse stem cell virus (MSCV) long terminal repeat. The gene of interest and EGFP are co-expressed in the transduced cells and retroviral infection can be monitored by detecting EGFP fluorescence in live cells using fluorescence microscopy (*see* **Note 3**).
7. HEK293T cells, mycoplasma free (*see* **Note 4**), maintained in exponential growth by continuous passage and incubation at 37°C in a humidified incubator in an atmosphere of 10% CO_2.
8. 0.45-µm Filters.

2.2. LIM1863-Mph Cell Culture and Retroviral Infection

1. LIM1863-*Mph* tissue culture medium ($RF\text{-}10^{+}$): RPMI-1640 (*see* **Note 5**) supplemented with 10% (v/v) heat-inactivated FCS, 20 mM HEPES, 100 U/mL penicillin, and 100 µg/mL streptomycin (RF-10), and the following additives ($^{+}$): 0.9 µg/mL insulin, 1 µg/mL hydrocortisone, 10^{-4} M 1-thioglycerol (Sigma, St. Louis, MO; #M6145; add 4.33 µL/500 mL medium). Combine, filter sterilize (*see* **Note 6**), and store at 4°C.
2. Trypsin/EDTA solution: 0.05% (w/v) Trypsin 1:250 and 0.02% (w/v) EDTA, pH 7.0.

3. Retrovirus particle-containing supernatant fluids: Thaw vial just before use, do not freeze and thaw repeatedly (*see* **Note 7**).
4. Polybrene (Millipore): 4 mg/mL stock, store small volume aliquots at –20°C. Polybrene is an infection/transfection reagent that improves retroviral infection of mammalian cells.

3. Methods

3.1. Preparing Retrovirus for Gene Transfer

1. Twenty-four hours before transfection, harvest exponentially growing HEK293T cells and plate them into 10-cm tissue culture petri dishes at 1×10^6 cells/plate in 10 mL of DF-10. The HEK293T cells are harvested after brief exposure to trypsin/EDTA.
2. Three hours before transfection, remove DF-10 and replace with a fresh 10 mL of DF-10 (pre-warmed to 37°C).
3. For each 10-cm monolayer of cells to be transfected, prepare the calcium phosphate–DNA coprecipitate as follows. Place 0.5 mL of 2×HEBS (warmed to room temperature and mixed well before use) in a 10-mL polypropylene test tube. Add the appropriate amount of plasmid DNA (*see* **Note 8**) to a 1.5-mL polypropylene Eppendorf tube and make the volume up to 250 µL with ddH_2O; then add 250 µL of 2 M $CaCl_2$. Slowly (drop wise, while bubbling air into the tube to gently mix the solutions) add the contents of the Eppendorf tube to the test tube.
4. Incubate at room temperature for 20 min.
5. At the end of the incubation, pipette the mixture up and down once, and add it drop wise to the cell monolayer, covering as much of the surface as possible. Return cells to the incubator.
6. After 16 hours of incubation, remove the medium from the petri dish and add fresh warm DF-10 (*see* **Note 9**). Incubate the cells for a further 48 hours.
7. Transfer the culture medium from each petri dish to a separate centrifuge tube and centrifuge at 800×*g* for 5 min. A fresh 10 mL of medium can be added to the HEK293T cells at this stage and another batch of virus collected, however we do not do this routinely.
8. Filter the retrovirus-containing supernatant fluid from each plate through a 0.45-µm filter, aliquot at ~1.5-mL per cryotube, and store at –80°C (*see* **Note 10**).

3.2. Establishing LIM1863-Mph Cultures

LIM1863 cells *(13)* form spheres or organoids that expand in culture by fission as the organoid cells divide and become too large in number for the structure to remain as one organoid. In addition to forming organoids, when LIM1863 cells are cultured in medium with additives (RF-10$^+$), the occasional organoid will spontaneously (i.e. without added cytokines or stimuli) anchor to the tissue culture plastic and form a monolayer patch (**Fig. 21.1a**) (*see* **Note 11**). The LIM1863-*Mph* culture system *(4, 12)* was established by adapting low-passage LIM1863 organoid cells *(13)* to efficiently form monolayer patches. We enriched for the monolayer patches by removing approximately two thirds of the spent medium and free-floating organoids and replenishing with freshly prepared RF-10$^+$. After repeating this several times, the tissue culture flasks were covered with numerous monolayer patches. Strikingly, after several days' incubation, starting at the centre of the monolayer patch (**Fig. 21.1b**), the cells lifted off the tissue culture plastic to assemble new organoids (**Fig. 21.1c**). Importantly, we discovered that the property to efficiently form monolayer patches was preserved when the newly formed free-floating organoids were transferred to a new flask. Thus to passage the cells and perpetuate the LIM1863-*Mph* phenotype, we routinely passage the cells by transferring newly formed organoids to a fresh flask. Within 3 to 4 days, the new flask is littered with monolayer patches, which then go on to generate organoids and so on (*see* **Note 12**). LIM1863-*Mph* cells are incubated at 37°C in a humidified incubator in an atmosphere of 5% CO_2.

3.3. Seeding LIM1863-Mph Cells for Infection with Retrovirus

Because monolayer formation is spontaneous, the number of monolayer patches will vary from culture to culture, which introduces variation between replicate flasks or tissue culture wells. To overcome this, we take advantage of our observation that unlike the mature organoids, which are resistant to enzymatic dispersal *(13)*, the monolayer cells can be dispersed into a single-cell suspension with trypsin/EDTA solution *(4)*. Subsequently, the cells

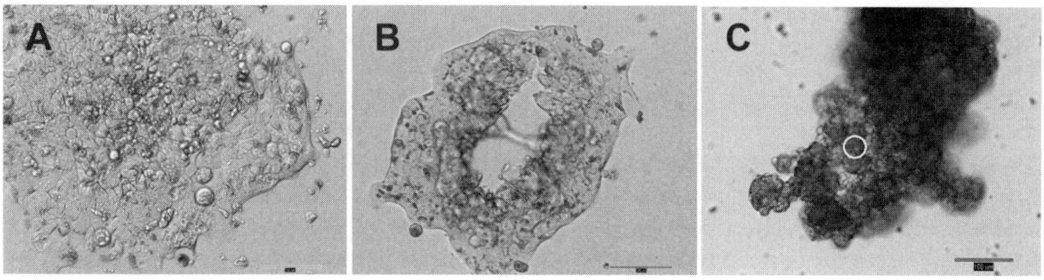

Fig. 21.1. The LIM1863-*Mph* culture system. **A** A spontaneously formed monolayer patch—the LIM1863-*Mph* cells spread out extensively on the tissue culture plastic forming monolayer patches of variable size. **B** Several days after monolayers are formed; the cells at the centre of the patch dissociate and begin to re-assemble organoid spheres with adjacent peripheral cells, generating a ring of organoid assembly. **C** Eventually all the monolayer cells are incorporated into organoids, which remain loosely attached to each other forming clusters of organoids. The *white circle* represents one organoid sphere in the cluster. The newly formed organoids will generate numerous new monolayer patches when transferred to a new tissue culture vessel and the whole cycle begins again. Note that each DIC image is at the same magnification; scale bars = 100 µm.

can be seeded at a density to either preferentially form monolayer patches or anchored organoids (**Fig. 21.2**).

1. Detach monolayer LIM1863-*Mph* cells with trypsin/EDTA. Aspirate medium and free-floating organoids from the tissue culture flask or well, and wash the monolayer cells with warm (37°C) PBS to remove any trace of medium (*see* **Note 13**).
2. Add sufficient trypsin/EDTA to cover the monolayer cells. Leave at room temperature for 2 to 3 minutes, and then aspirate excess trypsin/EDTA, leaving enough on the cells so that they do not dry out. Incubate in a 37°C humidified incubator for 5 min.
3. Check the cells on an inverted microscope. They should be rounding up and, depending on the batch of the trypsin, some of the monolayer patches should be starting to disperse.

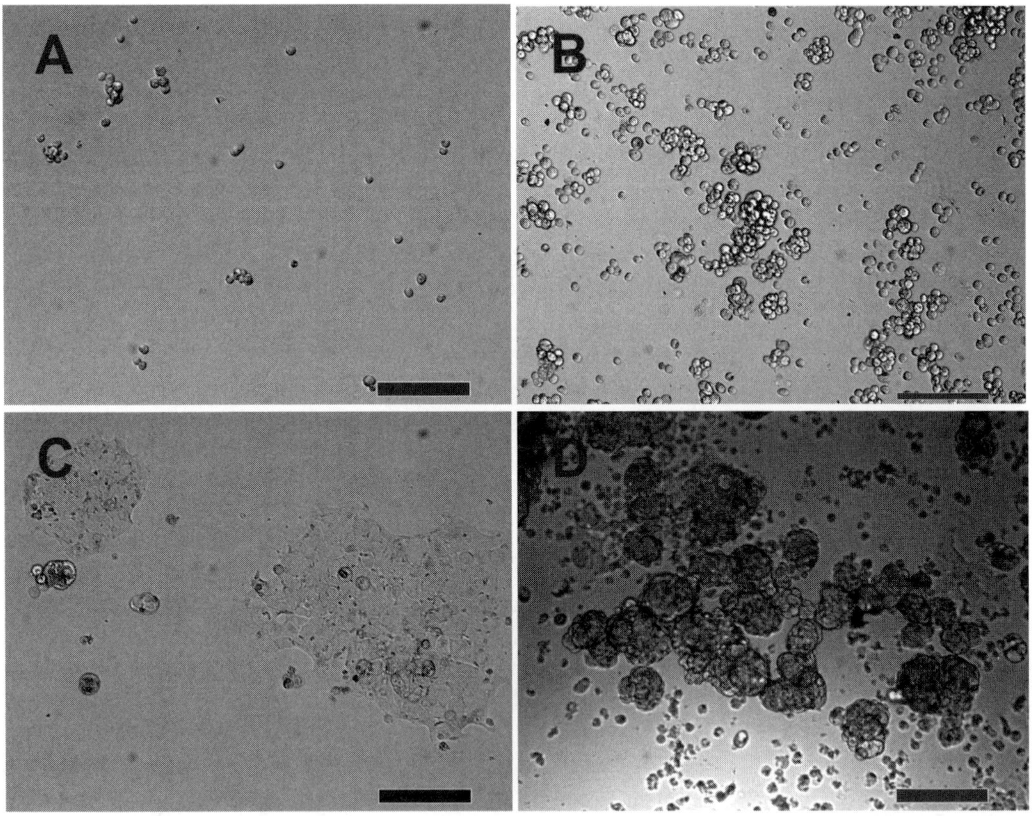

Fig. 21.2. Monolayer and organoid formation by LIM1863-*Mph* cells. LIM1863-*Mph* cells will preferentially form monolayers or organoids, depending on the density at which they are plated. LIM1863-*Mph* cells were added to tissue culture wells at (**A**) low density (2.5×10^4 cells/cm^2) or (**B**) high density (2.5×10^5 cells/cm^2) in RF-10$^+$ medium. After 4 days incubation at 37°C, the cells seeded at low density formed monolayer patches (**C**), while the cells seeded at high density formed organoids that were anchored to the tissue culture plastic (**D**). Note that each DIC image is at the same magnification; scale bars = 100 μm.

Tap the flask sharply to mechanically dislodge the cells and return the flask to the incubator for up to 5 min.

4. Add FCS-containing medium (RF-10$^+$) to "mop up" the trypsin and prevent further trypsinisation of the cells (*see* **Note 13**).
5. Remove any remaining organoids or cell clusters by passing the cell suspension through a 40-μm filter (sieve).
6. Count the cells and seed them in tissue culture vessels at less than 2.5×10^4 cells/cm^2 so that they preferentially form monolayer patches (*see* **Note 14**). Once the cells have formed the monolayer patches, which can take 4 to 10 days (**Fig. 21.2a, c**), aspirate the medium and any organoids that may have formed (*see* **Note 15**) and incubate overnight in fresh RF-10$^+$. The cells are now ready for infection with retrovirus. We routinely use 6-well tissue culture plates and infect duplicate cultures with each retrovirus (or pEQPAM3-E control, *see* **Note 8**) preparation.

3.4. Retroviral Infection

1. Thaw vials of retrovirus (or pEQPAM3-E control) (Section 3.1) just before use and add 7.5 μL of polybrene (5 μL of polybrene per mL of retrovirus-containing supernatant fluid). Aspirate the medium from the LIM1863-*Mph* monolayer cells and cover with retrovirus/polybrene (1.5 mL per well of a 6-well tissue culture plate).
2. Centrifuge the 6-well plates containing the LIM1863-*Mph* monolayer cells and the retroviral supernatant at low speed ($500 \times g$) for 1 hour in a bench top centrifuge fitted with buckets that hold tissue culture trays (*see* **Note 16**).
3. After centrifuging the plates, add an equal volume of RF-10$^+$ to the wells and incubate for 12 hours at 37°C in a humidified incubator.
4. Aspirate the retroviral solution and repeat steps 1 to 3 five times (i.e. two rounds of infection per day for three consecutive days).
5. After aspirating the final retroviral supernatant fluid, add 3 mL RF-10$^+$ and allow the cells to recover for 24 hours before proceeding with experiments.
6. The next steps depend on the retroviral plasmid used. For the MSCV-EGFP retroviral plasmid, the example cited earlier, the cells may be monitored for EGFP fluorescence using an inverted fluorescent microscope (**Fig. 21.3**) (*see* **Note 3**). If the plasmid encodes proteins that confer resistance to an antibiotic that would normally kill the cells, proceed with antibiotic selection (the concentration of antibiotic and length of antibiotic treatment required to kill uninfected cells must be pre-determined).

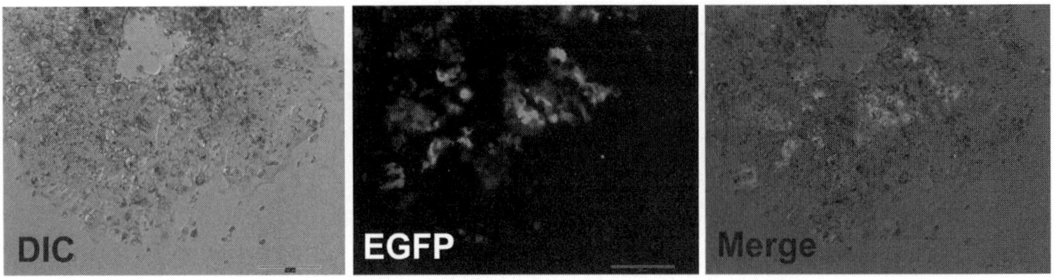

Fig. 21.3. LIM1863-*Mph* cells are susceptible to retroviral infection. LIM1863-*Mph* monolayer cells were infected with MSCV-EGFP retroviral particles. Expression of EGFP protein was detected in the infected cells by live cell fluorescence microscopy. Shown are the DIC (**A**) and fluorescence (**B**) images; and the "merge" when the two images are overlaid (**C**).

4. Notes

1. Retroviruses used as molecular tools for gene transfer do not have the full complement of genes that allow natural replication and cell infection. The retroviral particles are generated in cells that provide the envelope and structural proteins to assemble the retrovirus and allow one round of infection, however the genes for these are not encoded by the retroviral "genome" (the retroviral plasmid being delivered into the target cell) and hence the viruses are referred to as replication incompetent. However, these retroviruses can infect human cells and must be handled in accordance with PC2 guidelines (Class II cabinets etc.; also *see* **Chapter 8, Volume 1**).
2. One can generate stable continuous cells that constitutively express the "helper" plasmid products, which are referred to as packaging cells, so that only the retroviral plasmid (encoding the gene of interest) needs to be transfected for retrovirus production.
3. Once the cells are fixed however, EGFP fluorescence may be lost. To detect EGFP expression in fixed cells, use commercially available antibodies and immunofluorescence microscopy (or immunohistochemistry). Also, note that co-expression of the gene of interest cannot be assumed. In addition to detecting EGFP; the expression of the gene of interest must also be confirmed.
4. It is important to have mycoplasma-free cells for retrovirus production to avoid infecting the LIM1863 cells with mycoplasma.
5. To avoid possible contamination of media with endotoxin, etc., use single-strength commercially available RPMI (e.g. Gibco; Invitrogen, Carlsbad, CA; #11875 is relatively inexpensive if purchased in bulk i.e. 10× 500 mL) and 1 M HEPES solution.

6. The FCS, antibiotics and all the additives (apart from thioglycerol and HEPES, which are stored at 4°C) are stored as aliquots at –20°C, and added to the RPMI and filtered (filtration is an added precaution against contamination of aliquots, since the individual reagents are supplied sterile, and it ensures that any cryoprecipitate in the FCS for example, is removed from the liquid medium). Also note that the single-strength liquid RPMI contains glutamine; however glutamine is unstable at 4°C and needs to be replenished after ~1 month—store aliquots of glutamine at –20°C and add as required. It is also very important to ensure LIM1863-*Mph* cells are not infected with mycoplasma, which can alter their properties. We routinely test all continuously passaged cell lines twice a year for mycoplasma and have a strict policy of screening new cell lines before they are introduced into the "mycoplasma-free" tissue culture area; cells must be tested and confirmed to be mycoplasma-free.
7. The retroviral particles have a lipid bilayer envelope and repeated cycles of freeze–thaw will "lyse" this membrane. Retroviral particles must have an intact envelope that contains the *env* gene product in order to bind to the target cell surface (termed "adsorption"), which is the first step in viral infection.
8. To prepare MSCV-EGFP retrovirus for example, cells would be transfected with pRDF (viral envelope), pEQPAM3-E (GAG and POL), and MSCV-EGFP. We also routinely prepare supernatant fluid harvested from HEK293T cells transfected with the pEQPAM3-E plasmid alone to use for mock infection of target cells.
9. At this stage, the DF-10 medium can be replaced with the medium that the target cells grow in, which in this case would be RF-10^+.
10. It is critical to filter the culture supernatant fluids to ensure that HEK293T cells are not inadvertently transferred to the retrovirus preparation as these will be used to infect the target cells. If the fluids are not filtered, contamination of the target cell population with the rare surviving HEK293T cell is a real risk. Filtration may decrease the viral titre however; carry over of HEK293T cells to the target cell population must be avoided.
11. Formation of monolayer patches is not as efficient in medium without additives, and if cells are continuously passaged as organoids, this monolayer-forming property is selected against and the cells must then be induced with a cytokine to form monolayers *(15)*.
12. Monolayer formation appears to be density dependent; the cultures must be maintained relatively dense by frequent

low-volume passage rather than splitting the cells hard, so that the medium transferred with the cells into the passage flask is not diluted too far. Conditioned medium (supernatant harvested from LIM1863-*Mph* cultures) can be mixed with fresh RF-10⁺ and added to the passage flask to facilitate monolayer formation; this requirement for conditioned medium implicates endogenous cytokine release in this dynamic process. In addition, monolayer formation is most efficient when the organoids are small and grow as clusters of small organoids rather than larger spheres that can be several cells deep.

13. It is important to wash away the medium as it contains FCS, which contains proteins with trypsin cleavage sites that will exhaust ("mop up") the trypsin.
14. If the cells are plated at high density, i.e. 2.5×10^5 cells/cm^2, they preferentially form organoids. One ends up with a bed of single anchored organoids across the surface of the petri dish (**Figs. 21.2b, d**).
15. Note that some tissue culture dishes may not be perfectly flat and the cells tend to roll into the centre before they anchor to the tissue culture well, which increases the cell density and consequently, organoids will form (*see* **Note 14**).
16. This is an old technique that improves virus adsorption/infection, although the mechanism is not understood. This *g* force is insufficient to pellet virus particles, but may somehow alter cell shape or the surface of the cells, which facilitates the infection process.

References

1. Giles, R.H., J.H. van Es, and H. Clevers. (2003) Caught up in a Wnt storm: Wnt signaling in cancer. *Biochim Biophys Acta.* **1653**, 1–24.
2. Clevers, H. (2006) Wnt/beta-catenin signaling in development and disease. *Cell.* **127**, 469–480.
3. Vincan, E., P.K. Darcy, M.J. Smyth, E.W. Thompson, R.J. Thomas, W.A. Phillips, et al. (2005) Frizzled-7 receptor ectodomain expression in a colon cancer cell line induces morphological change and attenuates tumor growth. *Differentiation.* **73**, 142–153.
4. Vincan, E., P.K. Darcy, C.A. Farrelly, M.C. Faux, T. Brabletz, and R.G. Ramsay. (2007) Frizzled-7 dictates three-dimensional organization of colorectal cancer cell carcinoids. *Oncogene.* **26**, 2340–2352.
5. Caldwell, G.M., C. Jones, K. Gensberg, S. Jan, R.G. Hardy, P. Byrd, et al. (2004) The Wnt antagonist sFRP1 in colorectal tumorigenesis. *Cancer Res.* **64**, 883–888.
6. Suzuki, H., D.N. Watkins, K.W. Jair, K.E. Schuebel, S.D. Markowitz, W. Dong Chen, et al. (2004) Epigenetic inactivation of SFRP genes allows constitutive WNT signaling in colorectal cancer. *Nat Genet.* **36**, 417–422.
7. Aguilera, O., M.F. Fraga, E. Ballestar, M.F. Paz, M. Herranz, J. Espada, et al. (2006) Epigenetic inactivation of the Wnt antagonist DICKKOPF-1 (DKK-1) gene in human colorectal cancer. *Oncogene.* **25**, 4116–4121.
8 Homepage, W. http://www.stanford.edu/~rnusse/wntwindow.html.
9. Cadigan, K.M. and Y.I. Liu. (2006) Wnt signaling: complexity at the surface. *J Cell Sci.* **119**, 395–402.
10. Lee, G.Y., P.A. Kenny, E.H. Lee, and M.J. Bissell. (2007) Three-dimensional culture

models of normal and malignant breast epithelial cells. *Nat Methods.* **4**, 359–365.

11. Simpson, M.J., D.C. Zhang, M. Mariani, K.A. Landman, and D.F. Newgreen. (2007) Cell proliferation drives neural crest cell invasion of the intestine. *Dev Biol.* **302**, 553–568.
12. Vincan, E., T. Brabletz, M.C. Faux, and R.G. Ramsay. (2007) A human three-dimensional cell line model allows the study of dynamic and reversible epithelial-mesenchymal and mesenchymal-epithelial transition that underpins colorectal carcinogenesis. *Cells Tissues Organs.* **185**, 20–28.
13. Whitehead, R.H., J.K. Jones, A. Gabriel, and R.E. Lukies. (1987) A new colon carcinoma cell line (LIM1863) that grows as organoids with spontaneous differentiation into crypt-like structures in vitro. *Cancer Res.* **47**, 2683–2689.
14. Kelly, P.F., J. Vandergriff, A. Nathwani, A.W. Nienhuis, and E.F. Vanin. (2000) Highly efficient gene transfer into cord blood non-obese diabetic/severe combined immunodeficiency repopulating cells by oncoretroviral vector particles pseudotyped with the feline endogenous retrovirus (RD114) envelope protein. *Blood.* **96**, 1206–1214.
15. Bates, R.C. and A.M. Mercurio. (2003) Tumor necrosis factor-alpha stimulates the epithelial-to-mesenchymal transition of human colonic organoids. *Mol Biol Cell.* **14**, 1790–1800.

Chapter 22

Analysing Tissue and Gene Function in Intestinal Organ Culture

Helen E. Abud, Heather M. Young, and Donald F. Newgreen

Abstract

The study of growth, differentiation, and migration of different cell types within the developing intestine has been enhanced by the development of methods to grow intestinal tissue in organ culture. Here, we describe the innovative method of catenary culture where the tubular architecture of the intestine is maintained and normal cell differentiation occurs. Rapid analysis of gene function can be achieved using low voltage, square wave electroporation to introduce expression constructs into the epithelial cell layer of cultured explants. This whole-organ culture system allows cells, signalling pathways, and gene function to be analysed in intact explants of embryonic gut that are accessible for experimental manipulation and live cell imaging.

Key words: Organ culture, Electroporation, Intestine, Time-lapse imaging, Epithelium, Enteric nervous system.

1. Introduction

The developing gastrointestinal tract is composed of a lumen lined with epithelial cells and enclosed by multiple cellular layers including connective tissues, smooth muscle, and migratory neural precursors. We describe details of a method for culturing mouse embryonic gut in suspension by attachment to pieces of filter paper (catenary culture) *(1)*. Identical methods work equally well for explants of avian intestine *(2, 3)*. The overall three-dimensional structure of the intestine is maintained with all cell layers intact, which allows signalling between adjacent cell layers to be conserved. Importantly, most aspects of on-going

Elizabeth Vincan (ed.), *Wnt Signaling, Volume I: Pathway Methods and Mammalian Models, vol. 468*

Book doi: 10.1007/978-1-59745-249-6

development in cell differentiation and tissue organisation are maintained in these cultures *(1, 4)*. This permits tissues and signalling pathways, such as the Wnt signalling pathway, to be analysed in a system that accurately mimics intestinal differentiation in vivo but at the same time is amenable to systematic experimental intervention.

We describe details of using the method of low voltage, square wave electroporation to deliver expression constructs to the epithelial cell layer of cultured gut explants. Electroporation was initially developed in avian embryos for manipulating gene expression in the neural tube *(5, 6)*. We have adapted and optimised this technique for the introduction of expression constructs into the epithelial cell layer of intestinal explants *(7)*.

One of the advantages of working with mouse tissue is the repertoire of genetically manipulated animals that are available for analysis. Tissue from either wild-type or a wide range of mutant animals can be utilised in intestinal organ culture. Of particular interest is the ability to mark different cell types with markers such as green fluorescent protein (GFP) or its colour variants, which allows living and fixed cells to be visualised by fluorescence microscopy *(8, 9)*. For example, explants can be prepared from mice in which specific cell types within the embryonic gut express GFP *(10)*. If migratory cell populations such as neural crest cells are to be examined, co-cultures can be also established between explants of gut in which all cells express GFP and gut explants from embryonic wild-type mice; only the migratory GFP-positive cells will migrate into the explants of wild-type gut. Similar co-cultures with avian gut can utilise the chick–quail chimeric system for cell labelling *(3)*.

2. Materials

2.1. Dissecting Solutions and Tissue Culture Medium

All ingredients must be sterile.

1. Dissecting solution. For 1 L of dissecting solution, add 180 mL of 5× Dulbecco's Modified Eagle's Medium (DMEM) (Thermotrace, Noble Park, Australia) and 19 mL of HEPES buffer (Thermotrace). Make up to 1 L with dH_2O. Store at 4°C and use within 7 days.
2. Tissue Culture Medium. For 10 mL of culture medium, add 1.8 mL of 5× DMEM (Thermotrace), 100 µL of 7.5% (w/v) sodium bicarbonate dissolved in dH_2O, 100 µL of 200 mM l-glutamine (Thermotrace), 100 µL solution containing 5.4 mg/mL penicillin and 10 mg/mL streptomycin sulphate (both from Thermotrace) dissolved in dH_2O, and 1 mL of fetal bovine serum (Thermotrace). Make up to 10 mL with dH_2O. Store at 4°C and use within 7 days.

2.2. Support Paper for Gut Explants and Culture Dishes

1. Black Millipore paper (0.45-μm, Black gridded, type HABG; Millipore, Billerica, MA).
2. Tissue culture 60-well HL-A plate, 0.01-mL working volume (Lux 5260; Nunc Inc., Rochester, NY).

2.3. Culture of Explants

1. 37°C, 5% CO_2 incubator.
2. Plastic container with wet paper towels to provide a humidified environment.

2.4. Electroporation

1. EndoFree Plasmid Maxi Kit (Cat No. 12362; Qiagen, Hilden, Germany).
2. Micropipette Puller (P-97; Sutter, Novato, CA).
3. Glass Capillaries (GC100TF-10; SDR Clinical Technology, Sydney, Australia).
4. ECM830 Electroporator (BTX Molecular Delivery Systems, Harvard Instruments, Holliston, MA).
5. Genetrode, L-shape (gold tip) 3MM including holder and cables (514-kit; BTX Molecular Delivery Systems).
6. Scalpel blades.
7. Phosphate-buffered saline (PBS), sterile.

2.5. Imaging of Electroporated Explants

1. Fluorescent dissecting microscope (we use either Stereolumar; Zeiss, Berlin, Germany; or MZ16; Leica Microsystems, Heerbrugg, Switzerland) and digital camera.
2. Upright fluorescent microscope and confocal microscope and associated imaging software.

2.6. Co-culture of Explants

Ret^{TGM} mice in which all neural crest cells in the embryonic gut express GFP *(10)* or mice in which every cell expresses GFP (kind gift of A. Nagy, Mount Sinai Hospital, Toronto, Ontario, Canada)

2.7. Time-lapse Imaging

1. Multiwell cell culture plate (C-24767, Molecular Probes, Invitrogen, Carlsbad, CA).
2. Upright Zeiss Axioskop fixed-stage fluorescence microscope.

2.8. Immuno histochemistry

1. Microscope coverslips (22 × 22 and 22 × 50 mm).
2. 0.1 M phosphate buffer, pH 7.2.
3. Fixative: 4% (w/v) formaldehyde solution in 0.1 M PB, pH 7.2 (prepared from 40% formaldehyde solution) (Ajax Finechem, Taren Point, Asutralia).
4. Permeabilisation solution: 1% (v/v) Triton X-100 (Sigma, St. Louis, MO) in 0.1 M PB, pH 7.2.
5. Antibody dilution buffer: 0.1 M PB, pH 7.2.
6. Mounting medium: Fluorescent mounting medium (Code No. S3023; Dako, Glostrup, Denmark).

7. 15-mm Netwells, Cat No. 3478 (Corning, Lindfield, Australia).
8. Primary antibodies: Anti-β-catenin, (1:200, Cat No. 610154; BD Biosciences, San Jose, CA), anti-GFP (1:100; Rockland, Gilbertsville, PA), and anti-PGP9.5 (1:1,000; The Binding Site, Birmingham, UK). Dilute primary antibodies in 0.1% (w/v) bovine serum albumin (BSA) in PBS.
9. Secondary antibodies: Anti-mouse Alexa-Fluor 488 or 546, Anti-rabbit Alexa-Fluor 488 or 546 (Invitrogen), anti-sheep FITC (Jackson Immunoresearch, West Grove, PA). Dilute secondary antibodies 1:200 in 0.1% BSA in PBS.

3. Methods

In this technique, intact segments of gut are dissected from embryonic mice and suspended across a "V" cut in a piece of filter paper—the cut ends of the gut segments are attached to the filter paper by lightly applied pressure using forceps. The filter paper plus attached gut segment is floated on tissue culture medium in small wells so that the gut explant is suspended within the medium and is not in contact with any surfaces. The explants can then be grown in culture for over 1 week. We have grown the entire post-gastric gastrointestinal tract from E10.5 and E11.5 mice, the esophagus from E11.5 to E14.5 mice, and the entire large intestine and segments of small intestine from E12.5 to E14.5 mice in catenary culture. We have not explored catenary cultures as a method for growing the embryonic stomach. This basic culture method can then accommodate the addition of growth factors and chemical inhibitors, introduction of expression constructs by electroporation, and co-culture with other cells or tissues. The phenotypes can be analysed by live cell imaging (including time-lapse) of a variety of cell populations or the tissue can be analysed by immunohistochemistry after fixation.

3.1. Preparation of Explants

1. Euthanise time-mated pregnant mice and carefully remove the embryos to avoid damaging the gut, which is herniated from around E12.5.
2. Embryos are staged using the system of Theiler *(11)*.
3. Dissect the gastrointestinal tract region of interest. Perform the dissections in dissecting solution or tissue culture medium. The mesentery can be removed or left intact. For example, when growing the caudal small intestine, caecum, and entire large intestine from E11.5 mice (**Fig. 22.1a**), the mesentery between the gut regions is left intact (*see* **Note 1**).

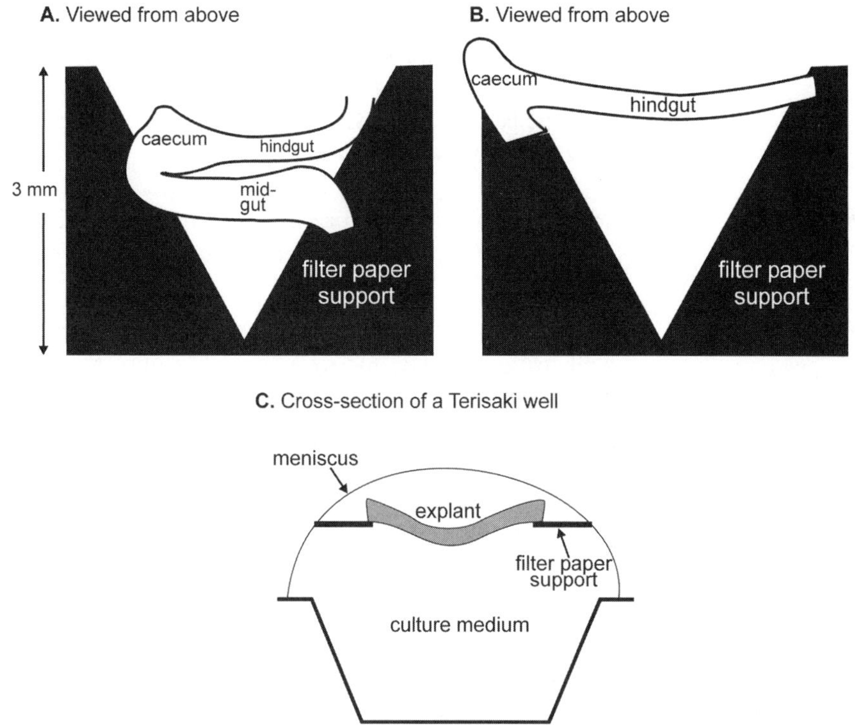

Fig. 22.1. Catenary culture of intestinal explants. Diagrams showing the placement of gut explants on filter paper supports (**A**, **B**) and the floating of the explants in culture medium in a Terisaki well (**C**). Explants of midgut plus hindgut with the mesentery intact from E11.5 mice (**A**), or the hindgut only from E11.5 to E14.5 mice (**B**) can be suspended across the "V" cut in the filter paper support. The filter paper support and attached gut explant is floated on the top of the medium in the Terisaki well. The explant is therefore suspended and not in contact with the surface of the culture dish (**C**).

3.2. Filter Paper Support and Attachment of Gut Segments

1. Place sheets of black filter paper in 70% ethanol for a minimum of 5 minutes to sterilise them and then transfer the sheets into the dissection solution or into tissue culture medium in a small Petri dish.
2. Cut 3 × 3-mm squares of paper (along the grid lines printed on the paper) using a scalpel. Cut a "V"-shaped notch from each square. The "V" shape enables gut segments of differing lengths to be attached—short segments are attached near the apex of the "V", and long segments are attached near the open end of the "V" (**Fig. 22.1a, b**).
3. Attach the cut ends of each gut segment to the paper using lightly applied pressure with forceps. The segments will adhere more easily to the paper if they are in dissecting solution than in tissue culture medium because of the absence of protein in the dissecting solution. Nonetheless, segments of gut will also adhere to filter paper supports in tissue culture medium. If the segments are attached to filter paper supports in dissecting medium, they are then transferred into a small Petri dish containing tissue culture medium for a minimum of 2–3 minutes. The suspended gut

segment needs to be supported during the transfer or else they will not remain attached to the filter paper supports, so each filter paper support and attached gut segment are transferred by sliding the flat side of the scalpel blade underneath them and lifting them out of one solution and into the next.

3.3. Culture of Explants

1. Add 25 μL of tissue culture medium to the required number of wells in a 60-well HL-A tissue culture plate, 0.01-mL working volume.
2. Carefully transfer each explant plus the filter paper support to a well using the flat side of a scalpel, where the explant will float on the surface of the culture medium in the well.
3. Add 25 μL of dissecting medium to some wells in which there are no explants so that the humidity within the plate is high. The cultures are maintained in a 5% CO_2 environment at 37°C for 1–10 days.
4. The medium is changed every day by removing 10 μL of medium with a pipette, adding 10 μL of fresh medium, and then repeating this process several times for each well. If the entire medium from each well is removed, the explants are likely to be dislodged from the filter paper supports or damaged.

3.4. Electroporation

Low-voltage square wave electroporation is used to introduce expression constructs to the epithelial cell layer of cultured intestinal explants of midgut and hindgut. A variety of plasmid expression vectors can be used. It is not necessary to incorporate tissue-specific promoters as constitutive expression vectors such as those using the EF1α promoter can be utilised. It is beneficial to use vectors that permit identification of cells expressing mutant proteins by co-expression of enhanced GFP (EGFP), either as a bicistronic reporter or a fusion protein *(7)*. Plasmid DNA of high quality is required for efficient electroporation (*see* **Note 2**).

1. Prepare DNA using a commercial kit such as the endo-free plasmid maxi prep from Qiagen according to the manufacturer's instructions. Resuspend the DNA at a concentration of 1 mg/mL and store it at –20°C in aliquots.
2. Gastrointestinal tracts from E12.5 to E14.5 embryos are dissected intact in sterile PBS.
3. For injection of DNA into the lumen, glass capillary tubes are pulled using a micropipette puller to produce needles (approximate conditions: heat 720, pull 100, velocity 50, and time 50). The tips of needles are then cut with a scalpel blade under a dissecting microscope to produce a needle of approximately 50 μm in diameter.
4. Flush DNA through the lumen (**Fig. 22.2a**) and attach small segments of intestine to filter supports as described above.

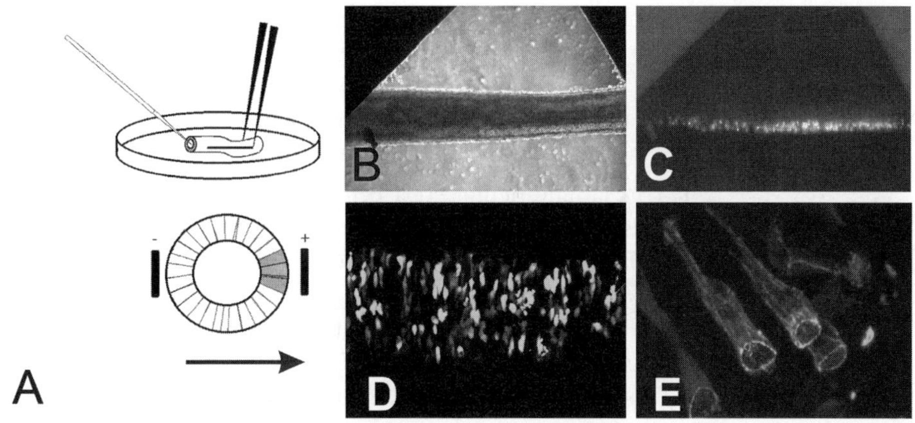

Fig. 22.2. Electroporation of embryonic gut explants. **A** Schematic representation of the electroporation procedure. DNA is injected into the lumen of E12.5 to E13.5 embryonic gut and electric pulses are applied across the segments using wire electrodes. The DNA is specifically electroporated into one side of the epithelial cell layer. **B** Bright-field image of electroporated explant attached to filter support 48 hours after electroporation with an EGFP construct. **C** Fluorescent image of the same explant showing many cells expressing EGFP on one side of the explant. **D** Confocal optical section of the epithelial cell layer showing many EGFP expressing cells. **E** Confocal maximum intensity projection of a live gut explant electroporated with EGFP–actin. Four electroporated cells show the formation of an actin ring at their apical surface. This figure has been modified from the original article in ref. *(7)*; AGA Institute.

It is important for this to be done rapidly with the ends of the segments attached to the filter paper essentially "trapping" the DNA solution in the lumen (*see* **Note 3**).

5. Immediately electroporate the prepared explants by applying pulses through the gut wall. The electrodes are adjusted to be separated by a 4-mm gap with the explant placed between. Electric pulses are delivered unidirectionally for three pulses of a duration of 50 ms. We obtain maximal transfection of epithelial cells using 40 to 45 V (*see* **Note 4**). The electrical pulses are applied unidirectionally to confine expression of constructs to the epithelium on one side of the gut tube, thereby providing an experimental and control region within one explant (**Fig. 22.2a**). Following electroporation, place the explants immediately into culture.

3.5. Imaging of Explants Following Electroporation

1. Following electroporation, expression of GFP can be visualised by 2 to 10 hours post-transfection with maximum expression observed 48 hours after electroporation (**Fig. 22.2b–e**). Explants can be viewed using a fluorescent stereodissecting microscope. In general, when conditions are optimal, most explants will exhibit transfected cells with an efficiency between 5 and 20%.
2. Sections of explants containing many transfected cells can be dissected by reference to GFP expression under the stereomicroscope and mounted for more detailed analysis using methods

such as live cell confocal microscopy. Alternatively, explants can be fixed and processed for immunohistochemistry.

3.6. Co-cultures

To examine interactions between tissues or the behaviour of different cell types, co-cultures can be used.

1. To examine neural crest cell migration, sources of neural crest cells can be placed on the filter paper support in contact with one or both cut ends of the gut explant before the start of the culture period (**Fig. 22.3a**). The sources of neural crest cells could include the neural tube containing pre-migratory neural crest cells or explants of gut from transgenic mice in which the neural crest cells express reporter genes such as GFP or lacZ *(10)*.
2. Co-cultures of gut explants with other tissues or cell types that express or secrete specific growth factors, viruses, and other proteins can also be examined.

3.7. Time-lapse Imaging

1. Explants in which sub-populations of the cells within the embryonic gut express fluorescent transgene markers such as GFP or Ds-Red can be imaged immediately or after growth in culture for several days.
2. Place the explants and filter paper supports in tissue culture medium in a chamber of a multi-well cell culture plate (C-24767; Molecular Probes). Wedge the filter paper support under the Sylgard template that forms the walls of the wells so there is

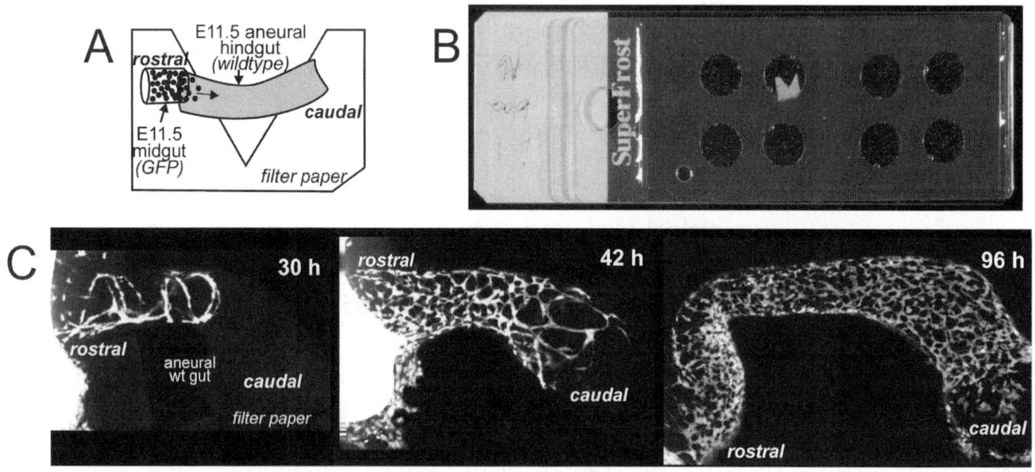

Fig. 22.3. Co-culture of two explants of embryonic gut. **A** Diagram of co-cultures between explants of gut containing GFP+ neural crest cells and wild-type gut. A segment of gut containing GFP+ neural crest cells is placed on the filter paper support abutting one end of a segment of wild-type gut that is suspended across a "V" cut in a piece of filter paper. During the culture period, the GFP+ neural crest cells migrate into the recipient wild-type gut explant. **B** Photograph of an explant mounted on filter paper in a multi-well cell culture plate for time-lapse imaging *(12)*. The explant is wedged under the sides of the wells to minimise movement. **C** Fluorescence micrographs of the co-cultures after 30, 42, and 96 hours. GFP+ neural crest cells from the donor gut colonise the explant of wild-type recipient gut.

minimal movement of the preparations during imaging (**Fig. 22.3b**). Time-lapse imaging can then be performed using conventional fluorescence or confocal microscopy using a microscope with a heated stage (**Fig. 22.3c**). Preparations containing GFP-positive neural crest cells have been imaged for up to 36 hours *(12)*. Time-lapse photography can also be performed on explants that have been electroporated with GFP-fusion constructs. In order to visualise the epithelial cell layer, the cultured explants can be transversely sliced into smaller segments ("doughnuts") using a scalpel blade and mounted on their cut sides on a slide overlaid with dissecting medium (see set up in **Fig. 22.4a**). Growth factors and inhibitors can be added and the effect on the cellular localisation of GFP-fusion constructs analysed *(4, 7)*. These experiments are only suitable for short-term analysis over several minutes (*see* **Note 5**).

3.8. Immuno histochemistry

Explants to be processed for immunohistochemistry can be processed in whole mount, either as intact segments on filter paper supports or in smaller segments, which is more suitable for the analysis of the epithelial cell layer. Explants can also be processed for frozen and paraffin sectioning with the immunohistochemistry performed subsequently on the sectioned tissue. In the latter case, note that exposure to organic solvents leaches the black dye out of the filters and this discolours the tissue, so the tissue should be freed from the support before processing.

1. For whole-mount immunohistochemistry, remove explants from the incubator, and fix them by removing 10 μL of medium with a pipette, and then adding 10 μL of 4% formaldehyde solution in 0.1 M PB and then repeating this process several times for each well. The explants are fixed overnight at 4 °C. *See* **Note 6**.
2. The explants can then either be processed while still attached to the filter paper supports or cut into small segments with a scalpel blade. The small segments of tissue can be processed through the immunohistochemistry steps using netwells.
3. After fixation, wash the explants in 0.1 M PB and expose them to 1% Triton X-100 for 30 min. Then incubate the tissue in the primary antibody overnight at 4 °C with gentle rocking. Wash the explants in 0.1 M PB for 6–8 hours and then incubate in secondary antibody overnight at 4 °C. The tissue is then washed for several hours in 0.1 M PB for several hours before mounting. *See* **Note 7**.
4. Mount explants and filter paper supports on pre-prepared slides on which two 22 × 22-mm coverslips have been stuck with clear nail polish so that there is a small gap between the two coverslips (**Fig. 22.4a**). Place the gut explants and filter paper supports in the gap between the two 22 × 22-mm cover-

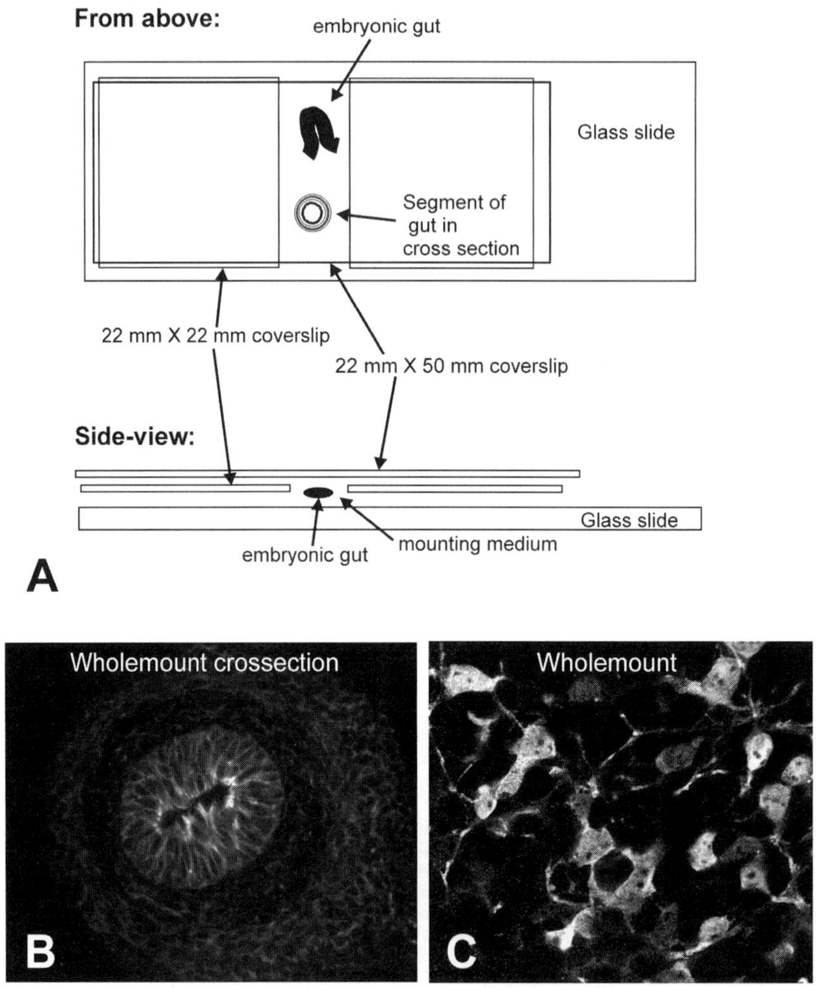

Fig. 22.4. Immunohistochemistry and imaging of gut explants. **A** Diagrams showing method of mounting whole-mount explants of embryonic gut so that the tissue does not get mechanically damaged. Two 22 × 22-mm coverslips are attached to a glass slide with nail polish at least 24 hours prior to use. A space of approximately 5 mm is left between the coverslips, which is where the tissue will be placed. The whole-mount preparations of tissue are placed in mounting medium between the two small coverslips and then a 22 × 50-mm coverslip is placed over the top. **B** Cross section of an explanted E13.5 gut stained with the β-catenin primary antibody and the Alexa-Fluor 488 secondary antibody visualised by confocal microscopy. **C** Whole-mount preparation of an explant of gut from an E11.5 wild-type mouse that had been grown in catenary culture for 3 days. The preparation was then fixed, processed for immunohistochemistry using primary antibodies against the pan-neuronal marker, PGP9.5 and Alexa-488-conjugated secondary antibodies, and examined using a confocal microscope. During the culture period, many PGP9.5+ cells have developed, which extend neurites.

slips, cover with mounting medium, and then place a 22 × 50-mm coverslip over the top. This prevents the explants from being squashed.

5. Segments of explants can be mounted in a similar way by carefully arranging the segments with forceps and adding the mounting medium gently.

6. Leave the preparations overnight to allow the mounting medium to gel, and then seal the edges with clear nail polish.
7. The expression pattern of individual proteins can then be visualised using fluorescence or confocal microscopy.

4. Notes

1. Rapid dissection of tissue from animals is essential to maintain viability of cultured explants. Prolonged dissection times may produce explants with significant necrosis. It is recommended that all materials be prepared in advance and the animals killed immediately prior to dissection in order to minimise dissection times.
2. Electroporation is very sensitive to the quality of construct DNA. Poor-quality plasmid DNA will not electroporate efficiently. DNA may have to be prepared again if electroporation efficiency is low.
3. When using the electroporation technique, it is useful to add a marker dye (such as Fast Green) to the DNA solution. After the lumen is flushed, it is then very easy to see how far the DNA has travelled down the intestinal tract. Any ruptures caused during dissection of the intestinal tract may result in DNA leaking from the lesion rather than remaining in the lumen. Ruptures of this kind and failure to flush the DNA sufficiently throughout the lumen will result in very inefficient electroporation.
4. Different instruments may need the electroporation conditions to be adjusted in order to maximise electroporation efficiency *(13)*. Optimisation can be achieved by varying the voltage in steps of 5 V.
5. Viability of small segments of gut tissue ("doughnuts") placed on their sides for imaging declines rapidly, making these studies only suitable for short-term experiments.
6. A number of different fixatives have been tested including PFA, methanol (–20°C), acetone, and Zamboni. As with other immunolabelling techniques, the best fixative varies with the particular antibody and antigen. However, for fixatives that include organic solvents, remove the tissue from the filter holder before fixation to avoid discolouring the tissue with the dye from the filter. Also, filter deformation, which can occur in organic solvents, hinders mounting of specimens.
7. Fixed intestine whole mounts may be subjected to antigen retrieval (e.g. citrate buffer, pH 6.0, 95°C, 30 min) to improve the binding of many antibodies.

Acknowledgments

The authors acknowledge support for this work from NHMRC project grants 191502 and 400251 to HA, 145628 to HY, and 237144 and 436971 to DFN. We also thank those who have originally worked on developing this technique, especially Lynne Hartley and Catherine Hearn.

References

1. Hearn, C. J., Young, H. M., Ciampoli, D., Lomax, A. E., and Newgreen, D. (1999) Catenary cultures of embryonic gastrointestinal tract support organ morphogenesis, motility, neural crest cell migration, and cell differentiation. *Dev Dyn* **214**, 239–247.
2. Druckenbrod, N. R. and Epstein, M. L. (2007) Behavior of enteric neural crest-derived cells varies with respect to the migratory wavefront. *Dev Dyn* **236**, 84–92.
3. Simpson, M. J., Zhang, D. C., Mariani, M., Landman, K. A., and Newgreen, D. F. (2007) Cell proliferation drives neural crest cell invasion of the intestine. *Dev Biol* **302**, 553–568.
4. Abud, H. E., Watson, N., and Heath, J. K. (2005) Growth of intestinal epithelium in organ culture is dependent on EGF signalling. *Exp Cell Res* **303**, 252–262.
5. Itasaki, N., Bel-Vialar, S., and Krumlauf, R. (1999) 'Shocking' developments in chick embryology: electroporation and in ovo gene expression. *Nat Cell Biol* **1**, E203–207.
6. Muramatsu, T., Mizutani, Y., Ohmori, Y., and Okumura, J. (1997) Comparison of three nonviral transfection methods for foreign gene expression in early chicken embryos in ovo. *Biochem Biophys Res Commun* **230**, 376–380.
7. Abud, H. E., Lock, P., and Heath, J. K. (2004) Efficient gene transfer into the epithelial cell layer of embryonic mouse intestine using low-voltage electroporation. *Gastroenterology* **126**, 1779–1787.
8. Chalfie, M., Tu, Y., Euskirchen, G., Ward, W. W., and Prasher, D. C. (1994) Green fluorescent protein as a marker for gene expression. *Science* **263**, 802–805.
9. Hadjantonakis, A. K. and Nagy, A. (2001) The color of mice: in the light of GFP-variant reporters. *Histochem Cell Biol* **115**, 49–58.
10. Enomoto, H., Crawford, P. A., Gorodinsky, A., Heuckeroth, R. O., Johnson, E. M., Jr., and Milbrandt, J. (2001) RET signaling is essential for migration, axonal growth and axon guidance of developing sympathetic neurons. *Development* **128**, 3963–3974.
11. Theiler, K. (ed.) (1989) *The house mouse: atlas of mouse development.* Springer-Verlag, New York.
12. Young, H. M., Bergner, A. J., Anderson, R. B., Enomoto, H., Milbrandt, J., Newgreen, D. F., and Whitington, P. M. (2004) Dynamics of neural crest-derived cell migration in the embryonic mouse gut. *Dev Biol* **270**, 455–473.
13. Swartz, M., Eberhart, J., Mastick, G. S., and Krull, C. E. (2001) Sparking new frontiers: using in vivo electroporation for genetic manipulations. *Dev Biol* **233**, 13–21.

Chapter 23

Genetics of Wnt Signaling During Early Mammalian Development

Terry P. Yamaguchi

Abstract

Proper cell–cell communication is necessary to orchestrate the cell fate determination, proliferation, movement, and differentiation that occurs during the development of a complex, multicellular organism. Members of the Wnt family of secreted signaling molecules regulate these processes in virtually every embryonic tissue and during the homeostatic maintenance of adult tissues. Mammalian genetic studies have been particularly useful in illustrating the specific roles that Wnt signaling pathways play in embryonic development, and in the etiology of diseases such as cancer. This chapter will largely focus on the functional roles that Wnts, signaling through the Wnt/ -catenin pathway, play during early mammalian development.

Key words: Wnt3a, Beta-catenin, Gastrulation, Mouse, Axis formation, Somitogenesis

1. Introduction

Wnt genes encode a large family of highly conserved secreted glycoproteins that play critical roles in development and disease *(1, 2)*. Starting with the discovery of the first Wnt gene, Wnt1 (Int-1), as a common site of integration for the Mouse Mammary Tumor Virus in breast cancers *(3)*, mice have played a central role in the study of Wnt function. Considerable effort has been devoted to the elucidation of the biochemical pathways that transduce Wnt signals, at least in part because of the role that Wnt pathways play in initiating cancer (see the Wnt homepage http://www.stanford.edu/~rnusse/wntwindow.html for details

Elizabeth Vincan (ed.), *Wnt Signaling, Volume I: Pathway Methods and Mammalian Models, vol. 468*

Book doi: 10.1007/978-1-59745-249-6

on Wnt genes, pathways, and links to the many reviews on Wnt signaling). These pathways include the Wnt/β-catenin pathway, and lesser-understood pathways such as the Wnt/PCP and the Wnt/Ca^{2+} pathways (*see* **Chapters 1** and **10** in this volume for detailed reviews of these pathways). Although studies in many different model organisms and cell lines have established the framework for investigations into the mechanisms of Wnt signaling, the ability to precisely manipulate the germline, coupled with the suitability of the mouse as a model mammalian organism for the study of human disease, makes the mouse a particularly attractive system to study Wnt function in vivo.

2. Brief Overview of the Wnt/β-Catenin Signaling Pathway

β-Catenin is the primary transducer of Wnt signals in the canonical Wnt/β-catenin signaling pathway. In the absence of Wnt ligand, cytoplasmic β-catenin is maintained at low levels due to the activity of the destruction complex. This multiprotein complex, which includes the tumor suppressors adenomatous polyposis coli (APC) and Axin, and members of the glycogen synthase kinase (GSK)-3 and casein kinase (CK)-1 families, binds β-catenin and phosphorylates it on N-terminal residues to target it for ubiquitylation and destruction by the proteasome *(4)*. The binding of Wnt ligands to the seven-pass transmembrane receptor Frizzled (Fz) and the single-pass transmembrane coreceptor lipoprotein receptor-related proteins 5 and 6 (Lrp5/6) promotes the recruitment of Axin to Lrp and Fz-bound Dishevelled, presumably affecting the stability or conformation of the destruction complex. Although the precise mechanisms are not clear, the reduced activity of the destruction complex is thought to be due to the inhibition of GSK3-mediated phosphorylation of β-catenin. Nonphosphorylated β-catenin is stable, accumulates, and translocates into the nucleus where it binds to T-cell factor (Tcf)/lymphoid enhancer factor (Lef) transcription factors. Tcf/Lef proteins bind to specific sequences in the regulatory elements of target genes. When the Wnt/β-catenin pathway is inactive, Tcf/Lef factors repress the transcription of target genes through interactions with the Groucho/TLE repressors. The binding of β-catenin to Tcf/Lef displaces Groucho and recruits transcriptional coactivators, converting Tcf/Lef proteins from repressors to activators of target genes *(5, 6)*. Thus stimulation of the Wnt/β-catenin signaling pathway results in the activation of transcriptional programs of gene expression.

The 19 known Wnt ligands have been traditionally classified as canonical Wnts if they have been shown to stimulate the

Wnt/β-catenin pathway, or as non-canonical Wnts if they have not. Assays for Wnt/β-catenin pathway activation are numerous and varied and include β-catenin stabilization, cell transformation, and β-catenin/Tcf-luciferase reporter assays in vitro, axis duplication assays in Xenopus embryos, and assessment of cells or embryos for the expression of characterized Wnt/β-catenin target genes. Fewer assays are available for the assessment of Wnt activity in alternative signaling pathways. Probably the most widely accepted assay for ligand activity in the vertebrate PCP pathway remains the axis extension assay in Xenopus and Zebrafish embryos. Overexpression of noncanonical Wnt ligands interferes with convergent-extension (CE) cell movements and results in shorter, broader embryos (reviewed in ref. *(7)*), a phenotype distinguishable from the axis duplications observed upon injection of canonical Wnts. Recent studies have blurred the line that distinguishes Wnt ligands from one another. For instance, Wnt5a and Wnt11, long considered non-canonical Wnts, can indeed signal via β-catenin if the appropriate receptor combination is present on the responding cell *(8, 9)*. These results led to the suggestion that pathway selection is not an intrinsic property of the ligand, but rather is determined by receptor context. Nevertheless, lipid modification of Wnts is necessary for canonical Wnt signaling, leaving open the possibility that differential palmitoylation of Wnts could contribute to pathway selectivity (discussed in ref. *(10)*).

Mammalian genetic studies have contributed substantially to our understanding of the function of Wnt ligands during development. *Wnt* genes are generally expressed in a highly localized and temporally regulated fashion in virtually all tissues throughout embryogenesis. The majority of *Wnt* genes have been mutated by conventional knockout studies *(11)*. Although many of these mutations lead to dramatic developmental defects, a significant number result in viable, fertile animals when homozygous null. This is presumably due to genetic redundancy, since many *Wnt* genes are expressed in overlapping spatial and temporal domains in numerous tissues. Surprisingly few of these viable *Wnt* mutants (discussed below) have been published, or examined directly in double mutant tests for redundancy, likely due to the insurmountable depression that the developmental biologist experiences following the discovery of a viable phenotype. Targeted null mutations of genes encoding central components of Wnt signaling pathways have led to interesting phenotypes that either manifest as early embryonic lethal, necessitating the use of conditional alleles to address function in later tissues, or as highly specific phenotypes that are only apparent at relatively late developmental stages due to genetic redundancy. This broad range of phenotypes has frequently made the assignment of a given Wnt to a specific signaling pathway somewhat ambiguous. This chapter reviews recent progress in our understanding of the genetic networks underlying Wnt signaling during early mammalian preimplantation and postimplantation development.

3. Wnt Signaling During Preimplantation Development

The fertilized egg is totipotent, that is, it possesses the capacity to generate all differentiated cell types. Shortly after fertilization, the conceptus undergoes a series of reductive cleavage divisions that lead to the first establishment of polarized cells, and to the first cellular differentiation events. These early events result in the formation of the blastocyst by embryonic day (E) 3.5 (reviewed in ref. *(12)*), which consists of an outer epithelial extraembryonic trophectoderm (TE), and an internal clump of inner cell mass (ICM) cells (**Fig. 23.1**). The ICM is asymmetrically located within the TE due to the formation of the blastocoel cavity, leading to a distinct proximal–distal axis defined by the ICM at the proximal pole. The TE interacts directly with the maternal uterine wall during implantation (E4.5), and subsequently contributes to the trophoblast lineages of the placenta. The TE that surrounds the blastocoelic cavity is termed the mural TE, while the TE directly in contact with the epiblast is known as the polar TE. The ICM gives rise to the epiblast (which produces all of the cells of the embryo proper) and the overlying primitive endoderm (PE)

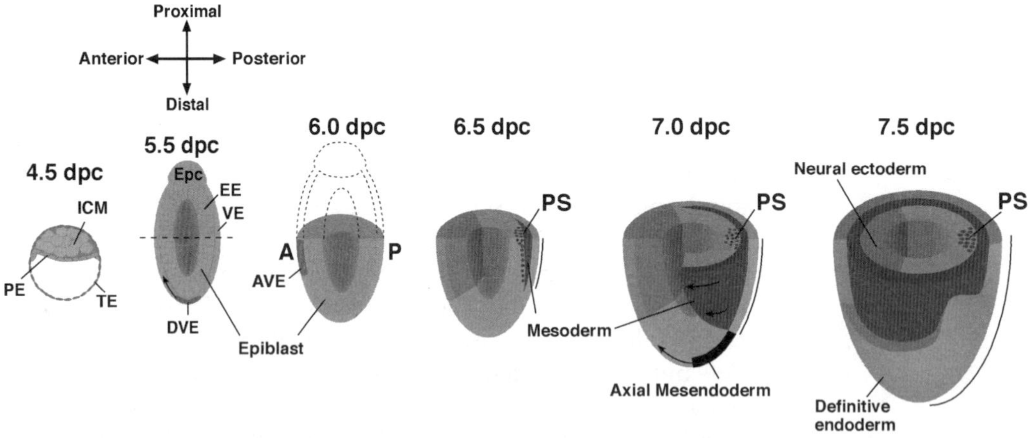

Fig. 23.1. Perimplantation and postimplantation stages of mammalian development. Schematic diagrams illustrating perimplantation (4.5 days post-coitum (dpc)) and postimplantation (5.5 dpc onward) mouse embryos. The *dashed line* overlying the 5.5-dpc embryo demarcates the extraembryonic ectoderm above, from the contiguous epiblast below. Please see the text for details. *PE*, primitive endoderm; *TE*, trophectoderm; *ICM*, inner cell mass; *Epc*, ectoplacental cone; *EE*, extraembryonic ectoderm; *VE*, visceral endoderm; *DVE*, distal visceral endoderm; *AVE*, anterior visceral endoderm; *PS*, primitive streak.

around the time of implantation, and the PE subsequently generates the visceral endoderm (VE) and parietal endoderm.

At least 6 *Wnt* genes are expressed in overlapping but tissue-specific patterns during preimplantation development *(13, 14)*. For instance, *Wnt1* expression appears restricted to the ICM, while *Wnt3a* messenger RNA (mRNA) is detected throughout the blastocyst. Although presumed null mutations have been generated in all 19 *Wnt* genes *(11)*, preimplantation phenotypes have not been observed in any single *Wnt* mutant. Genetic redundancy may explain the lack of an early phenotype given that many components of the Wnt/β-catenin pathway are expressed in the preimplantation embryo *(15, 16)*. However, nuclear localization of β-catenin has not been observed in preimplantation embryos *(17)*, nor has the expression of β-catenin/Tcf-*lacZ* reporter transgenes been detected *(17–21)*, suggesting that the Wnt/β-catenin pathway is not active in the embryo at these stages. Consistent with these studies, preimplantation phenotypes were not observed in either loss-of-function (LOF) or gain-of-function (GOF) studies of β-catenin, nor in single or double mutants of any other component of the Wnt/β-catenin or Wnt/PCP pathway examined to date, including *Lrp5*, *Lrp6*; *Dvl1*, *Dvl2*, or *Dvl3*; *Tcf1*, *Tcf3*, *Tcf4*; and *Lef1*, *Vangl2*, and *Celsr1*, among others *(11)*. These results strongly suggest that the mammalian embryo does not require the Wnt/β-catenin or Wnt/PCP signaling pathways for proper preimplantation development.

Interestingly, Wnt ligand expression by the blastocyst may serve to coordinate the embryo–uterus interactions necessary for implantation. A β-catenin/Tcf *lacZ* reporter is expressed in the maternal uterine epithelium at sites of embryo implantation, and expression is dependent upon the presence of the blastocyst *(22)*. This activation can be mimicked by injection of beads coated with Wnt7a- but not Wnt5a-expressing cells into the uterine lumen. Lumenal injections of the Wnt antagonist sFRP2 significantly inhibited blastocyst implantation, suggesting that a Wnt ligand(s) is necessary for successful implantation. While these results are intriguing, formal proof that β-catenin is required in the uterine epithelium for embryo implantation is lacking. Future studies aimed at the identification of relevant target genes in the epithelium will be necessary to understand the underlying mechanisms.

4. Wnt Signaling and the Establishment of the Body Plan

Wnt signaling pathways play crucial roles in the development of the postimplantation embryo. Since the role of the Wnt/β-catenin pathway in body plan formation has been reviewed *(23, 24)*, this section provides a brief overview of the important developmental processes that occur after blastocyst implantation, and then focuses on recent advances in our understanding of the role of Wnt signaling in these processes.

The embryo undergoes major morphogenetic changes after implantation as the spherical blastocyst transforms into an elongated, cup-shaped structure known as the egg cylinder. The egg cylinder can be considered as two regions, an embryonic and extraembryonic region that is composed of three distinct tissues, the embryonic epiblast, the extraembryonic ectoderm located proximally to the epiblast, and an overlying sheet of VE that cocoons the epiblast and extraembryonic ectoderm (**Fig. 23.1**). Recent studies suggest that morphogenetic rearrangements of the polar TE occurring shortly after implantation lead to the folding and formation of the extraembryonic ectoderm *(25)*. This morphogenetic folding, together with increased proliferation of polar TE *(26)*, drives the growth of the epiblast and extraembryonic ectoderm in a distal direction, resulting in their envelopment by the VE.

Although the extraembryonic ectoderm and VE contribute only to extraembryonic tissues, both tissues are in direct contact with the epiblast and are crucial for the establishment of the anterior–posterior (AP) body axis. The epiblast is an epithelium of pluripotent stem cells that gives rise to all the cells of the embryo proper including the germ line. The imparting of positional information to the epiblast is crucial for the establishment of the body plan. The formation of the AP axis is a fundamental first step in the establishment of this plan as the anterior pole defines where the head will form, while the posterior pole defines the site of primitive streak (PS) and ultimately tail formation. The VE plays a particularly important role in the specification of the AP axis. Around E5.5, VE cells located at the distal end of the egg cylinder (DVE; **Fig. 23.1**) migrate proximally and asymmetrically to become the anterior VE (AVE) *(27, 28)*. The AVE defines the prospective anterior pole, as surgical removal of the AVE results in the loss of anterior epiblast and neurectoderm fates *(29)*. Concomitantly, the expression of powerful developmental signaling molecules including the Tgf-β family member, *Nodal*, *Wnt3*, and *Wnt8a* in the proximal epiblast shifts posteriorly to define the prospective site of PS formation (and consequently, mesoderm formation) at the posterior pole (**Fig. 23.1**). The restriction of the AVE to the distal tip of the embryo, AVE migration, and the expression of proximal posterior epiblast markers such as *Nodal* appears to be under the control of signals secreted by the extraembryonic ectoderm *(30)*. Thus interactions between extraembryonic ectoderm, VE, and the epiblast, are responsible for the formation of the AVE and the posterior localization of gene expression in the epiblast, thereby converting proximal–distal polarity into AP polarity. This phenomenon is frequently referred to as axis conversion or rotation.

The Wnt/β-catenin pathway plays an important role in the initial establishment of the AP axis. Embryos homozygous for a null allele of β-catenin do not initiate AP axis formation due to the abnormal specification of the DVE, and the failure of the DVE to rotate anteriorly *(31)*. Chimera studies demonstrate that β-catenin is required in epiblast, and not in VE cells, suggesting that β-catenin regulates the specification and movement of VE cells in a non-cell autonomous fashion. One way that the β-catenin pathway may indirectly regulate DVE specification and movement may be via the Nodal signaling pathway. Nodal signaling from the epiblast is essential for the formation of the DVE *(32)*. The Nodal co-receptor *Cripto*, which is also required for the anterior movement of the DVE *(33)*, is a direct target gene of the Wnt/β-catenin pathway and is not expressed in β-catenin null mutants *(34)*. These and other results suggest that β-catenin regulates DVE migration by facilitating Nodal signaling in the epiblast through the activation of *Cripto*.

Wnt/β-catenin signaling also has important roles in DVE migration and AP axis formation that are independent of the Nodal pathway. The paired-type homeobox gene *Otx2* is another important regulator of AP axis formation. The DVE forms in *Otx2*$^{-/-}$ embryos but it does not rotate anteriorly *(35, 36)*. Expression of ***dickkopf1*** (*Dkk1*), which encodes a secreted negative regulator of Wnt signaling, is normally found in a proximal subset of wild-type DVE cells and then becomes asymmetrically localized as DVE cells migrate anteriorly. Although several DVE markers were expressed in *Otx2*$^{-/-}$ embryos, *Dkk1* expression was not initiated, suggesting that it might play a role in directional migration of DVE cells as a downstream target of Otx2 *(37)*. Gel shift assays and promoter–reporter analyses in transgenic mice demonstrated that Otx2 indeed directly activates the *Dkk1* promoter. Remarkably, the forced expression of *Dkk1* in the *Otx2* spatial domain, by "knocking-in" a *Dkk1* complementary DNA (cDNA) into the *Otx2* locus, led to the partial rescue of axis rotation defects. These results suggest that Otx2 regulates anterior specification, at least in part, by activating *Dkk1* and thereby inhibiting the Wnt/β-catenin pathway. This is supported by the observation that genetically reducing Wnt/β-catenin signaling in *Otx2*$^{-/-}$ mutants by deleting one copy of the β-catenin gene (*Ctnnb1*), also partially rescues the anterior migration defect of DVE cells. Moreover, cultures of whole E5.5 embryos with beads coated with recombinant protein demonstrated that Dkk1 attracted migrating DVE cells, while Wnt3a repelled them *(37)*. Clearly, *Dkk1* plays an essential role in anterior specification since a targeted mutation of *Dkk1* leads to the loss of anterior head structures *(38)*. Together, these results suggest that asymmetrical Wnt/β-catenin signaling promotes the anteriorly directed migration of DVE cells necessary for the specification of the AP body axis.

5. Gastrulation and the Regulation of Mesoderm Formation by Wnt Signaling

Gastrulation is a complex morphogenetic process that is responsible for the formation of the mesoderm and definitive endoderm germ layers and for the elaboration of the body plan. Gastrulation begins at E6.5 with the formation of the PS, a transient developmental structure that forms at the posterior terminus (**Fig. 23.1**) and converts pluripotent epiblast stem cells into mesoderm and endoderm progenitors. Many powerful inducing factors, including members of the Wnt, Fgf, and Tgfb/Bmp families, are expressed in the PS where they play important roles in the regulation of cell fates. Epiblast cells fated to give rise to mesoderm cells undergo an epithelial-to-mesenchymal transition and generally ingress and migrate away from the PS to generate the mesodermal germ layer.

The Wnt/β-catenin pathway plays a major role in the formation of the PS during gastrulation. Both *Wnt3* and *Wnt8a* are radially expressed in the proximal epiblast prior to gastrulation and become posteriorly localized, forecasting the site of PS formation, after the DVE migrates anteriorly to orient the AP axis *(37, 39)*. *Wnt3* but not *Wnt8a* plays an essential role in PS formation since *Wnt3*$^{-/-}$ embryos lack a PS and mesoderm *(39)*, while *Wnt8a* mutants appear normal (T. Yamaguchi and A. McMahon, unpublished observations). The lack of a phenotype in the *Wnt8a* mutants is presumably due to redundancy with *Wnt3*, although this has not been tested directly. Analyses of AVE markers indicate that *Wnt3–/–* mutants are able to establish a correctly oriented AP axis, however the overlying ectoderm does not acquire anterior neural identity. Thus the absence of both anterior neural identity and posterior structures demonstrates that Wnt3 is necessary for patterning of the epiblast along the entire AP axis. Wnts appear to function as posteriorizing signals since ectopic expression of Wnt8a leads to the enhanced formation of posterior structures while reducing the formation of anterior neuroectoderm *(40)*.

Recent experiments have shed light on the molecular signals responsible for the activation of *Wnt3* in the proximal posterior epiblast and for the positioning of the PS in the posterior embryo. It has been known for some time that *Nodal* is required in the epiblast for gastrulation and PS formation *(41, 42)*. Nodal activity likely determines the site of PS formation since combined mutations in the Nodal antagonists *Cerberus-like* and *Lefty1*, which are expressed in the AVE, lead to ectopic Nodal signaling and the formation of multiple PS *(43)*. Nodal appears to regulate PS formation by regulating *Wnt3*, but does so indirectly through the activation of *Bmp4* in the extraembryonic ectoderm, which in turn signals back to the epiblast to induce *Wnt3 (32, 44)*. Wnt3 completes a positive feedback loop by stimulating *Nodal* expression in the epiblast. It has been suggested that *Nodal* is a direct target gene of Wnt3 since Tcf binding sites were identified in the proximal epiblast enhancer *(44)*, however the functional relevance of these sites has not yet been tested by mutational analyses. Nevertheless, these studies clearly demonstrate that reciprocal inductive interactions between the epiblast and contiguous extraembryonic tissues are important for the transcriptional activation of *Wnt3* and the subsequent formation of the PS.

Although not formally demonstrated through genetic means, in vitro studies and comparisons of phenotypes and gene expression profiles of *Wnt3*-null and β-catenin-null mutants are consistent with Wnt3 signaling via β-catenin *(31, 34, 45, 46)*. In general, antibody reagents detect active β-catenin protein at, or near, sites of *Wnt3* and *Wnt8a* expression in the pregastrulation embryo, however the precise locations are often inconsistent with reported sites of β-catenin/Tcf-lacZ reporter activity. For instance, cytoplasmic dephosphorylated β-catenin is most strongly expressed in the extraembryonic VE at pregastrulation and early gastrulation stages, and is poorly detected, if at all, in the nucleus and cytoplasm of epiblast cells at these stages, despite the strong expression of Wnt ligands there *(17, 37)*. In contrast, β-catenin/Tcf *lacZ* reporters (see ref. *(18)* for review) are only weakly expressed in extraembryonic tissues, but are strongly activated in the posterior epiblast as the PS forms *(17, 19, 20)*. These reporter studies, together with the previously mentioned chimera analyses demonstrating that β-catenin function is required in the epiblast and not in the VE *(31)*, call into question the relevance of the β-catenin expression detected in the VE.

While it is clear that *Wnt3* is essential for the formation of the earliest mesoderm to arise in the PS, *Wnt3* expression is down regulated in the PS by E7.5 *(47)*, 6 days before mesoderm formation ceases in the PS-derived tailbud at E13.5. Several other *Wnt* genes are expressed in the PS at E7.5, including *Wnt2b*, *Wnt3a*, *Wnt5a*, *Wnt5b*, and *Wnt11* (**Fig. 23.2**) *(48)*, and presumably compensate for the absence of *Wnt3* expression. *Wnt3a* is first expressed in the PS at E7.5, and appears to function as the primary mesoderm inducer at this stage since *Wnt3a* null embryos die at mid-gestation with no apparent posterior mesoderm, PS, or tailbud, and instead display ectopic neural tubes *(48, 49)*. The construction of a genetic series of *Wnt3a* alleles, through the breeding of animals carrying null and hypomorphic mutations, clearly demonstrate that *Wnt3a* plays a central role in the formation of all trunk and tail mesoderm *(50)*.

Multiple lines of genetic evidence convincingly demonstrate that Wnt3a regulates mesoderm fates by signaling through the Wnt/β-catenin pathway. Conditional inactivation of *Ctnnb1* (β-catenin) in the PS at E7.5 (the stage at which *Wnt3a* is first

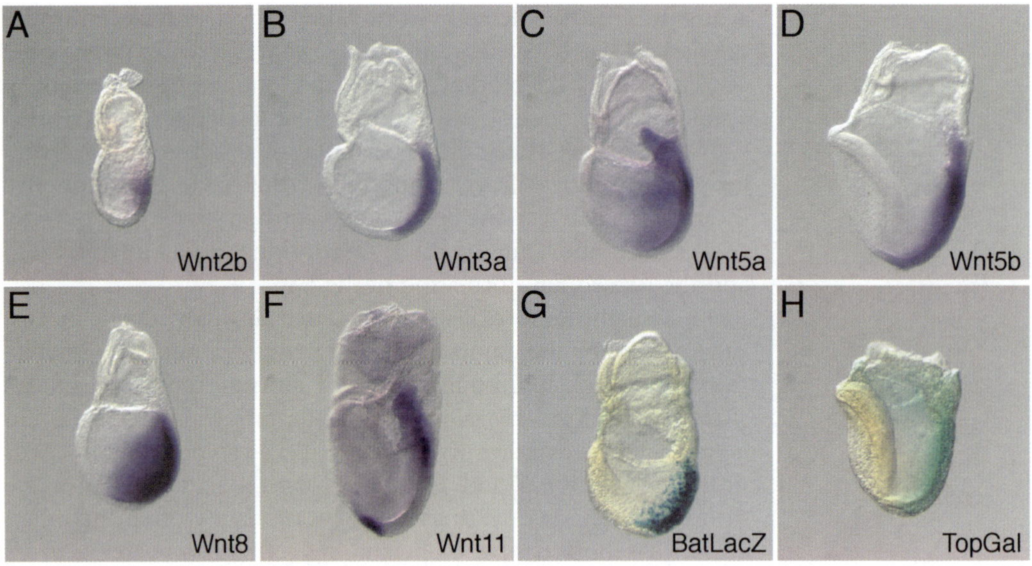

Fig. 23.2. Expression of Wnt genes and β-catenin/Tcf–lacZ reporter transgenes during gastrulation. Whole-mount in situ hybridization demonstrates that the expression of *Wnt2b*, *Wnt3a*, *Wnt5a*, *Wnt5b*, *Wnt8*, and *Wnt11* is posteriorly localized and largely centered on the primitive streak. The expression of the BATlacZ and TOPgal reporters indicate that the Wnt/β-catenin signaling pathway is active in the primitive streak.

expressed there) results in a loss of posterior mesoderm that is similar to the mesoderm deficit observed in *Wnt3a* mutants *(51)*. Importantly, expression of a stabilized form of β-catenin in *Wnt3a*$^{-/-}$ embryos was sufficient to rescue the posterior mesoderm deficit, providing formal genetic proof that β-catenin mediates the mesoderm-inducing activity of Wnt3a. Numerous studies provide additional support: a constitutively active form of Lef1 rescues the phenotypes observed in the hypomorphic *Wnt3a* mutant *vestigial tail* (*vt*) *(52)*, *Tcf1;Lef1* double mutants phenocopy the *Wnt3a* mutants *(53)*, *Lrp6* mutants genetically interact with *vt (54, 55)*, and, finally, β-catenin/Tcf-responsive β-gal reporters and endogenous β-catenin/Tcf target genes (see below) are downregulated by LOF alleles of *Wnt3a* and *Ctnnb1*, and upregulated by GOF alleles of *Ctnnb1 (21, 51, 52, 56–60)*. Together, the data unequivocally demonstrates that Wnt3a utilizes the canonical β-catenin/Tcf-Lef pathway to induce the formation of all streak-derived mesoderm.

With the exception of *Wnt5a*, null mutations in the remaining *Wnt* genes expressed during gastrulation have not yielded gastrulation phenotypes. For instance, embryos homozygous for a targeted disruption of *Wnt2b* are viable and fertile (T. Tsukiyama and T. Yamaguchi, unpublished data). Again, the lack of an apparent phenotype is likely due to genetic redundancy with canonical Wnts such as *Wnt3a*. Although *Wnt5a* expression in the PS overlaps considerably with other Wnts (**Fig. 23.2**), embryos homozygous for a targeted null mutation of *Wnt5a* display shortened trunks and do not form tails, resulting in a posterior truncation that is milder than that observed in *Wnt3a*$^{-/-}$ mutants *(61)*. Molecular and morphological analyses of *Wnt5a* mutants demonstrate that this posterior truncation phenotype is not due to an aberrant specification of posterior mesodermal fates but instead likely arises from defects in CE cell movements in or near the PS (ref. *(61)* and unpublished data).

Despite recent observations in vitro that suggest that Wnt5a can signal through β-catenin *(9)*, little evidence exists to support a role for β-catenin in transducing Wnt5a signals in vivo. For instance, the expression of Wnt/β-catenin target genes and β-catenin/Tcf–lacz reporters in the PS is unaffected by the absence of *Wnt5a* (ref. *(61)* and unpublished data). Moreover, genetic experiments demonstrate that *Wnt5a* interacts with core components of the Wnt/PCP pathway such as *Vangl2/Looptail* during axis extension *(62)*. Thus Wnt5a appears to control PS morphogenesis by regulating the Wnt/PCP pathway, however the precise cellular and molecular mechanisms underlying Wnt5a activity in the PS remain largely unknown. It is intriguing to note that embryos lacking the Wnt antagonists *sFRP1* and *sFRP2* display a phenotype remarkably similar to the *Wnt5a* mutant phenotype *(63)*. Wnt5a has been reported to inhibit the activation of the Wnt/β-catenin pathway *(7)*, leading to the expectation that this pathway will be hyperactivated in *Wnt5a* mutants, however no supporting evidence for this was found in the *Wnt5a* mutants nor in the *sFRP1;sFRP2* double mutants. The relationship between Wnt5a and sFRP1/2 may be complex since Wnt5a does not appear to simply bind directly to sFRP1 *(64, 65)*.

6. The Wnt3a/β-Catenin Pathway and Mesodermal Segmentation

Fate mapping, transplantation, and gene expression studies of the E7.5 gastrulating embryo indicate that different mesodermal lineages arise from different AP regions of the PS *(66, 67)*. Paraxial presomitic mesoderm (PSM) progenitors arise in the anterior PS, while intermediate, lateral, and extraembryonic mesoderm cells arise from successively posterior regions of the streak. Transplantation studies performed in the chick suggest that while these mesodermal fates are determined in the PS, mesodermal progenitors are not irreversibly committed to these fates until after they have left the PS *(68, 69)*. Different mesodermal fates likely arise from the combinatorial activities of multiple secreted signaling molecules, however the mechanism controlling the commitment and differentiation of mesodermal progenitors to mature cell types is not well understood.

The continuous formation of new mesodermal progenitors in the PS, together with the morphogenetic movements of gastrulation and CE, drive the anteriorly directed movement of mesodermal cells out of the PS. Once immature PSM cells, for example, get a prescribed distance from the PS, they undergo a dramatic transition in gene expression, initiating a poorly understood maturation process that results in the termination of early mesoderm specification genes, and the activation of a segment boundary formation program that cleaves the anterior mesoderm every 2 hours, resulting in the periodic formation of somites.

Remarkably, the Wnt3a/β-catenin pathway directly, and indirectly, controls the majority of these developmental events. In addition to the roles already described for Wnt3a in the specification of mesoderm fates, Wnt3a also plays a major role in the subsequent segmentation of mesoderm. Although it was originally observed that only the first 7–10 (of the approximately 60–65 expected) pairs of somites form in *Wnt3a* null mutants, the lack of posterior somites was primarily attributed to the impaired formation of mesoderm in the PS and tailbud *(48)*. It was not until it was shown that the direct Wnt3a target gene, *Axin2*, is expressed in a graded and oscillating fashion in the PSM that Wnt3a was directly implicated in segmentation *(57)*. This led to the development of the clock and gradient model, modified from preexisting clock and wavefront models, that proposed that the Wnt3a/β-catenin (and Notch) pathways are core components of a molecular oscillator or segmentation clock that controls periodic somite formation, and, together with Fgf8, establishes a morphogen gradient that sets the segment boundary position (reviewed in refs. *(70, 71)*). Recent studies using *Wnt3a* nulls, and conditional *Ctnnb1* LOF and GOF alleles refutes the proposed role for Wnt3a/β-catenin signaling in the segmentation clock, suggesting instead that the Wnt3a gradient maintains posterior PSM cells in an immature progenitor state that is permissive but not instructive for the cycling segmentation clock *(51)*. Conditional stabilization of β-catenin in the PSM did not disrupt molecular oscillations but led to a dramatic expansion of immature PSM and a delay in the formation of segment boundaries, clearly supporting the proposed role for Wnt3a in setting the position of segment boundaries. Thus Wnt3a plays a central role in eliciting proper posterior development, functioning to coordinate PSM maturation with the segmentation clock and boundary formation.

7. Complex Transcriptional Networks Activated by Wnt3a/β-Catenin Signaling

Attempts to elucidate the transcriptional networks activated by the Wnt3a/β-catenin pathway during early mammalian development have led to the identification of several important direct target genes. The T-box transcription factor and early pan-mesodermal marker, *T* (also known as *Brachyury*), was one of the first developmentally relevant direct target genes to be identified *(52, 56, 60)*. Embryos homozygous for this classic mouse mutation display a posterior truncation phenotype that is remarkably similar to the *Wnt3a*$^{-/-}$ phenotype, suggesting that *Wnt3a* and *T* may function in the same pathway. Indeed, *T* is downstream of Wnt3a/β-catenin signaling since its expression is downregulated in the PS of *Wnt3a* null mutants, completely absent in embryos conditionally lacking β-catenin in the PS, and upregulated in embryos expressing a stabilized form of β-catenin in the PS *(51, 60)*. Molecular analyses demonstrated that Tcf/Lef binding sites found *in* the *T* promoter bind Lef1 and are essential for the promoter to drive expression in reporter assays in vitro, and in the PS in transgenic animals *(52, 56, 60)*. These results indicate that T is a major transcriptional effector of Wnt3a/β-catenin signaling during posterior development. Surprisingly little is known about the target genes of T itself.

Targeted mutations in several other developmental genes lead to phenotypes that are similar to the *Wnt3a* and *T* mutant phenotypes, suggesting their potential involvement in this pathway. For instance, the lack of posterior somites and the ectopic neural tubes observed in embryos lacking *Tbx6*, a PSM-specific Tbox transcription factor closely related to *T*, closely resembles the *Wnt3a*$^{-/-}$ phenotype *(72)*. However, in contrast to *Wnt3a* and *T* mutants that lack tailbuds, *Tbx6* mutants display a grossly enlarged tailbud, indicating that *T* and *Tbx6* have divergent functions. Analysis of *Tbx6* expression in *Wnt3a* and *T* mutants, coupled with genetic analyses of *T* and *Tbx6* mutants, suggests that *Tbx6* is likely a direct target of T, and therefore an indirect target of the Wnt3a/β-catenin pathway *(59, 60, 73, 74)*. *Tbx6* may play an important role in integrating inputs from multiple signaling pathways since *Tbx6* transcription is also directly regulated by the Notch pathway *(75)*.

Interestingly, many of the target genes of Tbx6 that have been described to date are also directly co-regulated by the Wnt/β-catenin or Notch pathways, i.e., the target gene promoters possess functional binding sites for both Tbx6 and Tcf or RBPJκ (the transcriptional effector of the Notch pathway). For example, Tbx6 and the Wnt/β-catenin pathway synergistically activate the expression of the bHLH transcription factor *Mesogenin1* in the PSM (ref. *(76)*; W.C. Dunty and T. Yamaguchi, unpublished).

Mesogenin1 and *Tbx6* are expressed in similar domains in the PSM and targeted mutations of *Mesogenin1* reveal a mutant phenotype that is remarkably similar to the *Tbx6* mutant phenotype *(77)*. Analysis of gene expression in the enlarged tailbuds of *Tbx6* and *Mesogenin* mutants indicates that the tailbuds are largely composed of immature mesodermal progenitors. This suggests that Tbx6 and Mesogenin, under the control of Wnt3a, promote the progression or maturation of posterior mesodermal progenitors to anterior, segmented somites.

Several other Wnt3a/β-catenin target genes help to illustrate the diverse activities of this pathway during posterior development. The *Delta-like1* (*Dll1*) gene encodes a Notch ligand that is necessary for the activation of the segmentation clock and proper segmentation *(78, 79)*. *Dll1* expression is dependent upon Tbx6 and the Wnt3a/β-catenin pathway, indicating that Wnt3a controls segmentation in part through the activation of *Dll1*, and therefore, Notch activity *(21, 58, 59)*. Wnt3a also controls segmentation through the repression of the segment boundary determination gene *Ripply2 (51)*. In addition, Wnt3a plays important roles in imparting positional information to the PSM. *Wnt3a*$^{-/-}$ mice display homeotic transformations of the vertebrae that are similar to transformations observed in *Hox* mutants *(80)*. Indeed, Wnt3a indirectly regulates *Hox* gene expression through the activation of members of the *Cdx* family of homeodomain encoding genes, which are known regulators of *Hox* gene activity (ref. *(81)*, and references therein). Remarkably, Wnt3a also participates in the determination of the left–right (LR) body axis, indirectly regulating the LR determinant and Dll1/Notch target gene, *Nodal*, through the activation of *Dll1 (21, 82, 83)*. This suggests that Dll1 links Wnt3a to segmentation and the orientation of the LR axis.

8. Summary

The Wnt/β-catenin pathway regulates multiple events crucial for early mammalian embryogenesis, including the specification of the AP and LR body axes, PS and mesoderm formation, mesodermal patterning, and segmentation. Once gastrulation has been initiated and the body axes have been established, Wnt3a plays a particularly important role in the coordination of the many developmental events that are necessary for proper posterior development. Wnt3a functions at the top of an interacting transcriptional network of genes that starts with the induction of the transcription factors *T* and *Tbx6*, both of which play central roles in integrating the inputs from multiple signaling pathways. Wnt3a subsequently activates numerous other developmental regulator genes in cooperation with T and Tbx6. Although significant progress has been made in the identification of direct and indirect target genes of the Wnt3a/β-catenin pathway, relatively little is known about how these molecular targets control cellular behavior. Recently, Wnts have emerged as important homeostatic regulators of adult tissues, stimulating the self-renewal of hematopoietic, skin, and gastrointestinal stem cells *(1, 2)*. Genetic mutations that cause aberrant activation of Wnt/β-catenin signaling lead to hematopoietic and gastrointestinal malignancies, suggesting that these cancers arise from the co-opting of normal, physiological regulators of stem cell self-renewal. It stands to reason then that achieving a mechanistic understanding of Wnt/β-catenin signaling in normal embryonic and adult tissues will provide valuable insights into the formation of cancers caused by dysregulated Wnt signaling. Future studies addressing how Wnt/β-catenin signals and their target genes elicit mesoderm progenitors from pluripotent epiblast stem cells during gastrulation will be particularly enlightening in this regard.

Acknowledgments

I apologize to the many authors whose work was omitted due to time and space constraints. I thank Kristin Biris for assistance with manuscript and figure preparation, and members of the laboratory for comments on the manuscript. Work originating in my laboratory was supported by the Intramural Research Program of the National Institutes of Health (NIH), the National Cancer Institute, and the Center for Cancer Research.

References

1. Clevers, H. (2006) Wnt/beta-catenin signaling in development and disease. *Cell* **127**, 469–480.
2. Reya, T. and Clevers, H. (2005) Wnt signalling in stem cells and cancer. *Nature* **434**, 843–850.
3. Nusse, R. and Varmus, H. E. (1982) Many tumors induced by the mouse mammary tumor virus contain a provirus integrated in the same region of the host genome. *Cell* **31**, 99–109.
4. Kimelman, D. and Xu, W. (2006) beta-catenin destruction complex: insights and questions from a structural perspective. *Oncogene* **25**, 7482–7491.
5. Stadeli, R., Hoffmans, R., and Basler, K. (2006) Transcription under the control of nuclear Arm/beta-catenin. *Curr Biol* **16**, R378–385.
6. Willert, K. and Jones, K. A. (2006) Wnt signaling: is the party in the nucleus? *Genes Dev* **20**, 1394–1404.
7. Veeman, M. T., Axelrod, J. D., and Moon, R. T. (2003) A second canon. Functions and mechanisms of beta-catenin-independent Wnt signaling. *Dev Cell* **5**, 367–377.
8. Tao, Q., Yokota, C., Puck, H., Kofron, M., Birsoy, B., Yan, D., Asashima, M., Wylie, C. C., Lin, X., and Heasman, J. (2005) Maternal wnt11 activates the canonical wnt signaling pathway required for axis formation in Xenopus embryos. *Cell* **120**, 857–871.
9. Mikels, A. J. and Nusse, R. (2006) Purified Wnt5a protein activates or inhibits beta-catenin-TCF signaling depending on receptor context. *PLoS Biol* **4**, e115.

10. Mikels, A. J. and Nusse, R. (2006) Wnts as ligands: processing, secretion and reception. *Oncogene* **25**, 7461–7468.
11. van Amerongen, R. and Berns, A. (2006) Knockout mouse models to study Wnt signal transduction. *Trends Genet* **22**, 678–689.
12. Rossant, J. and Tam, P. P. (2004) Emerging asymmetry and embryonic patterning in early mouse development. *Dev Cell* **7**, 155–164.
13. Mohamed, O. A., Dufort, D., and Clarke, H. J. (2004) Expression and estradiol regulation of Wnt genes in the mouse blastocyst identify a candidate pathway for embryomaternal signaling at implantation. *Biol Reprod* **71**, 417–424.
14. Kemp, C., Willems, E., Abdo, S., Lambiv, L., and Leyns, L. (2005) Expression of all Wnt genes and their secreted antagonists during mouse blastocyst and postimplantation development. *Dev Dyn* **233**, 1064–1075.
15. Hamatani, T., Carter, M. G., Sharov, A. A., and Ko, M. S. (2004) Dynamics of global gene expression changes during mouse preimplantation development. *Dev Cell* **6**, 117–131.
16. Wang, Q. T., Piotrowska, K., Ciemerych, M. A., Milenkovic, L., Scott, M. P., Davis, R. W., and Zernicka-Goetz, M. (2004) A genome-wide study of gene activity reveals developmental signaling pathways in the preimplantation mouse embryo. *Dev Cell* **6**, 133–144.
17. Mohamed, O. A., Clarke, H. J., and Dufort, D. (2004) Beta-catenin signaling marks the prospective site of primitive streak formation in the mouse embryo. *Dev Dyn* **231**, 416–424.
18. Barolo, S. (2006) Transgenic Wnt/TCF pathway reporters: all you need is Lef? *Oncogene* **25**, 7505–7511.
19. Maretto, S., Cordenonsi, M., Dupont, S., Braghetta, P., Broccoli, V., Hassan, A. B., Volpin, D., Bressan, G. M., and Piccolo, S. (2003) Mapping Wnt/beta-catenin signaling during mouse development and in colorectal tumors. *Proc Natl Acad Sci U S A* **100**, 3299–3304.
20. Merrill, B. J., Pasolli, H. A., Polak, L., Rendl, M., Garcia-Garcia, M. J., Anderson, K. V., and Fuchs, E. (2004) Tcf3: a transcriptional regulator of axis induction in the early embryo. *Development* **131**, 263–274.
21. Nakaya, M. A., Biris, K., Tsukiyama, T., Jaime, S., Rawls, J. A., and Yamaguchi, T. P. (2005) Wnt3a links left-right determination with segmentation and anteroposterior axis elongation. *Development* **132**, 5425–5436.
22. Mohamed, O. A., Jonnaert, M., Labelle-Dumais, C., Kuroda, K., Clarke, H. J., and Dufort, D. (2005) Uterine Wnt/beta-catenin signaling is required for implantation. *Proc Natl Acad Sci U S A* **102**, 8579–8584.
23. Marikawa, Y. (2006) Wnt/beta-catenin signaling and body plan formation in mouse embryos. *Semin Cell Dev Biol* **17**, 175–184.
24. Yamaguchi, T. P. (2001) Heads or tails: Wnts and anterior-posterior patterning. *Curr Biol* **11**, R713–724.
25. Perea-Gomez, A., Meilhac, S. M., Piotrowska-Nitsche, K., Gray, D., Collignon, J., and Zernicka-Goetz, M. (2007) Regionalization of the mouse visceral endoderm as the blastocyst transforms into the egg cylinder. *BMC Dev Biol* 7, 96.
26. Copp, A. J. (1979) Interaction between inner cell mass and trophectoderm of the mouse blastocyst. II. The fate of the polar trophectoderm. *J Embryol Exp Morphol* **51**, 109–120.
27. Srinivas, S., Rodriguez, T., Clements, M., Smith, J. C., and Beddington, R. S. (2004) Active cell migration drives the unilateral movements of the anterior visceral endoderm. *Development* **131**, 1157–1164.
28. Thomas, T., Yamagishi, H., Overbeek, P. A., Olson, E. N., and Srivastava, D. (1998) The bHLH factors, dHAND and eHAND, specify pulmonary and systemic cardiac ventricles independent of left-right sidedness. *Dev Biol* **196**, 228–236.
29. Thomas, P. and Beddington, R. (1996) Anterior primitive endoderm may be responsible for patterning the anterior neural plate in the mouse embryo. *Curr Biol* **6**, 1487–1496.
30. Rodriguez, T. A., Srinivas, S., Clements, M. P., Smith, J. C., and Beddington, R. S. (2005) Induction and migration of the anterior visceral endoderm is regulated by the extra-embryonic ectoderm. *Development* **132**, 2513–2520.
31. Huelsken, J., Vogel, R., Brinkmann, V., Erdmann, B., Birchmeier, C., and Birchmeier, W. (2000) Requirement for beta-catenin in anterior-posterior axis formation in mice. *J Cell Biol* **148**, 567–578.
32. Brennan, J., Lu, C. C., Norris, D. P., Rodriguez, T. A., Beddington, R. S., and Robertson, E. J. (2001) Nodal signalling in the epiblast patterns the early mouse embryo. *Nature* **411**, 965–969.
33. Ding, J., Yang, L., Yan, Y. T., Chen, A., Desai, N., Wynshaw-Boris, A., and Shen, M. M. (1998) Cripto is required for correct orientation of the anterior-posterior axis in the mouse embryo. *Nature* **395**, 702–707.

34. Morkel, M., Huelsken, J., Wakamiya, M., Ding, J., van de Wetering, M., Clevers, H., Taketo, M. M., Behringer, R. R., Shen, M. M., and Birchmeier, W. (2003) Beta-catenin regulates Cripto- and Wnt3-dependent gene expression programs in mouse axis and mesoderm formation. *Development* **130**, 6283–6294.
35. Kimura, C., Yoshinaga, K., Tian, E., Suzuki, M., Aizawa, S., and Matsuo, I. (2000) Visceral endoderm mediates forebrain development by suppressing posteriorizing signals. *Dev Biol* **225**, 304–321.
36. Perea-Gomez, A., Lawson, K. A., Rhinn, M., Zakin, L., Brulet, P., Mazan, S., and Ang, S. L. (2001) Otx2 is required for visceral endoderm movement and for the restriction of posterior signals in the epiblast of the mouse embryo. *Development* **128**, 753–765.
37. Kimura-Yoshida, C., Nakano, H., Okamura, D., Nakao, K., Yonemura, S., Belo, J. A., Aizawa, S., Matsui, Y., and Matsuo, I. (2005) Canonical Wnt signaling and its antagonist regulate anterior-posterior axis polarization by guiding cell migration in mouse visceral endoderm. *Dev Cell* **9**, 639–650.
38. Mukhopadhyay, M., Shtrom, S., Rodriguez-Esteban, C., Chen, L., Tsukui, T., Gomer, L., Dorward, D. W., Glinka, A., Grinberg, A., Huang, S. P., Niehrs, C., Belmonte, J. C., and Westphal, H. (2001) Dickkopf1 is required for embryonic head induction and limb morphogenesis in the mouse. *Dev Cell* **1**, 423–434.
39. Liu, P., Wakamiya, M., Shea, M. J., Albrecht, U., Behringer, R. R., and Bradley, A. (1999) Requirement for Wnt3 in vertebrate axis formation. *Nat Genet* **22**, 361–365.
40. Popperl, H., Schmidt, C., Wilson, V., Hume, C. R., Dodd, J., Krumlauf, R., and Beddington, R. S. (1997) Misexpression of Cwnt8C in the mouse induces an ectopic embryonic axis and causes a truncation of the anterior neuroectoderm. *Development* **124**, 2997–3005.
41. Zhou, X., Sasaki, H., Lowe, L., Hogan, B. L., and Kuehn, M. R. (1993) Nodal is a novel TGF-beta-like gene expressed in the mouse node during gastrulation. *Nature* **361**, 543–547.
42. Conlon, F. L., Lyons, K. M., Takaesu, N., Barth, K. S., Kispert, A., Herrmann, B., and Robertson, E. J. (1994) A primary requirement for nodal in the formation and maintenance of the primitive streak in the mouse. *Development* **120**, 1919–1928.
43. Perea-Gomez, A., Vella, F. D., Shawlot, W., Oulad-Abdelghani, M., Chazaud, C., Meno, C., Pfister, V., Chen, L., Robertson, E., Hamada, H., Behringer, R. R., and Ang, S. L. (2002) Nodal antagonists in the anterior visceral endoderm prevent the formation of multiple primitive streaks. *Dev Cell* **3**, 745–756.
44. Ben-Haim, N., Lu, C., Guzman-Ayala, M., Pescatore, L., Mesnard, D., Bischofberger, M., Naef, F., Robertson, E. J., and Constam, D. B. (2006) The nodal precursor acting via activin receptors induces mesoderm by maintaining a source of its convertases and BMP4. *Dev Cell* **11**, 313–323.
45. Haegel, H., Larue, L., Ohsugi, M., Fedorov, L., Herrenknecht, K., and Kemler, R. (1995) Lack of beta-catenin affects mouse development at gastrulation. *Development* **121**, 3529–3537.
46. Shimizu, H., Julius, M. A., Giarre, M., Zheng, Z., Brown, A. M., and Kitajewski, J. (1997) Transformation by Wnt family proteins correlates with regulation of beta-catenin. *Cell Growth Differ* **8**, 1349–1358.
47. Liu, T., Liu, X., Wang, H., Moon, R. T., and Malbon, C. C. (1999) Activation of rat frizzled-1 promotes Wnt signaling and differentiation of mouse F9 teratocarcinoma cells via pathways that require Galpha(q) and Galpha(o) function. *J Biol Chem* **274**, 33539–33544.
48. Takada, S., Stark, K. L., Shea, M. J., Vassileva, G., McMahon, J. A., and McMahon, A. P. (1994) Wnt-3a regulates somite and tailbud formation in the mouse embryo. *Genes Dev* **8**, 174–189.
49. Yoshikawa, Y., Fujimori, T., McMahon, A. P., and Takada, S. (1997) Evidence that absence of Wnt-3a signaling promotes neuralization instead of paraxial mesoderm development in the mouse. *Dev Biol* **183**, 234–242.
50. Greco, T. L., Takada, S., Newhouse, M. M., McMahon, J. A., McMahon, A. P., and Camper, S. A. (1996) Analysis of the vestigial tail mutation demonstrates that Wnt-3a gene dosage regulates mouse axial development. *Genes Dev* **10**, 313–324.
51. Dunty, W. C., Jr., Biris, K. K., Chalamalasetty, R. B., Taketo, M. M., Lewandoski, M., and Yamaguchi, T. P. (2008) Wnt3a/ -catenin signaling controls posterior body development by coordinating mesoderm formation and segmentation. *Development* **135**, 85–94.
52. Galceran, J., Hsu, S. C., and Grosschedl, R. (2001) Rescue of a Wnt mutation by an activated form of LEF-1: regulation of maintenance but not initiation of Brachyury expression. *Proc Natl Acad Sci U S A* **98**, 8668–8673.

53. Galceran, J., Farinas, I., Depew, M. J., Clevers, H., and Grosschedl, R. (1999) Wnt3a–/–-like phenotype and limb deficiency in Lef1(–/–)Tcf1(–/–) mice. *Genes Dev* **13**, 709–717.

54. Pinson, K. I., Brennan, J., Monkley, S., Avery, B. J., and Skarnes, W. C. (2000) An LDL-receptor-related protein mediates Wnt signalling in mice. *Nature* **407**, 535–538.

55. Kokubu, C., Heinzmann, U., Kokubu, T., Sakai, N., Kubota, T., Kawai, M., Wahl, M. B., Galceran, J., Grosschedl, R., Ozono, K., and Imai, K. (2004) Skeletal defects in ringelschwanz mutant mice reveal that Lrp6 is required for proper somitogenesis and osteogenesis. *Development* **131**, 5469–5480.

56. Arnold, S. J., Stappert, J., Bauer, A., Kispert, A., Herrmann, B. G., and Kemler, R. (2000) Brachyury is a target gene of the Wnt/beta-catenin signaling pathway. *Mech Dev* **91**, 249–258.

57. Aulehla, A., Wehrle, C., Brand-Saberi, B., Kemler, R., Gossler, A., Kanzler, B., and Herrmann, B. G. (2003) Wnt3a plays a major role in the segmentation clock controlling somitogenesis. *Dev Cell* **4**, 395–406.

58. Galceran, J., Sustmann, C., Hsu, S. C., Folberth, S., and Grosschedl, R. (2004) LEF1-mediated regulation of Delta-like1 links Wnt and Notch signaling in somitogenesis. *Genes Dev* **18**, 2718–2723.

59. Hofmann, M., Schuster-Gossler, K., Watabe-Rudolph, M., Aulehla, A., Herrmann, B. G., and Gossler, A. (2004) WNT signaling, in synergy with T/TBX6, controls Notch signaling by regulating Dll1 expression in the presomitic mesoderm of mouse embryos. *Genes Dev* **18**, 2712–2717.

60. Yamaguchi, T. P., Takada, S., Yoshikawa, Y., Wu, N., and McMahon, A. P. (1999) T (Brachyury) is a direct target of Wnt3a during paraxial mesoderm specification. *Genes Dev* **13**, 3185–3190.

61. Yamaguchi, T. P., Bradley, A., McMahon, A. P., and Jones, S. (1999) A Wnt5a pathway underlies outgrowth of multiple structures in the vertebrate embryo. *Development* **126**, 1211–1223.

62. Qian, D., Jones, C., Rzadzinska, A., Mark, S., Zhang, X., Steel, K. P., Dai, X., and Chen, P. (2007) Wnt5a functions in planar cell polarity regulation in mice. *Dev Biol* **306**, 121–133.

63. Satoh, W., Gotoh, T., Tsunematsu, Y., Aizawa, S., and Shimono, A. (2006) Sfrp1 and Sfrp2 regulate anteroposterior axis elongation and somite segmentation during mouse embryogenesis. *Development* **133**, 989–999.

64. Xu, Q., D'Amore, P. A., and Sokol, S. Y. (1998) Functional and biochemical interactions of Wnts with FrzA, a secreted Wnt antagonist. *Development* **125**, 4767–4776.

65. Dennis, S., Aikawa, M., Szeto, W., d'Amore, P. A., and Papkoff, J. (1999) A secreted frizzled related protein, FrzA, selectively associates with Wnt-1 protein and regulates wnt-1 signaling. *J Cell Sci* **112 (Pt 21)**, 3815–3820.

66. Tam, P. P. and Beddington, R. S. (1987) The formation of mesodermal tissues in the mouse embryo during gastrulation and early organogenesis. *Development* **99**, 109–126.

67. Kinder, S. J., Tsang, T. E., Quinlan, G. A., Hadjantonakis, A. K., Nagy, A., and Tam, P. P. (1999) The orderly allocation of mesodermal cells to the extraembryonic structures and the anteroposterior axis during gastrulation of the mouse embryo. *Development* **126**, 4691–4701.

68. Garcia-Martinez, V. and Schoenwolf, G. C. (1992) Positional control of mesoderm movement and fate during avian gastrulation and neurulation. *Dev Dyn* **193**, 249–256.

69. Schoenwolf, G. C., Garcia-Martinez, V., and Dias, M. S. (1992) Mesoderm movement and fate during avian gastrulation and neurulation. *Dev Dyn* **193**, 235–248.

70. Aulehla, A. and Herrmann, B. G. (2004) Segmentation in vertebrates: clock and gradient finally joined. *Genes Dev* **18**, 2060–2067.

71. Pourquie, O. (2003) The segmentation clock: converting embryonic time into spatial pattern. *Science* **301**, 328–330.

72. Chapman, D. L. and Papaioannou, V. E. (1998) Three neural tubes in mouse embryos with mutations in the T-box gene Tbx6. *Nature* **391**, 695–697.

73. Chapman, D. L., Garvey, N., Hancock, S., Alexiou, M., Agulnik, S. I., Gibson-Brown, J. J., Cebra-Thomas, J., Bollag, R. J., Silver, L. M., and Papaioannou, V. E. (1996) Expression of the T-box family genes, Tbx1-Tbx5, during early mouse development. *Dev Dyn* **206**, 379–390.

74. Chapman, D. L., Cooper-Morgan, A., Harrelson, Z., and Papaioannou, V. E. (2003) Critical role for Tbx6 in mesoderm specification in the mouse embryo. *Mech Dev* **120**, 837–847.

75. White, P. H., Farkas, D. R., and Chapman, D. L. (2005) Regulation of Tbx6 expression by Notch signaling. *Genesis* **42**, 61–70.

76. Wittler, L., Shin, E. H., Grote, P., Kispert, A., Beckers, A., Gossler, A., Werber, M.,

and Herrmann, B. G. (2007) Expression of Msgn1 in the presomitic mesoderm is controlled by synergism of WNT signalling and Tbx6. *EMBO Rep* **8**, 784–789.

77. Yoon, J. K. and Wold, B. (2000) The bHLH regulator pMesogenin1 is required for maturation and segmentation of paraxial mesoderm. *Genes Dev* **14**, 3204–3214.
78. Hrabe de Angelis, M., McIntyre, J., 2nd, and Gossler, A. (1997) Maintenance of somite borders in mice requires the Delta homologue DII1. *Nature* **386**, 717–721.
79. Barrantes, I. B., Elia, A. J., Wunsch, K., Hrabe de Angelis, M. H., Mak, T. W., Rossant, J., Conlon, R. A., Gossler, A., and de la Pompa, J. L. (1999) Interaction between Notch signalling and Lunatic fringe during somite boundary formation in the mouse. *Curr Biol* **9**, 470–480.
80. Ikeya, M. and Takada, S. (2001) Wnt-3a is required for somite specification along the anteroposterior axis of the mouse embryo and for regulation of cdx-1 expression. *Mech Dev* **103**, 27–33.
81. Pilon, N., Oh, K., Sylvestre, J. R., Savory, J. G., and Lohnes, D. (2007) Wnt signaling is a key mediator of Cdx1 expression in vivo. *Development* **134**, 2315–2323.
82. Krebs, L. T., Iwai, N., Nonaka, S., Welsh, I. C., Lan, Y., Jiang, R., Saijoh, Y., O'Brien, T. P., Hamada, H., and Gridley, T. (2003) Notch signaling regulates left-right asymmetry determination by inducing Nodal expression. *Genes Dev* **17**, 1207–1212.
83. Raya, A., Kawakami, Y., Rodriguez-Esteban, C., Buscher, D., Koth, C. M., Itoh, T., Morita, M., Raya, R. M., Dubova, I., Bessa, J. G., de la Pompa, J. L., and Belmonte, J. C. (2003) Notch activity induces Nodal expression and mediates the establishment of left-right asymmetry in vertebrate embryos. *Genes Dev* **17**, 1213–1218.

Chapter 24

Tissue-Specific Transgenic, Conditional Knockout and Knock-In Mice of Genes in the Canonical Wnt Signaling Pathway

Koji Aoki and Makoto M.Taketo

Abstract

The Wnt signaling pathway plays key roles in the development and homeostasis of a number of organs such as the brain, lung, liver, heart, gastrointestinal tract, mammary gland, skin, and bone, as well as of the immune system. Studies on conventional knockout mice of the genes in the Wnt signaling pathway have revealed its essential roles in these tissues; however, most of these knockout mice die during embryogenesis or soon after birth. Through more advanced techniques such as Cre/*loxP* and tetracycline-inducible systems, a gene of interest can be expressed or inactivated in a tissue-specific and time-controlled manner. Here we review recent papers on the tissue-specific transgenic, conditional knockout and knock-in mice of the genes in the Wnt signaling pathway. In addition to such engineered mice, we also list reporter mice that have been generated to determine the activity of the canonical Wnt signaling pathway in mouse tissues.

Key words: Transgenic, Conditional, Knockout, Knock-in, Tissue-specific, Wnt signaling, Cre/*loxP*, Tet, β-Catenin, APC

1. Introduction

The Wnt signaling pathway is divided into the canonical and non-canonical pathways *(1)*. While the former activates T-cell factor (TCF)/β-catenin-dependent transcription, the latter is involved in several pathways including Rho-dependent cytoskeletal rearrangements *(1)*. The mutations in the *Adenomatous polyposis coli* (*APC*) gene have been found in 60% of human colorectal cancer *(2)*. Loss of APC results in the accumulation of β-catenin, which activates the canonical Wnt signal transcription *(1)*. Because

Elizabeth Vincan (ed.), *Wnt Signaling, Volume I: Pathway Methods and Mammalian Models, vol. 468*

Book doi: 10.1007/978-1-59745-249-6

suppression of the canonical Wnt pathway by dominant negative TCF inhibits the proliferation of colon cancer cells (e.g., ref. *[3]*), these reports suggest that activation of the canonical Wnt pathway is essential for tumorigenesis. In addition to cancer cell proliferation, it has been indicated by a number of conventional knockout mouse studies that the canonical Wnt pathway also plays key roles in regulating the proliferation of progenitor cells in developing tissues *(4)*. With more advanced techniques in mouse genetic engineering, it has been demonstrated that additional important roles are played by the canonical Wnt pathway, not only in the developing but also in adult tissues. In this review, we summarize the mutant mice in which the canonical Wnt pathway is activated or inhibited conditionally.

2. General Strategies for Construction of Transgenic and Conditional Gene Knockout and Knock-In Mice

2.1. Transgenic Mice

In the conventional strategy for transgenic mouse construction, a gene of interest is expressed under the control of a tissue-specific promoter (**Fig. 24.1a**, "transgenic-(1)") *(5)*. Transgene expression can be regulated more tightly through Cre-mediated site-specific recombination at *loxP* (locus of X-over of P1; **Fig. 24.1a**, "transgenic-(2)"). As shown in "transgenic-(2)", Cre placed under the control of a tissue-specific promoter deletes the Stop cassette, sandwiched by *loxP*, that inhibits expression of the transgene, which induces the transgene expression in the target tissue. In addition, transgene expression can be regulated by an inducible promoter. Through the use of the *Mx1* promoter, transgene expression can be controlled by interferon-α, interferon-β, or synthetic double-stranded RNA polyinosinic–polycytidylic acid (polyI–polyC), although this strategy is limited to expression in the liver and immune system (**Fig. 24.1a**, "transgenic-(3)") *(5)*. Using a tetracycline-controlled system ("Tet-on" or "Tet-off"), transgene expression can be induced by tetracycline or its analog, doxycycline (DOX) (**Fig. 24.1a**, "transgenic-(4)") *(5)*. Tetracycline-controlled transactivator rtTA (Tet-on) or rTA (Tet-off) under the control of a tissue-specific promoter induces transgene expression in the tissue through tet-responsive element (TRE). In the Tet system, addition (Tet-on) or removal (Tet-off) of tetracycline can induce the transgene expression.

2.2. Conditional Gene Knockout and Knock-In Mice

Cre and *loxP* have been used widely to excise the target segment from a gene of interest, which induces conditional gene knockout or knock-in (**Fig. 24.1b**) *(5)*. Expression of Cre under the control of a tissue-specific promoter allows deletion of the segment flanked

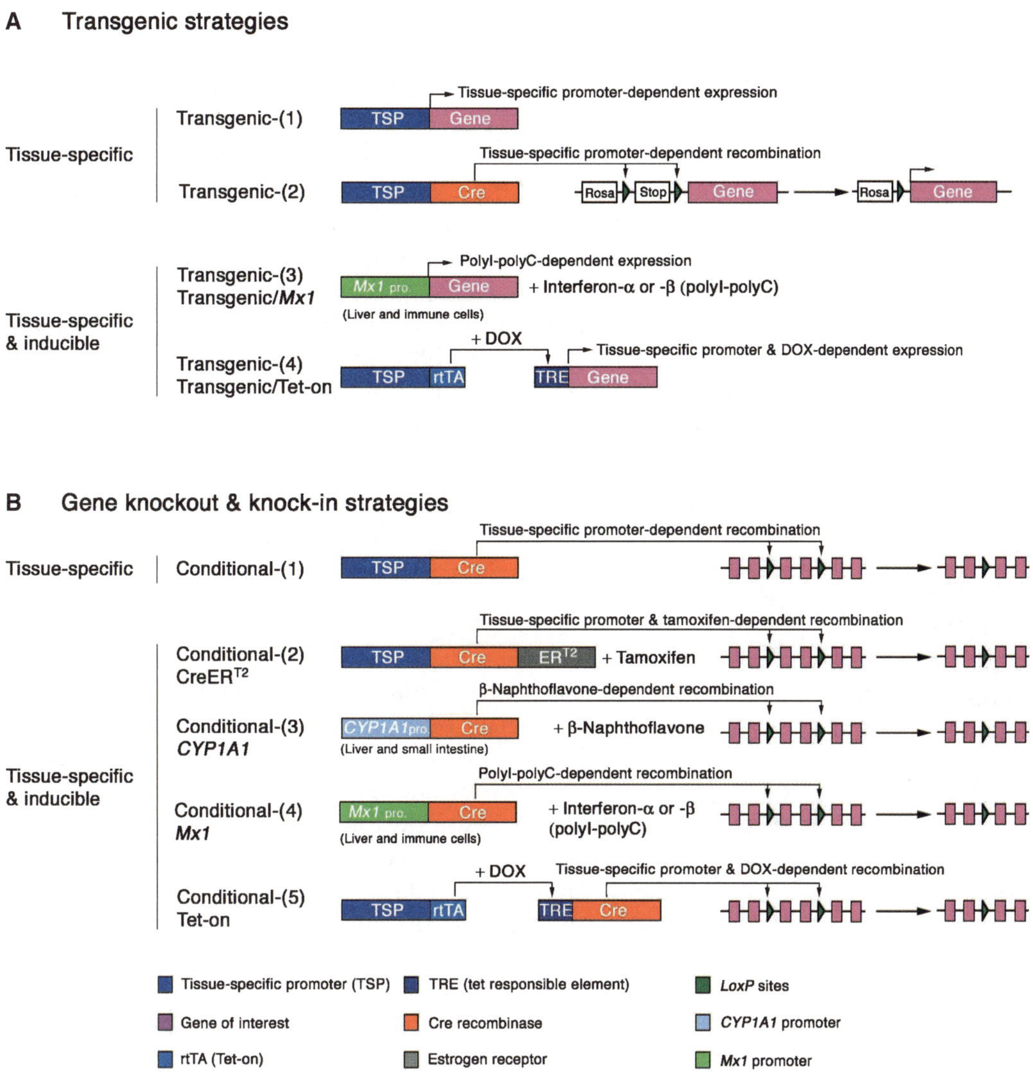

Fig. 24.1. General strategies for generation of transgenic and conditional gene knockout and knock-in mice. **A** Transgenic strategies. In the design "transgenic-(1)," a transgene is expressed under the control of a tissue-specific promoter (TSP). In "transgenic-(2)," a transgene is expressed following Cre-mediated recombination. In "transgenic-(3)," expression of a transgene is induced by addition of interferon or polyI–polyC. In "transgenic-(4)," transgene expression is induced by addition of tetracycline or DOX. **B** Conditional gene knockout and knock-in strategies. In the design "conditional-(1)," expression of Cre driven by a tissue-specific promoter induces knockout or knock-in of a gene of interest conditionally. In "conditional-(2)", a Cre-mediated gene knockout is induced by injection of tamoxifen into the mutant mice. In "conditional-(3)," a Cre-mediated gene knockout is induced by injection of β-naphthoflavone into the mice. In "conditional-(4)," a Cre-mediated gene knockout is induced by polyI–polyC. In "conditional-(5)," a Cre-mediated gene knockout is induced by addition (Tet-on) or removal (Tet-off) of DOX.

by *loxP* sites and inactivation of the target gene in the tissue (**Fig. 24.1b**, "conditional-(1)"). The recombination activity of Cre can be regulated by an estrogen analog, tamoxifen, when fused with a mutant form of the estrogen receptor binding domain (Cre-ER^{T2}, tamoxifen-inducible system; **Fig. 24.1b**, "conditional-(2)") *(5)*.

In this tamoxifen-inducible system, administration of tamoxifen or its analog, 4-hydroxy-tamoxifen (4-OHT), into mice activates Cre and induces recombination at the *loxP* sites. The Cre activity can also be controlled through the use of an inducible promoter. Addition of β-naphthoflavone induces expression of Cre placed under the control of the *Cyp1a1* promoter and induces recombination at *loxP* sites, although this strategy is limited to expression in the liver and gastrointestinal tract (**Fig. 24.1b**, "conditional-(3)") *(6)*. As explained in the "transgenic strategies," Cre expression can be regulated through the *Mx1* promoter (**Fig. 24.1b**, "conditional-(4)") or the Tet system as well (**Fig. 24.1b**, "conditional-(5)"). Using the *Mx1* promoter or the Tet system, the Cre-mediated recombination at *loxP* sites can be regulated in an inducible manner by interferon (or polyI–polyC) or tetracycline.

3. Strategies for Activation or Inhibition of the Canonical Wnt Signaling Pathway

3.1. Activation of the Canonical Wnt Signaling Pathway (Table 24.1)

For activation of the canonical Wnt signaling pathway, the following strategies have been reported. 1) Expression of WNT ligands such as WNT1 *(7)*. 2) Expression of stable β-catenin mutants: because its N-terminal region encodes amino acids that include all four phosphorylation targets for ubiquitylation, N-terminally deleted (β-cateninΔN) *(8, 9)* or exon 3-deleted β-catenin (β-cateninΔEx3) *(10)* is stable and activates TCF-dependent transcription. β-Catenin with mutations in the phosphorylation target amino acids are also stable (e.g., β-catenin S33Y) *(11)*. In addition, the stable β-catenin mutant fused with LEF1 (β-cateninΔN-LEF1) activates the transcription strongly *(12)*. The activity of β-catenin can be manipulated by fusing it with the estrogen receptor-binding domain (β-cateninΔN-ER) *(13)*. 3) Inactivation of the scaffolding proteins such as APC and Axin: because the scaffolding complex stimulates glycogen synthase kinase-3β (GSK3β)-mediated phosphorylation of unphosphorylated β-catenin, the absence of intact APC or Axin protein causes accumulation of β-catenin and activation of canonical Wnt signaling *(14, 15)*. 5) Expression of kinase-inactive GSK3β: because β-catenin is degraded following GSK3β-mediated phosphorylation, kinase-inactive GSK3β inhibits the degradation by dominantly antagonizing the endogenous activity of GSK3β and consequently activates canonical Wnt signaling *(16)*.

Table 24.1
Strategy for activation or inhibition of the canonical Wnt signaling pathway

Category	Gene	Transgene, knockout and knock-in gene products	Engineering type	Wnt signal transcription	References
Ligand	*Wnt*	WNT1	Transgenic	Activation	*(7)*
		WNT4	Transgenic	Activation	*(69)*
		WNT10b	Transgenic	Activation	*(83)*
		WNT14	Transgenic	Activation	*(76)*
	Dkk	DKK1	Transgenic	Inhibition	*(17)*
		DKK2	Transgenic	Inhibition	*(18)*
Receptor	*Lrp*	LRP5 G171V (dominant negative)	Transgenic	Inhibition	*(19)*
	Frizzled	Frizzled-5	Conditional knockout	Inhibition	*(20)*
β-catenin	*Catnb*	β-CateninΔN (stable mutant)	Transgenic	Activation	*(8, 9)*
		β-CateninΔN-ER (stable mutant)	Transgenic	Activation	*(13)*
		β-CateninS33Y-ER (stable mutant)	Transgenic	Activation	*(11)*
		β-CateninΔN fused with Lef1 (stable mutant)	Transgenic	Activation	*(12)*
		β-CateninΔEx3 (stable mutant)	Conditional knock-in	Activation	*(10)*
		β-CateninΔEx3-6 (no functional protein)	Conditional knockout	Inhibition	*(21)*
		β-CateninΔEx2-6 (no functional protein)	Conditional knockout	Inhibition	*(22)*

(continued)

Table 24.1 (continued)

Category	Gene	Transgene, knockout and knock-in gene products	Engineering type	Wnt signal transcription	References
LEF/TCF	*Lef1*	Mouse LEF1ΔN32 (dominant negative)	Transgenic	Inhibition	*(24)*
	Tcf3	TCF3 with mutation L383P and P407I (dominant negative)	Transgenic	Inhibition	*(25)*
Scaffolding protein	*Apc*	APCΔEx14 (truncated protein APCΔ580)	Conditional knockout	Activation	*(14, 15)*
		APCΔEx11-12 (truncated protein APCΔ468)	Conditional knockout	Activation	*(51)*
		APC	Transgenic	Inhibition	*(26)*
	Axin	Axin	Transgenic	Inhibition	*(27)*
	GSK3β	Kinase-inactive GSK3β	Transgenic	Activation	*(16)*
		GSK3β	Transgenic	Inhibition	*(28)*

3.2. Inhibition of the Canonical Wnt Signaling Pathway (Table 24.1)

For inhibition of the canonical Wnt signaling pathway, the following strategies have been employed. 1) Expression of Dickkopf (DKK) proteins such as DKK1 *(17)* and DKK2 *(18)*: because DKK is an inhibitor of Wnt ligands, expression of DKK inhibits the downstream signal activation. 2) Expression of a mutant low-density lipoprotein receptor-related protein (LRP), such as LRP5G171V, that cannot bind to Frizzled *(19)*, or inactivation of Frizzled *(20)*: because LRP and Frizzled are receptor subunits for the Wnt ligands, their inactivation shuts off the downstream signal activation. 3) Inactivation of β-catenin: because β-catenin binds to LEF/TCF and activates them, loss of β-catenin inhibits the Wnt signal transcription completely *(21, 22)*. Because β-catenin also plays key roles in the cadherin-dependent cell adhesion, however, it should be noted that the loss of β-catenin can disrupt the adhesion as well as TCF-dependent transcription *(23)*. 4) Expression of dominant negative LEF *(24)* or TCF *(25)*: LEF or TCF whose N-terminal region responsible for β-catenin binding is deleted inhibits the TCF/β-catenin transcription. 5) Expression of APC *(26)*, Axin *(27)*, or GSK3β *(28)*: expression of APC, Axin, or GSK3β stimulates the degradation of β-catenin and suppresses Wnt signal-mediated transcription.

4. Transgenic, Conditional Knockout and Knock-In Mice of the Genes in the Wnt Signaling Pathway

A number of gene-engineered mice have been generated to study the roles of the Wnt signaling pathway in various tissues, making use of tissue-specific promoters. Currently available mice are listed below.

4.1. Central Nervous System and Related Tissues (Table 24.2)

To analyze the development and functions of the central nervous system (CNS), the promoters of engrailed1 (*En1*) *(29)*, prion protein (*Prnp*) *(30, 31)*, Ca^{2+}–calmodulin-dependent protein kinase II (*Camk2a*) *(19, 28)*, brain 4 (*Pou3f4*) *(32)*, *Wnt1 (22, 33)*, nestin 11 (*Nes11*) *(34)*, nestin 8 (*Nes8*) *(34)*, forkhead box G1 (*Foxg1*) *(35)*, and *Axin2 (36)* have been used. To analyze the development of the pituitary gland and the optic and otic nerves, the promoters of *Pitx1* and Pit1 (*Pou1f1*) *(37)*, *Pax6 (38, 39)*, and *Pax2 (40)* have been used, respectively.

4.2. Lung and Liver (Table 24.2)

To analyze the development, homeostasis, and growth of the lung and liver, the promoters of surfactant-associated protein C (*Sftpc*) *(12, 41)* and albumin 1 (*Alb1*) *(42)* have been used, respectively. A Tet-off system with tTA under the control of the rat Clara cell secretory protein (*Scgb1a1*) promoter has been used for expression in the lung *(43)*. For expression in the liver, $CreER^{T2}$

Table 24.2
Transgenic, conditional knockout and knock-in mice of the genes in the Wnt signaling pathway

Tissue	Gene	Transgene, knockout and knock-in gene products	Engineering type	Wnt signal transcription	Promoter driving transgene	Promoter-Cre (CreER)	Phenotype	References
CNS	*Wnt1*	WNT1	Knock-in	Activation	*Engrailed-1*		Overproliferation	*(29)*
	Catnb	Human β-cateninS37F (stable mutant)	Transgenic	Activation	*Prion protein*		No obvious abnormality	*(30)*
		β-CateninΔEx3 (stable mutant)	Conditional knock-in	Activation		*CamKII*-Cre, *Brn4*-Cre	Synaptic vesicle localization, increase in CNS mass	*(19, 32)*
		β-CateninΔEx3 (stable mutant)	Conditional knock-in	Activation		*Wnt-1*-Cre	Induction of the sensory neural cell formation	*(33)*
		β-CateninΔEx3 (stable mutant)	Conditional knock-in	Activation		*Nes11*-Cre, *Nes8*-Cre	Dorso-ventral fate shift	*(34)*
		β-CateninΔEx2-6 (no functional protein)	Conditional knockout	Inhibition		*Wnt-1*-Cre	Brain malformation	*(22)*
		β-CateninΔEx2-6 (no functional protein)	Conditional knockout	Inhibition		*FoxG1*-Cre	Disruption of neuro-epithelial structures	*(35)*
		β-CateninΔEx2-6 (no functional protein)	Conditional knockout	Inhibition		*Nes11*-Cre, *Nes8*-Cre	Dorso-ventral fate shift	*(34)*
		β-CateninΔEx3-6 (no functional protein)	Conditional knockout	Inhibition		*Brn4*-Cre	Decrease in CNS mass	*(32)*
	Axin	Axin	Transgenic/Tet-on	Inhibition	*Axin2*-rtTA		Defects in brain development	*(36)*
	GSK3-β	GSK3-β	Transgenic/Tet-on	Inhibition	*CamKII*-tTA		Neurodegeneration	*(28)*
Pituitary	*Catnb*	β-CateninΔEx3 (stable mutant)	Conditional knock-in	Activation		*Pitx1*-Cre, *Pit*-Cre	Defects in differentiation	*(37)*

		β-CateninΔEx2-6 (no functional protein)	Conditional knockout	Inhibition		*Pitx1*-Cre, *Pit*-Cre	Defects in differentiation	*(37)*
Optic	*Catnb*	β-CateninΔEx3 (stable mutant)	Conditional knock-in	Activation		*Pax6*-Cre	Inappropriate optic cup pattern, absence of lens development	*(38, 39)*
		β-CateninΔEx2-6 (no functional protein)	Conditional knockout	Inhibition		*Pax6*-Cre	Disorganized lens epithelial cells	*(39)*
Otic	*Catnb*	β-CateninΔEx3 (stable mutant)	Conditional knock-in	Activation		*Pax2*-Cre	Induction of the otic placode	*(40)*
		β-CateninΔEx2-6 (no functional protein)	Conditional knockout	Inhibition		*Pax2*-Cre	Defects in otic placode induction	*(40)*
Lung	*Wnt5a*	Wnt5a	Transgenic	Activation	*Surfactant protein C*		Developmental abnormalities	*(41)*
	Catnb/ Lef1	β-CateninΔN fused with mouse full-length Lef1 (stable mutant)	Transgenic	Activation	*Surfactant protein C*		Developmental abnormalities	*(12)*
	Catnb	β-CateninΔEx3 (stable mutant)	Conditional knock-in/ Tet-off	Activation	*CCSP*-rtTA	Tet-off-Cre	Pulmonary tumors	*(43)*
Liver	*Catnb*	β-CateninΔEx3 (stable mutant)	Conditional knock-in	Activation		AdenovirusCMV-Cre	Hepatomegaly	*(45)*
		β-CateninΔEx2-6 (no functional protein)	Conditional knockout	Inhibition		*Albumin*-Cre	Developmental abnormalities	*(42)*
	Apc	APCΔEx14 (truncated protein APCΔ580)	Conditional knock-in/ tamoxifen inducible	Activation		*TTR*-CreERT2	Liver zonation	*(44)*
Heart	*Catnb*	β-CateninΔEx3-6 (no functional protein)	Conditional knockout /tamoxifen inducible	Inhibition		α-*MHC*-CreERT2	Defects in cardiomyocyte growth	*(46)*
		β-CateninΔEx2-6 (no functional protein)	Conditional knockout	Inhibition			Multiple heart formation	*(47)*

(continued)

Table 24.2 (continued)

Tissue	Gene	Transgene, knockout and knock-in gene products	Engineering type	Wnt signal transcription	Promoter driving transgene	Promoter-Cre (CreER)	Phenotype	References
	Lef1	LEF1ΔN20 (dominant negative)	Transgenic	Inhibition	α-*MHC*		Defects in cardiomyocyte growth	*(46)*
Vascular system	*Catnb*	β-CateninΔEx2-6 (no functional protein)	Conditional knockout	Inhibition		*Tie2*-Cre	Defects in vascular pattern	*(48)*
Immune system	*Catnb*	Human β-cateninΔN87 (stable mutant)	Transgenic	Activation	*Lck*		Increase in mature thymocytes	*(49)*
		Human β-cateninS33A (stable mutant)	Transgenic/ interferon- inducible	Activation	*Mx1*		Block of differentiation	*(52)*
		β-CateninΔEx3 (stable mutant)	Conditional knock-in /interferon inducible	Activation		*Mx1*-Cre	Block of differentiation	*(53)*
		β-CateninΔEx3 (stable mutant)	Conditional knock-in	Activation		*Lck*-Cre	Induction of thymocyte differentiation	*(50)*
		β-CateninΔEx2-6 (no functional protein)	Conditional knockout /interferon inducible	Inhibition		*Mx1*-Cre	No obvious phenotype	*(54)*
	Apc	APCΔEx11-12 (truncated protein APCΔ468)	Conditional knock-in	Activation		*Lck*-Cre	Defects in T cell differentiation	*(51)*

controlled by the transthyretin (*Ttr*) promoter has been reported *(44)*, and Cre expression by a recombinant adenovirus injected through the tail vein has been employed as well *(45)*.

4.3. Heart and Vascular System (Table 24.2)

To analyze the development and growth of the heart, the promoter of α-myosin heavy chain (*Myh6*) has been used *(46, 47)*, and the promoter of Tie2 (*Tek*) has been employed for expression in the vascular analysis *(48)*.

4.4. Immune System (Table 24.2)

To analyze the differentiation and functions of the immune cells, the promoters of *Lck (49–51)* and *Mx1 (50, 52–54)* have been used. As described above, the *Mx1* promoter activity can be regulated by interferon or polyI–polyC.

4.5. Gastrointestinal Tract (Table 24.3)

To study the development of the pancreas, the promoter of *Pdx1* has been used *(23, 55)*; and K19 (*Krt19*) has been used for investigation of tumorigenesis in gastric mucosa *(56)*. To analyze the homeostasis, differentiation, and tumorigenesis of the intestines, the following promoters have been employed: villin 1 (*Vil1*) *(57, 58)*, *Krt19 (10, 20)*, liver fatty acid-binding protein (*Fabpl*) *(8, 10, 26)*, calbindin-D9K (*S100g*) *(59)*, and *Cyp1a1 (6, 15)*. Direct injection of a recombinant adenovirus encoding Cre has also allowed Cre expression in the colonic mucosa *(14)*, and DKK1 protein has been delivered to the intestines through tail vein injection of a recombinant adenovirus encoding DKK1 *(60)*.

4.6. Mammary Gland (Table 24.3)

To investigate the development, homeostasis and tumorigenesis of the mammary gland, mouse mammary tumor virus long terminal repeat (*MMTV-LTR*) has been used widely *(7, 16, 61–63)*. The uses of promoters of K5 (*Krt5*) *(64)* and whey acidic protein (*Wap*) *(65)* have also been reported. A Tet-on system with rtTA under the control of *MMTV-LTR* has been used *(27, 66)*.

4.7. Reproductive Systems (Table 24.3)

The promoters of *MMTV-LTR* for expression in the prostate *(67, 68)*, *Wnt4* for expression in the testis *(69)*, tissue nonspecific alkaline phosphatase (*Akp2*) for expression in the germ cells *(70)*, Amh receptor 2 (*Amhr2*) for expression in the ovary *(71)*, and Müllerian-inhibiting substance type II receptor gene that is identical to *Amhr2* for expression in the uterus *(72)* have been used.

4.8. Skin (Table 24.4)

To analyze the development, homeostasis, and tumorigenesis of the skin, the promoter of K14 (*Krt14*) has been used widely *(9, 13, 17, 21, 24, 25, 73)*, and those of forkhead box N1 (*Foxn1*) *(18)*, K5 (*Krt5*) *(11)*, and *En1 (74)* have also been reported. The Tet-on system using *Krt14*-rtTA has been reported as well *(75)*.

4.9. Bone (Table 24.4)

For expression in the bone, the promoter of Collagen 2a1 (*Col2a1*) has been used widely *(76, 77)*, and the promoters of

Table 24.3
Transgenic, conditional knockout and knock-in mice of the genes in the Wnt signaling pathway

Tissue	Gene	Transgene, knockout and knock-in gene products	Engineering type	Wnt signal transcription	Promoter driving transgene	Promoter-Cre (CreER)	Phenotype	References
Pancreas	*Catnb*	β-CateninΔEx3 (stable mutant)	Conditional knock-in	Activation		*Pdx1*-Cre (early) *Pdx1*-Cre (late)	Larger pancreas size	*(55)*
		β-CateninΔEx2-6 (no functional protein)	Conditional knock-out	Inhibition		*Pdx1*-Cre	Developmental abnormalities (independent of the Wnt signal)	*(23)*
Gastric mucosa	*Wnt1*	WNT1	Transgenic	Activation	*K19*		Gastric tumor	*(56)*
Intestine	*Dkk1*	DKK1	Transgenic	Inhibition	*Villin*		Crypt ablation	*(57)*
	Frizzled-5	Frizzled-5	Conditional knock-out	Inhibition		*K19*-Cre	Defects in Paneth cell differentiation	*(20)*
	Catnb	Human β-cateninΔN 89 (stable mutant)	Transgenic	Activation	*Fabpl*		Villus branching	*(8)*
		β-CateninΔN131 (stable mutant)	Transgenic	Activation	*Calbindin-D9K*		Adenoma formation	*(59)*
		β-CateninΔEx3 (stable mutant)	Conditional knock-in	Activation		*K19*-Cre, *Fabpl*-Cre	Adenoma formation	*(10)*
		β-CateninΔEx2-6 (no functional protein)	Conditional knock-out	Inhibition		*CYP1A1*-Cre	Crypt ablation	*(6)*
	Apc	APCΔEx14 (truncated protein ApcΔ580)	Conditional knock-out/ tamoxifen inducible/ β-naphthoflavone inducible	Activation		AdenovirusCMV-Cre *CYP1A1*-Cre, *Villin*-CreERT2	Adenoma formation, defects in proliferation, migration, and differentiation	*(14, 15, 58)*

		APC	Transgenic	Inhibition	*Fabpl*		Disordered cell migration	*(26)*
Mammary gland	*Wnt1*	WNT1	Transgenic	Activation	*MMTV-LTR*		Mammary gland tumor	*(7)*
			Transgenic/Tet-On	Activation	*MMTV-LTR-*rtTA		Mammary gland tumor	*(66)*
	Wnt-10b	WNT10b mouse isoform I	Transgenic	Activation	*MMTV-LTR*		Mammary gland tumor	*(61)*
		WNT10b mouse isoform II	Transgenic	Activation	*MMTV-LTR*		Mammary gland tumor	*(61)*
	Catnb	Xenopus β-cateninΔN89 (stable mutant)	Transgenic	Activation	*MMTV-LTR*		Mammary gland tumor	*(62)*
		Mouse β-cateninΔN90 (stable mutant)	Transgenic	Activation	*MMTV-LTR*		Mammary gland tumor	*(63)*
		Mouse β-cateninΔN57 (stable mutant)	Transgenic	Activation	*K5*		Defects in normal development	*(64)*
		β-CateninΔEx3 (stable mutant)	Conditional knock-in	Activation		*MMTV-LTR*-Cre, *WAP*-Cre	Mammary gland tumor, squamous differentiation	*(65)*
	GSK3-β	Kinase-inactive GSK3β	Transgenic	Activation	*MMTV-LTR*		Mammary gland tumor	*(16)*
	Axin	Axin	Transgenic	Inhibition	*MMTV-LTR-*rtTA		Defects in lactation	*(27)*
Prostate	*Catnb*	β-CateninΔEx3 (stable mutant)	Conditional knock-in	Activation		*MMTV-LTR*-Cre	Prostate intraepithelial neoplasia	*(67, 68)*

(continued)

Table 24.3 (continued)

Tissue	Gene	Transgene, knockout and knock-in gene products	Engineering type	Wnt signal transcription	Promoter driving transgene	Promoter-Cre (CreER)	Phenotype	References
Testis	*Wnt4*	WNT4	Transgenic	Inhibition	Human endogenous *Wnt-4* promoter		Defects in testicular structure	*(69)*
Germ cell	*Catnb*	β-CateninΔEx3 (stable mutant)	Conditional knock-in	Activation		Tissue nonspecific alkaline phosphatase-Cre	Loss of germ cells	*(70)*
Ovary	*Catnb*	β-CateninΔEx3 (stable mutant)	Conditional knock-in	Activation		*Amhr2*-Cre	Ovarian tumor	*(71)*
Uterus	*Catnb*	β-CateninΔEx3-6 (no functional protein)	Conditional knockout	Inhibition		Müllerian Inhibiting Substance type II receptor (*MisrII*)-Cre	Developmental abnormalities	*(72)*

Table 24.4
Transgenic, conditional knockout and knock-in mice of the genes in the Wnt signaling pathway

Tissue	Gene	Transgene, knockout and knock-in gene products	Engineering type	Wnt signal transcription	Promoter driving transgene	Promoter-Cre (CreER)	Phenotype	References
Skin	*Dkk1*	DKK1	Transgenic	Inhibition	Human *K14*		Developmental abnormalities	*(17)*
	Dkk2	DKK2	Transgenic	Inhibition	*Foxn1*		Reduction in hair follicle density	*(18)*
	Catnb	Human β-cateninΔN87 (stable mutant)	Transgenic	Activation	*K14*		Hair follicle tumor	*(9)*
		β-CateninΔN-ER (stable mutant)	Transgenic/tamoxifen inducible	Activation	*K14*		Hair follicle tumor	*(13)*
		β-CateninS33Y-ER (stable mutant)	Transgenic/tamoxifen inducible	Activation	*K5*		Increase in follicle growth	*(11)*
		β-CateninΔEx3 (stable mutant)	Conditional knock-in/ tamoxifen inducible	Activation		*En1*-CreERT2	Dorsal dermal differentiation	*(74)*
		β-CateninΔEx2-6 (no functional protein)	Conditional knockout /tamoxifen inducible	Inhibition		*En1*-CreERT2	Dorsal dermal differentiation	*(74)*
		β-CateninΔEx3-6 (no functional protein)	Conditional knockout	Inhibition		*K14*-Cre	Stem cell maintenance	*(21)*
	Lef1	LEF1	Transgenic		*K14*		Normal skin functions	*(25)*
		Mouse LEF1ΔN32 (dominant negative)	Transgenic	Inhibition	*K14*		Progressive hair loss	*(24)*
	Tcf3	TCF3	Transgenic		*K14*		Defects in skin functions	*(25)*

(continued)

Table 24.4 (continued)

Tissue	Gene	Transgene, knockout and knock-in gene products	Engineering type	Wnt signal transcription	Promoter driving transgene	Promoter-Cre (CreER)	Phenotype	References
		TCF3 with mutation L383P and P407I	Transgenic	Inhibition	*K14*		Normal skin functions	*(25)*
		TCF3	Transgenic/Tet-on		*K14*-rtTA		Stem cell maintenance	*(75)*
	Apc	APCΔEx14 (truncated protein ApcΔ580)	Conditional knockout	Activation		*K14*-Cre	Aberrant growth in ectodermally derived epithelia	*(73)*
Bone	*Wnt14*	WNT4	Transgenic	Activation	*Collagen 2a1*		Ectopic joint formation	*(76)*
	DKK1	DKK1	Transgenic	Inhibition	*Collagen 1a1*		Osteopenia	*(79)*
	Lrp5	LRP5 G171V (dominant negative)	Transgenic	Inhibition	*Collagen 1a1*		Increase in bone mass	*(81)*
	Catnb	β-CateninΔN (stable mutant)	Transgenic	Activation	*Collagen 2a1*		Ectopic joint formation	*(76)*
		β-Catenin (695–781) fused with LEF1	Transgenic	Activation	*Collagen 2a1*		Severe skeletal defects	*(77)*
		β-CateninΔEx3 (stable mutant)	Conditional knock-in	Activation		*Prx1*-Cre, *Collagen 1a1*-Cre	Abnormal differentiation	*(78, 80)*
		β-CateninΔEx3-6 (no functional protein)	Conditional knockout	Inhibition		Adenovirus CMV-Cre	Inhibition of chondrogenesis and osteogenesis	*(82)*
		β-CateninΔEx2-6 (no functional protein)	Conditional knockout	Inhibition		*Prx1*-Cre, *Collagen 1a1*-Cre	Abnormal differentiation	*(78, 80)*
		β-CateninΔEx3-6 (no functional protein)	Conditional knockout	Inhibition		*Prion*-Cre	Chondrocyte differentiation	*(76)*

Adipose tissue	*Wnt10b*	WNT10b	Transgenic	Activation	*FABP4*		Inhibition of differentiation	*(83)*
Taste buds	*Dkk1*	DKK1	Transgenic/Tet-on	Inhibition	*K5*-rtTA		Disruption of fungiform papilla morphogenesis	*(84)*
	Catnb	β-CateninΔEx3 (stable mutant)	Conditional knock-in	Activation		*K14*-Cre	Enlarged fungiform papilla and taste buds	*(84)*
		β-CateninΔEx2-6 (no functional protein)	Conditional knockout	Inhibition		*K14*-Cre	Disruption of fungiform papilla morphogenesis	*(84)*
Tooth	*Catnb*	β-CateninΔEx3 (stable mutant)	Conditional knock-in	Activation		*K14*-Cre	Activation of tooth regeneration	*(85)*
Limb	*Catnb*	β-CateninΔEx3 (stable mutant)	Conditional knock-in	Activation		*Prx1*-Cre	Truncated limbs	*(86)*
		β-CateninΔEx3 (stable mutant)	Conditional knock-in	Activation		*BMP4*-Cre	Expanded apical ectodermal ridge	*(87)*
		β-CateninΔEx3-6 (no functional protein)	Conditional knockout	Inhibition		*Prx1*-Cre	Severe agenesis of the scapula	*(86)*
Kidney	*Apc*	APCΔEx14 (truncated protein ApcΔ580)	Conditional knockout	Activation		*Kidney-specific cadherin*-Cre	Cystic renal neoplasia	*(88)*
Embryo	*Catnb*	β-CateninΔEx3 (stable mutant)	Conditional knock-in	Activation		*Zona pellucida3*-Cre	Disorganized embryonic ectoderm	*(89)*
Muscle	*Catnb*	β-CateninΔEx2-6 (no functional protein)	Conditional knockout	Inhibition		AdenovirusCMV-Cre	Defects in fiber growth	*(90)*

paired related homeobox 1 (*Prrx1*) *(78)*, Collagen 1a1 (*Col1a1*) *(79–81)*, and prion protein (*Prnp*) *(76)* have been reported as well. Direct injection of a recombinant adenovirus encoding Cre has also been employed to express Cre in the bone *(82)*.

4.10. Other Tissues

The promoter of fatty acid biding protein 4 (*Fabp4*) has been used for expression in the adipose tissues *(83)*, and promoters of K5 (*Krt5*) and K14 (*Krt14*) have been employed for expression in the taste buds *(84)* and the tooth *(85)*, respectively. For expression in the limb, the promoters of *Prrx1 (86)* and *BMP4 (87)* have been reported, and the promoter of kidney-specific cadherin (*Cdh16*) has been used for expression in the kidney *(88)*. The promoter of zona pellucida glycoprotein 3 (*Zp3*) has been reported for expression in the ectoderm layer of early post-implantation embryos *(89)*. Direct injection of a recombinant adenovirus encoding Cre has also been useful for expression of Cre in the skeletal muscle *(90)*.

5. Reporter Mice for Detecting Active Wnt Signaling (Table 24.5)

About ten reporter mouse strains have been generated to monitor the activity of the canonical Wnt signaling pathway *(91)* (**Table 24.5**). Because *Axin2* is one of the transcriptional target genes of the Wnt signaling pathway, the knock-in mice that express the β-galactosidase gene controlled by the *Axin2* promoter can be used as reporter mice *(92)*. In cell culture experiments, transfection with TOPFLASH that contains six repeats of the LEF/TCF binding site has been used widely as a readout for the Wnt signaling activity *(93)*. Transgenic mice that express the β-galactosidase gene driven by three to eight repeats of the LEF/TCF binding site have been generated, i.e., BAT-gal mice *(94, 95)* and TOPGAL mice *(96–98)*. A reporter mouse strain that expresses enhanced green fluorescent protein (EGFP) controlled by such binding sites has been reported as well *(99)*. Modified TOPGAL and TOPEGFP mice that express the β-galactosidase or EGFP genes flanked by insulators have been generated recently *(100)*. The insulators can reduce the chromosome positional effects on the reporter gene activities.

6. Summary

In this review, we described conditional mutant mice of the genes in the Wnt signaling pathway, and the tissue-specific promoters used for the mutant strains. Additionally, tissue-specific promoters and

Table 24.5
Reporter mice for detecting active Wnt signaling

Reporter mouse	Target sequence	Minimal promoter	Reporter gene	Modification	References
Conductin-lacZ	Endogenous conductin promoter		β-*Galactosidase*		*(92)*
BAT-gal	7 X TCF consensus sites	TATA box of siamois	β-*Galactosidase*		*(94)*
BATlacZ	8 X TCF consensus sites	TATA box of siamois	β-*Galactosidase*		*(95)*
TOPGAL	3 X TCF consensus sites	c-Fos minimal promoter	β-*Galactosidase*		*(96)*
LEF/TCF-LacZ	6 X TCF consensus sites	Hsp68 minimal promoter	β-*Galactosidase*		*(97)*
TCF-LacZ	6 X TCF consensus sites	Thymidine kinase promoter	β-*Galactosidase*		*(98)*
LEF/TCF-EGFP	6 X TCF consensus sites	Thymidine kinase promoter	*EGFP*		*(99)*
ins-TOPGAL	6 X TCF consensus sites	Thymidine kinase promoter	β-*Galactosidase*	Insulators flank the reporter gene	*(100)*
ins-TOPEGFP	6 X TCF consensus sites	Thymidine kinase promoter	*EGFP*	Insulators flank the reporter gene	*(100)*

expression systems such as Flp/*FRT* are available and have been reviewed elsewhere *(5, 101, 102)*. Given that whole-body deletion or transgenic expression of Wnt pathway genes often results in embryonic lethality, the ability to control gene expression in a spatiotemporal manner has dramatically advanced our understanding of the consequences of modulating Wnt signaling in development, homeostasis, and tumorigenesis.

Acknowledgments

The research programs in our laboratory have been supported by grants from MESSC, Japan; OPSR, Japan; the University of Tokyo-Banyu Pharmaceutical Co. Joint Fund; the Takeda Foundation; the Mitsubishi Foundation; and the Sagawa Cancer Research Fund.

References

1. Barker, N. and Clevers, H. (2006) Mining the Wnt pathway for cancer therapeutics. *Nat Rev Drug Discov* **5**, 997–1014.
2. Powell, S. M., Zilz, N., Beazer-Barclay, Y., Bryan, T. M., Hamilton, S. R., Thibodeau, S. N., Vogelstein, B., and Kinzler, K. W. (1992) APC mutations occur early during colorectal tumorigenesis. *Nature* **359**, 235–237.
3. van de Wetering, M., Sancho, E., Verweij, C., de Lau, W., Oving, I., Hurlstone, A., van der Horn, K., Batlle, E., Coudreuse, D., Haramis, A. P., Tjon-Pon-Fong, M., Moerer, P., van den Born, M., Soete, G., Pals, S., Eilers, M., Medema, R., and Clevers, H. (2002) The β-catenin/TCF-4 complex imposes a crypt progenitor phenotype on colorectal cancer cells. *Cell* **111**, 241–250.
4. van Amerongen, R. and Berns, A. (2006) Knockout mouse models to study Wnt signal transduction. *Trends Genet* **22**, 678–689.
5. Lewandoski, M. (2001) Conditional control of gene expression in the mouse. *Nat Rev Genet* **2**, 743–755.
6. Ireland, H., Kemp, R., Houghton, C., Howard, L., Clarke, A. R., Sansom, O. J., and Winton, D. J. (2004) Inducible Cre-mediated control of gene expression in the murine gastrointestinal tract: effect of loss of β-catenin. *Gastroenterology* **126**, 1236–1246.
7. Tsukamoto, A. S., Grosschedl, R., Guzman, R. C., Parslow, T., and Varmus, H. E. (1988) Expression of the int-1 gene in transgenic mice is associated with mammary gland hyperplasia and adenocarcinomas in male and female mice. *Cell* **55**, 619–625.
8. Wong, M. H., Rubinfeld, B., and Gordon, J. I. (1998) Effects of forced expression of an NH2-terminal truncated β-Catenin on mouse intestinal epithelial homeostasis. *J Cell Biol* **141**, 765–777.
9. Gat, U., DasGupta, R., Degenstein, L., and Fuchs, E. (1998) De Novo hair follicle morphogenesis and hair tumors in mice expressing a truncated β-catenin in skin. *Cell* **95**, 605–614.
10. Harada, N., Tamai, Y., Ishikawa, T., Sauer, B., Takaku, K., Oshima, M., and Taketo, M. M. (1999) Intestinal polyposis in mice with a dominant stable mutation of the β-catenin gene. *EMBO J* **18**, 5931–5942.
11. Van Mater, D., Kolligs, F. T., Dlugosz, A. A., and Fearon, E. R. (2003) Transient activation of β-catenin signaling in cutaneous keratinocytes is sufficient to trigger the active growth phase of the hair cycle in mice. *Genes Dev* **17**, 1219–1224.
12. Okubo, T. and Hogan, B. L. (2004) Hyperactive Wnt signaling changes the developmental potential of embryonic lung endoderm. *J Biol* **3**, 11.
13. Lo Celso, C., Prowse, D. M., and Watt, F. M. (2004) Transient activation of β-catenin signalling in adult mouse epidermis is sufficient to induce new hair follicles but con-

tinuous activation is required to maintain hair follicle tumours. *Development* **131**, 1787–1799.

14. Shibata, H., Toyama, K., Shioya, H., Ito, M., Hirota, M., Hasegawa, S., Matsumoto, H., Takano, H., Akiyama, T., Toyoshima, K., Kanamaru, R., Kanegae, Y., Saito, I., Nakamura, Y., Shiba, K., and Noda, T. (1997) Rapid colorectal adenoma formation initiated by conditional targeting of the Apc gene. *Science* **278**, 120–123.
15. Sansom, O. J., Reed, K. R., Hayes, A. J., Ireland, H., Brinkmann, H., Newton, I. P., Batlle, E., Simon-Assmann, P., Clevers, H., Nathke, I. S., Clarke, A. R., and Winton, D. J. (2004) Loss of Apc in vivo immediately perturbs Wnt signaling, differentiation, and migration. *Genes Dev* **18**, 1385–1390.
16. Farago, M., Dominguez, I., Landesman-Bollag, E., Xu, X., Rosner, A., Cardiff, R. D., and Seldin, D. C. (2005) Kinase-inactive glycogen synthase kinase 3β promotes Wnt signaling and mammary tumorigenesis. *Cancer Res* **65**, 5792–5801.
17. Andl, T., Reddy, S. T., Gaddapara, T., and Millar, S. E. (2002) WNT signals are required for the initiation of hair follicle development. *Dev Cell* **2**, 643–653.
18. Sick, S., Reinker, S., Timmer, J., and Schlake, T. (2006) WNT and DKK determine hair follicle spacing through a reaction-diffusion mechanism. *Science* **314**, 1447–1450.
19. Bamji, S. X., Shimazu, K., Kimes, N., Huelsken, J., Birchmeier, W., Lu, B., and Reichardt, L. F. (2003) Role of β-catenin in synaptic vesicle localization and presynaptic assembly. *Neuron* **40**, 719–731.
20. van Es, J. H., Jay, P., Gregorieff, A., van Gijn, M. E., Jonkheer, S., Hatzis, P., Thiele, A., van den Born, M., Begthel, H., Brabletz, T., Taketo, M. M., and Clevers, H. (2005) Wnt signalling induces maturation of Paneth cells in intestinal crypts. *Nat Cell Biol* **7**, 381–386.
21. Huelsken, J., Vogel, R., Erdmann, B., Cotsarelis, G., and Birchmeier, W. (2001) β-Catenin controls hair follicle morphogenesis and stem cell differentiation in the skin. *Cell* **105**, 533–545.
22. Brault, V., Moore, R., Kutsch, S., Ishibashi, M., Rowitch, D. H., McMahon, A. P., Sommer, L., Boussadia, O., and Kemler, R. (2001) Inactivation of the β-catenin gene by Wnt1-Cre-mediated deletion results in dramatic brain malformation and failure of craniofacial development. *Development* **128**, 1253–1264.
23. Dessimoz, J., Bonnard, C., Huelsken, J., and Grapin-Botton, A. (2005) Pancreas-specific deletion of β-catenin reveals Wnt-dependent and Wnt-independent functions during development. *Curr Biol* **15**, 1677–1683.
24. Niemann, C., Owens, D. M., Hulsken, J., Birchmeier, W., and Watt, F. M. (2002) Expression of ΔNLef1 in mouse epidermis results in differentiation of hair follicles into squamous epidermal cysts and formation of skin tumours. *Development* **129**, 95–109.
25. Merrill, B. J., Gat, U., DasGupta, R., and Fuchs, E. (2001) Tcf3 and Lef1 regulate lineage differentiation of multipotent stem cells in skin. *Genes Dev* **15**, 1688–1705.
26. Wong, M. H., Hermiston, M. L., Syder, A. J., and Gordon, J. I. (1996) Forced expression of the tumor suppressor adenomatosis polyposis coli protein induces disordered cell migration in the intestinal epithelium. *Proc Natl Acad Sci U S A* **93**, 9588–9593.
27. Hsu, W., Shakya, R., and Costantini, F. (2001) Impaired mammary gland and lymphoid development caused by inducible expression of Axin in transgenic mice. *J Cell Biol* **155**, 1055–1064.
28. Lucas, J. J., Hernandez, F., Gomez-Ramos, P., Moran, M. A., Hen, R., and Avila, J. (2001) Decreased nuclear β-catenin, tau hyperphosphorylation and neurodegeneration in GSK-3β conditional transgenic mice. *EMBO J* **20**, 27–39.
29. Panhuysen, M., Vogt Weisenhorn, D. M., Blanquet, V., Brodski, C., Heinzmann, U., Beisker, W., and Wurst, W. (2004) Effects of Wnt1 signaling on proliferation in the developing mid-/hindbrain region. *Mol Cell Neurosci* **26**, 101–111.
30. Kratz, J. E., Stearns, D., Huso, D. L., Slunt, H. H., Price, D. L., Borchelt, D. R., and Eberhart, C. G. (2002) Expression of stabilized β-catenin in differentiated neurons of transgenic mice does not result in tumor formation. *BMC Cancer* **2**, 33.
31. Borchelt, D. R., Davis, J., Fischer, M., Lee, M. K., Slunt, H. H., Ratovitsky, T., Regard, J., Copeland, N. G., Jenkins, N. A., Sisodia, S. S., and Price, D. L. (1996) A vector for expressing foreign genes in the brains and hearts of transgenic mice. *Genet Anal* **13**, 159–163.
32. Zechner, D., Fujita, Y., Hulsken, J., Muller, T., Walther, I., Taketo, M. M., Crenshaw, E. B., 3rd, Birchmeier, W., and Birchmeier, C. (2003) β-Catenin signals regulate cell growth and the balance between progenitor cell expansion and differentiation in the nervous system. *Dev Biol* **258**, 406–418.

33. Lee, H. Y., Kleber, M., Hari, L., Brault, V., Suter, U., Taketo, M. M., Kemler, R., and Sommer, L. (2004) Instructive role of Wnt/β-catenin in sensory fate specification in neural crest stem cells. *Science* **303**, 1020–1023.
34. Backman, M., Machon, O., Mygland, L., van den Bout, C. J., Zhong, W., Taketo, M. M., and Krauss, S. (2005) Effects of canonical Wnt signaling on dorso-ventral specification of the mouse telencephalon. *Dev Biol* **279**, 155–168.
35. Junghans, D., Hack, I., Frotscher, M., Taylor, V., and Kemler, R. (2005) β-Catenin-mediated cell-adhesion is vital for embryonic forebrain development. *Dev Dyn* **233**, 528–539.
36. Yu, H. M., Liu, B., Costantini, F., and Hsu, W. (2007) Impaired neural development caused by inducible expression of Axin in transgenic mice. *Mech Dev* **124**, 146–156.
37. Olson, L. E., Tollkuhn, J., Scafoglio, C., Krones, A., Zhang, J., Ohgi, K. A., Wu, W., Taketo, M. M., Kemler, R., Grosschedl, R., Rose, D., Li, X., and Rosenfeld, M. G. (2006) Homeodomain-mediated β-catenin-dependent switching events dictate cell-lineage determination. *Cell* **125**, 593–605.
38. Miller, L. A., Smith, A. N., Taketo, M. M., and Lang, R. A. (2006) Optic cup and facial patterning defects in ocular ectoderm β-catenin gain-of-function mice. *BMC Dev Biol* **6**, 14.
39. Smith, A. N., Miller, L. A., Song, N., Taketo, M. M., and Lang, R. A. (2005) The duality of β-catenin function: a requirement in lens morphogenesis and signaling suppression of lens fate in periocular ectoderm. *Dev Biol* **285**, 477–489.
40. Ohyama, T., Mohamed, O. A., Taketo, M. M., Dufort, D., and Groves, A. K. (2006) Wnt signals mediate a fate decision between otic placode and epidermis. *Development* **133**, 865–875.
41. Li, C., Hu, L., Xiao, J., Chen, H., Li, J. T., Bellusci, S., Delanghe, S., and Minoo, P. (2005) Wnt5a regulates Shh and Fgf10 signaling during lung development. *Dev Biol* **287**, 86–97.
42. Tan, X., Behari, J., Cieply, B., Michalopoulos, G. K., and Monga, S. P. (2006) Conditional deletion of β-catenin reveals its role in liver growth and regeneration. *Gastroenterology* **131**, 1561–1572.
43. Mucenski, M. L., Nation, J. M., Thitoff, A. R., Besnard, V., Xu, Y., Wert, S. E., Harada, N., Taketo, M. M., Stahlman, M. T., and Whitsett, J. A. (2005) β-Catenin regulates differentiation of respiratory epithelial cells in vivo. *Am J Physiol Lung Cell Mol Physiol* **289**, L971–979.
44. Benhamouche, S., Decaens, T., Godard, C., Chambrey, R., Rickman, D. S., Moinard, C., Vasseur-Cognet, M., Kuo, C. J., Kahn, A., Perret, C., and Colnot, S. (2006) Apc tumor suppressor gene is the "zonation-keeper" of mouse liver. *Dev Cell* **10**, 759–770.
45. Harada, N., Miyoshi, H., Murai, N., Oshima, H., Tamai, Y., Oshima, M., and Taketo, M. M. (2002) Lack of tumorigenesis in the mouse liver after adenovirus-mediated expression of a dominant stable mutant of β-catenin. *Cancer Res* **62**, 1971–1977.
46. Chen, X., Shevtsov, S. P., Hsich, E., Cui, L., Haq, S., Aronovitz, M., Kerkela, R., Molkentin, J. D., Liao, R., Salomon, R. N., Patten, R., and Force, T. (2006) The β-catenin/T-cell factor/lymphocyte enhancer factor signaling pathway is required for normal and stress-induced cardiac hypertrophy. *Mol Cell Biol* **26**, 4462–4473.
47. Lickert, H., Kutsch, S., Kanzler, B., Tamai, Y., Taketo, M. M., and Kemler, R. (2002) Formation of multiple hearts in mice following deletion of β-catenin in the embryonic endoderm. *Dev Cell* **3**, 171–181.
48. Cattelino, A., Liebner, S., Gallini, R., Zanetti, A., Balconi, G., Corsi, A., Bianco, P., Wolburg, H., Moore, R., Oreda, B., Kemler, R., and Dejana, E. (2003) The conditional inactivation of the β-catenin gene in endothelial cells causes a defective vascular pattern and increased vascular fragility. *J Cell Biol* **162**, 1111–1122.
49. Mulroy, T., Xu, Y., and Sen, J. M. (2003) β-Catenin expression enhances generation of mature thymocytes. *Int Immunol* **15**, 1485–1494.
50. Gounari, F., Aifantis, I., Khazaie, K., Hoeflinger, S., Harada, N., Taketo, M. M., and von Boehmer, H. (2001) Somatic activation of β-catenin bypasses pre-TCR signaling and TCR selection in thymocyte development. *Nat Immunol* **2**, 863–869.
51. Gounari, F., Chang, R., Cowan, J., Guo, Z., Dose, M., Gounaris, E., and Khazaie, K. (2005) Loss of adenomatous polyposis coli gene function disrupts thymic development. *Nat Immunol* **6**, 800–809.
52. Kirstetter, P., Anderson, K., Porse, B. T., Jacobsen, S. E., and Nerlov, C. (2006) Activation of the canonical Wnt pathway leads to loss of hematopoietic stem cell repopulation and multilineage differentiation block. *Nat Immunol* **7**, 1048–1056.

53. Scheller, M., Huelsken, J., Rosenbauer, F., Taketo, M. M., Birchmeier, W., Tenen, D. G., and Leutz, A. (2006) Hematopoietic stem cell and multilineage defects generated by constitutive β-catenin activation. *Nat Immunol* 7, 1037–1047.
54. Cobas, M., Wilson, A., Ernst, B., Mancini, S. J., MacDonald, H. R., Kemler, R., and Radtke, F. (2004) β-Catenin is dispensable for hematopoiesis and lymphopoiesis. *J Exp Med* **199**, 221–229.
55. Heiser, P. W., Lau, J., Taketo, M. M., Herrera, P. L., and Hebrok, M. (2006) Stabilization of β-catenin impacts pancreas growth. *Development* **133**, 2023–2032.
56. Oshima, H., Matsunaga, A., Fujimura, T., Tsukamoto, T., Taketo, M. M., and Oshima, M. (2006) Carcinogenesis in mouse stomach by simultaneous activation of the Wnt signaling and prostaglandin E2 pathway. *Gastroenterology* **131**, 1086–1095.
57. Pinto, D., and Clevers, H. (2005) Wnt control of stem cells and differentiation in the intestinal epithelium. *Exp Cell Res* **306**, 357–363.
58. Andreu, P., Colnot, S., Godard, C., Gad, S., Chafey, P., Niwa-Kawakita, M., Laurent-Puig, P., Kahn, A., Robine, S., Perret, C., and Romagnolo, B. (2005) Crypt-restricted proliferation and commitment to the Paneth cell lineage following Apc loss in the mouse intestine. *Development* **132**, 1443–1451.
59. Romagnolo, B., Berrebi, D., Saadi-Keddoucci, S., Porteu, A., Pichard, A. L., Peuchmaur, M., Vandewalle, A., Kahn, A., and Perret, C. (1999) Intestinal dysplasia and adenoma in transgenic mice after overexpression of an activated β-catenin. *Cancer Res* **59**, 3875–3879.
60. Kuhnert, F., Davis, C. R., Wang, H. T., Chu, P., Lee, M., Yuan, J., Nusse, R., and Kuo, C. J. (2004) Essential requirement for Wnt signaling in proliferation of adult small intestine and colon revealed by adenoviral expression of Dickkopf-1. *Proc Natl Acad Sci U S A* **101**, 266–271.
61. Lane, T. F. and Leder, P. (1997) Wnt-10b directs hypermorphic development and transformation in mammary glands of male and female mice. *Oncogene* **15**, 2133–2144.
62. Imbert, A., Eelkema, R., Jordan, S., Feiner, H., and Cowin, P. (2001) ΔN89β-catenin induces precocious development, differentiation, and neoplasia in mammary gland. *J Cell Biol* **153**, 555–568.
63. Michaelson, J. S. and Leder, P. (2001) β-Catenin is a downstream effector of Wnt-mediated tumorigenesis in the mammary gland. *Oncogene* **20**, 5093–5099.
64. Teuliere, J., Faraldo, M. M., Deugnier, M. A., Shtutman, M., Ben-Ze'ev, A., Thiery, J. P., and Glukhova, M. A. (2005) Targeted activation of β-catenin signaling in basal mammary epithelial cells affects mammary development and leads to hyperplasia. *Development* **132**, 267–277.
65. Miyoshi, K., Rosner, A., Nozawa, M., Byrd, C., Morgan, F., Landesman-Bollag, E., Xu, X., Seldin, D. C., Schmidt, E. V., Taketo, M. M., Robinson, G. W., Cardiff, R. D., and Hennighausen, L. (2002) Activation of different Wnt/β-catenin signaling components in mammary epithelium induces transdifferentiation and the formation of pilar tumors. *Oncogene* **21**, 5548–5556.
66. Gunther, E. J., Moody, S. E., Belka, G. K., Hahn, K. T., Innocent, N., Dugan, K. D., Cardiff, R. D., and Chodosh, L. A. (2003) Impact of p53 loss on reversal and recurrence of conditional Wnt-induced tumorigenesis. *Genes Dev* **17**, 488–501.
67. Gounari, F., Signoretti, S., Bronson, R., Klein, L., Sellers, W. R., Kum, J., Siermann, A., Taketo, M. M., von Boehmer, H., and Khazaie, K. (2002) Stabilization of β-catenin induces lesions reminiscent of prostatic intraepithelial neoplasia, but terminal squamous transdifferentiation of other secretory epithelia. *Oncogene* **21**, 4099–4107.
68. Bierie, B., Nozawa, M., Renou, J. P., Shillingford, J. M., Morgan, F., Oka, T., Taketo, M. M., Cardiff, R. D., Miyoshi, K., Wagner, K. U., Robinson, G. W., and Hennighausen, L. (2003) Activation of β-catenin in prostate epithelium induces hyperplasias and squamous transdifferentiation. *Oncogene* **22**, 3875–3887.
69. Jordan, B. K., Shen, J. H., Olaso, R., Ingraham, H. A., and Vilain, E. (2003) Wnt4 overexpression disrupts normal testicular vasculature and inhibits testosterone synthesis by repressing steroidogenic factor 1/β-catenin synergy. *Proc Natl Acad Sci U S A* **100**, 10866–10871.
70. Kimura, T., Nakamura, T., Murayama, K., Umehara, H., Yamano, N., Watanabe, S., Taketo, M. M., and Nakano, T. (2006) The stabilization of β-catenin leads to impaired primordial germ cell development via aberrant cell cycle progression. *Dev Biol* **300**, 545–553.
71. Boerboom, D., Paquet, M., Hsieh, M., Liu, J., Jamin, S. P., Behringer, R. R., Sirois, J., Taketo, M. M., and Richards, J. S. (2005) Misregulated Wnt/β-catenin signaling leads to ovarian granulosa cell tumor development. *Cancer Res* **65**, 9206–9215.

72. Arango, N. A., Szotek, P. P., Manganaro, T. F., Oliva, E., Donahoe, P. K., and Teixeira, J. (2005) Conditional deletion of β-catenin in the mesenchyme of the developing mouse uterus results in a switch to adipogenesis in the myometrium. *Dev Biol* **288**, 276–283.
73. Kuraguchi, M., Wang, X. P., Bronson, R. T., Rothenberg, R., Ohene-Baah, N. Y., Lund, J. J., Kucherlapati, M., Maas, R. L., and Kucherlapati, R. (2006) Adenomatous polyposis coli (APC) is required for normal development of skin and thymus. *PLoS Genet* **2**, e146.
74. Atit, R., Sgaier, S. K., Mohamed, O. A., Taketo, M. M., Dufort, D., Joyner, A. L., Niswander, L., and Conlon, R. A. (2006) β-Catenin activation is necessary and sufficient to specify the dorsal dermal fate in the mouse. *Dev Biol* **296**, 164–176.
75. Nguyen, H., Rendl, M., and Fuchs, E. (2006) Tcf3 governs stem cell features and represses cell fate determination in skin. *Cell* **127**, 171–183.
76. Guo, X., Day, T. F., Jiang, X., Garrett-Beal, L., Topol, L., and Yang, Y. (2004) Wnt/β-catenin signaling is sufficient and necessary for synovial joint formation. *Genes Dev* **18**, 2404–2417.
77. Tamamura, Y., Otani, T., Kanatani, N., Koyama, E., Kitagaki, J., Komori, T., Yamada, Y., Costantini, F., Wakisaka, S., Pacifici, M., Iwamoto, M., and Enomoto-Iwamoto, M. (2005) Developmental regulation of Wnt/β-catenin signals is required for growth plate assembly, cartilage integrity, and endochondral ossification. *J Biol Chem* **280**, 19185–19195.
78. Hill, T. P., Spater, D., Taketo, M. M., Birchmeier, W., and Hartmann, C. (2005) Canonical Wnt/β-catenin signaling prevents osteoblasts from differentiating into chondrocytes. *Dev Cell* **8**, 727–738.
79. Li, J., Sarosi, I., Cattley, R. C., Pretorius, J., Asuncion, F., Grisanti, M., Morony, S., Adamu, S., Geng, Z., Qiu, W., Kostenuik, P., Lacey, D. L., Simonet, W. S., Bolon, B., Qian, X., Shalhoub, V., Ominsky, M. S., Zhu Ke, H., Li, X., and Richards, W. G. (2006) Dkk1-mediated inhibition of Wnt signaling in bone results in osteopenia. *Bone* **39**, 754–766.
80. Glass, D. A., 2nd, Bialek, P., Ahn, J. D., Starbuck, M., Patel, M. S., Clevers, H., Taketo, M. M., Long, F., McMahon, A. P., Lang, R. A., and Karsenty, G. (2005) Canonical Wnt signaling in differentiated osteoblasts controls osteoclast differentiation. *Dev Cell* **8**, 751–764.
81. Babij, P., Zhao, W., Small, C., Kharode, Y., Yaworsky, P. J., Bouxsein, M. L., Reddy, P. S., Bodine, P. V., Robinson, J. A., Bhat, B., Marzolf, J., Moran, R. A., and Bex, F. (2003) High bone mass in mice expressing a mutant LRP5 gene. *J Bone Miner Res* **18**, 960–974.
82. Chen, Y., Whetstone, H. C., Youn, A., Nadesan, P., Chow, E. C., Lin, A. C., and Alman, B. A. (2007) β-Catenin signaling pathway is crucial for bone morphogenetic protein 2 to induce new bone formation. *J Biol Chem* **282**, 526–533.
83. Longo, K. A., Wright, W. S., Kang, S., Gerin, I., Chiang, S. H., Lucas, P. C., Opp, M. R., and MacDougald, O. A. (2004) Wnt10b inhibits development of white and brown adipose tissues. *J Biol Chem* **279**, 35503–35509.
84. Liu, F., Thirumangalathu, S., Gallant, N. M., Yang, S. H., Stoick-Cooper, C. L., Reddy, S. T., Andl, T., Taketo, M. M., Dlugosz, A. A., Moon, R. T., Barlow, L. A., and Millar, S. E. (2007) Wnt-β-catenin signaling initiates taste papilla development. *Nat Genet* **39**, 106–112.
85. Jarvinen, E., Salazar-Ciudad, I., Birchmeier, W., Taketo, M. M., Jernvall, J., and Thesleff, I. (2006) Continuous tooth generation in mouse is induced by activated epithelial Wnt/β-catenin signaling. *Proc Natl Acad Sci U S A* **103**, 18627–18632.
86. Hill, T. P., Taketo, M. M., Birchmeier, W., and Hartmann, C. (2006) Multiple roles of mesenchymal β-catenin during murine limb patterning. *Development* **133**, 1219–1229.
87. Soshnikova, N., Zechner, D., Huelsken, J., Mishina, Y., Behringer, R. R., Taketo, M. M., Crenshaw, E. B., 3rd, and Birchmeier, W. (2003) Genetic interaction between Wnt/β-catenin and BMP receptor signaling during formation of the AER and the dorsal-ventral axis in the limb. *Genes Dev* **17**, 1963–1968.
88. Qian, C. N., Knol, J., Igarashi, P., Lin, F., Zylstra, U., Teh, B. T., and Williams, B. O. (2005) Cystic renal neoplasia following conditional inactivation of Apc in mouse renal tubular epithelium. *J Biol Chem* **280**, 3938–3945.
89. Kemler, R., Hierholzer, A., Kanzler, B., Kuppig, S., Hansen, K., Taketo, M. M., de Vries, W. N., Knowles, B. B., and Solter, D. (2004) Stabilization of β-catenin in the mouse zygote leads to premature epithelial-mesenchymal transition in the epiblast. *Development* **131**, 5817–5824.
90. Armstrong, D. D., Wong, V. L., and Esser, K. A. (2006) Expression of β-catenin is necessary

for physiological growth of adult skeletal muscle. *Am J Physiol Cell Physiol* **291**, C185–188.

91. Barolo, S. (2006) Transgenic Wnt/TCF pathway reporters: all you need is Lef? *Oncogene* **25**, 7505–7511.
92. Lustig, B., Jerchow, B., Sachs, M., Weiler, S., Pietsch, T., Karsten, U., van de Wetering, M., Clevers, H., Schlag, P. M., Birchmeier, W., and Behrens, J. (2002) Negative feedback loop of Wnt signaling through upregulation of conductin/axin2 in colorectal and liver tumors. *Mol Cell Biol* **22**, 1184–1193.
93. Korinek, V., Barker, N., Morin, P. J., van Wichen, D., de Weger, R., Kinzler, K. W., Vogelstein, B., and Clevers, H. (1997) Constitutive transcriptional activation by a β-catenin-Tcf complex in APC-/- colon carcinoma. *Science* **275**, 1784–1787.
94. Maretto, S., Cordenonsi, M., Dupont, S., Braghetta, P., Broccoli, V., Hassan, A. B., Volpin, D., Bressan, G. M., and Piccolo, S. (2003) Mapping Wnt/β-catenin signaling during mouse development and in colorectal tumors. *Proc Natl Acad Sci U S A* **100**, 3299–3304.
95. Nakaya, M. A., Biris, K., Tsukiyama, T., Jaime, S., Rawls, J. A., and Yamaguchi, T. P. (2005) Wnt3a links left-right determination with segmentation and anteroposterior axis elongation. *Development* **132**, 5425–5436.
96. DasGupta, R., and Fuchs, E. (1999) Multiple roles for activated LEF/TCF transcription complexes during hair follicle development and differentiation. *Development* **126**, 4557–4568.
97. Mohamed, O. A., Clarke, H. J., and Dufort, D. (2004) β-Catenin signaling marks the prospective site of primitive streak formation in the mouse embryo. *Dev Dyn* **231**, 416–424.
98. Staal, F. J., Meeldijk, J., Moerer, P., Jay, P., van de Weerdt, B. C., Vainio, S., Nolan, G. P., and Clevers, H. (2001) Wnt signaling is required for thymocyte development and activates Tcf-1 mediated transcription. *Eur J Immunol* **31**, 285–293.
99. Hochedlinger, K., Yamada, Y., Beard, C., and Jaenisch, R. (2005) Ectopic expression of Oct-4 blocks progenitor-cell differentiation and causes dysplasia in epithelial tissues. *Cell* **121**, 465–477.
100. Moriyama, A., Kii, I., Sunabori, T., Kurihara, S., Takayama, I., Shimazaki, M., Tanabe, H., Oginuma, M., Fukayama, M., Matsuzaki, Y., Saga, Y., and Kudo, A. (2007) GFP transgenic mice reveal active canonical Wnt signal in neonatal brain and in adult liver and spleen. *Genesis* **45**, 90–100.
101. Jonkers, J. and Berns, A. (2002) Conditional mouse models of sporadic cancer. *Nat Rev Cancer* **2**, 251–265.
102. Maddison, K. and Clarke, A. R. (2005) New approaches for modelling cancer mechanisms in the mouse. *J Pathol* **205**, 181–193.

INDEX

Printed In The United States Of America